Lecture Notes in Computer Science   9708

Commenced Publication in 1973
Founding and Former Series Editors:
Gerhard Goos, Juris Hartmanis, and Jan van Leeuwen

More information about this series at http://www.springer.com/series/7409

Kevin S. Xu · David Reitter
Dongwon Lee · Nathaniel Osgood (Eds.)

Social, Cultural, and Behavioral Modeling

9th International Conference, SBP-BRiMS 2016
Washington, DC, USA, June 28 – July 1, 2016
Proceedings

 Springer

Editors
Kevin S. Xu
University of Toledo
Toledo, OH
USA

David Reitter
Penn State University
University Park, PA
USA

Dongwon Lee
Penn State University
University Park, PA
USA

Nathaniel Osgood
University of Saskatchewan
Saskatoon, SK
Canada

ISSN 0302-9743   ISSN 1611-3349   (electronic)
Lecture Notes in Computer Science
ISBN 978-3-319-39930-0   ISBN 978-3-319-39931-7   (eBook)
DOI 10.1007/978-3-319-39931-7

Library of Congress Control Number: 2016940349

LNCS Sublibrary: SL3 – Information Systems and Applications, incl. Internet/Web, and HCI

Printed on acid-free paper

This Springer imprint is published by Springer Nature
The registered company is Springer International Publishing AG Switzerland

Preface

Improving the human condition requires understanding, forecasting, and impacting sociocultural behavior both in the digital and nondigital worlds. Increasing amounts of digital data, embedded sensors collecting human information, rapidly changing communication media, changes in legislation concerning digital rights and privacy, and the spread of 4G technology to developing countries are creating a new cyber-mediated world where the very precepts of why, when, and how people interact and make decisions are being called into question. For example, Uber took a deep understanding of human behavior vis-à-vis commuting, developed software to support this behavior, ended up saving human time (and thus capital) and reducing stress, and so indirectly created the opportunity for humans with more time and less stress to evolve new behaviors. Scientific and industrial pioneers in this area are relying on both social science and computer science to help make sense of and have an impact on this new frontier. To be successful a true merger of social science and computer science are needed. Solutions that rely only on the social science or only on the computer science are doomed to failure. For example, Anonymous developed an approach for identifying members of terror groups such as ISIS on the social media platform Twitter using state-of-the-art computational techniques. These accounts were then suspended. This was a purely technical solution. The result was that those individuals with suspended accounts just moved to new platforms, and resurfaced on Twitter under new IDs. In this case, failure to understand basic social behavior resulted in an ineffective solution.

The goal of the International Conference on Social Computing, Behavioral–Cultural Modeling, and Prediction and Behavior Representation in Modeling and Simulation (SBP-BRiMS) is to build this new community of social cyber scholars by bringing together and fostering interaction between members of the scientific, corporate, government, and military communities interested in understanding, forecasting, and impacting human sociocultural behavior. It is the charge of this community to build this new science, its theories, methods, and its scientific culture in a way that does not give priority to either social science or computer science, and to embrace change as the cornerstone of the community. Despite decades of work in this area, this new scientific field is still in its infancy. To meet this charge, to move this science to the next level, this community must meet the following three challenges: deep understanding, sociocognitive reasoning, and re-usable computational technology. Fortunately, as the papers in this volume illustrate, this community is poised to answer these challenges. But what does meeting these challenges entail?

Deep understanding refers to the ability to make operational decisions and theoretical arguments on the basis of an empirically based deep and broad understanding of the complex sociocultural phenomena of interest. Today, although more data are available digitally than ever before, we are still plagued by anecdotal-based arguments. For example, in social media, despite the wealth of information available, most analysts focus on small samples, which are typically biased and cover only a small time period, and use that to explain all events and make future predictions. The analyst finds the magic tweet or the unusual tweeter and uses that to prove their point. Tools that can help the analyst to reason using more data or less biased data are not widely used, are often more complex than the average analyst wants to use, or take more time than the analyst wants to spend to generate results. Not only are more scalable technologies needed, but so too is a better understanding of the biases in the data and ways to overcome them, and a cultural change to not accept anecdotes as evidence.

Sociocognitive reasoning refers to the ability of individuals to make sense of the world and to interact with it in terms of groups and not just individuals. Today most social-behavioral models either focus on (1) strong cognitive models of individuals engaged in tasks and thus model a small number of agents with high levels of cognitive accuracy but with little if any social context, or (2) light cognitive models and strong interaction models and thus model massive numbers of agents with high levels of social realism and little cognitive realism. In both cases, as realism is increased in the other dimension the scalability of the models fail, and their predictive accuracy on one of the two dimensions remains low. By contrast, as agent models are built where the agents are not just cognitive but socially cognitive, we find that the scalability increases and the predictive accuracy increases. Not only are agent models with sociocognitive reasoning capabilities needed, but so, too, is a better understanding of how individuals form and use these social cognitions.

More software solutions that support behavioral representation, modeling, data collection, bias identification, analysis, and visualization support human sociocultural behavioral modeling and prediction than ever before. However, this software is generally just piling up in giant black holes on the Web. Part of the problem is the fallacy of open source; the idea that if you merely make code open source others will use it. By contrast, most of the tools and methods available in Git or R are only used by the developer, if that. Reasons for lack of use include lack of documentation, lack of interfaces, lack of interoperability with other tools, difficulty of linking to data, and increased demands on the analyst's time due to a lack of tool-chain and workflow optimization. Part of the problem is the not-invented-here syndrome. For social scientists and computer scientists alike, it is simply more fun to build a quick and dirty tool for your own use than to study and learn tools built by others. And, part of the problem is the insensitivity of people from one scientific or corporate culture toward the reward and demand structures of the other cultures that impact what information can or should be shared and when. A related problem is double standards in sharing where universities are expected to share and companies are not; but increasingly universities are relying on intellectual property as a source of funding just like other companies. While common standards and representations would help, a cultural shift from a focus on sharing to a focus on re-use is as critical for moving this area to the next scientific level.

In this volume, and in all the work presented at the SBP-BRiMS 2016 conference, you will see suggestions of how to address the challenges just described. SBP-BRiMS 2016 continued the scholarly tradition of the past conferences out of which it has emerged like a phoenix: the Social Computing, Behavioral–Cultural Modeling, and Prediction (SBP) Conference and the Behavioral Representation in Modeling and Simulation (BRiMS) Society's conference. A total of 78 documents were submitted as full papers. Of these, 38 were accepted, for an acceptance rate of 49 %. Additionally there were a large number of papers describing emergent ideas and late-breaking results, or responses to the challenge problem were submitted and accepted. Finally there were nine tutorials covering a diversity of topics. This is an international group with papers submitted by authors from 13 countries.

The conference has a strong multidisciplinary heritage. As the papers in this volume show, people, theories, methods, and data from a wide number of disciplines are represented including computer science, psychology, sociology, communication science, public health, bioinformatics, political science, and organizational science. Numerous types of computational methods are used, including, but not limited to, machine learning, language technology, social network analysis and visualization, agent-based simulation, and statistics. Based on the author's self-selected area for each paper, the breakdown is as follows:

- Behavioral and social sciences: 17 submissions, nine accepted
- Health sciences: eight submissions, three accepted
- Information, systems, and network sciences: 27 submissions, 10 accepted
- Methodology: 12 submissions, eight accepted
- Military and intelligence applications: 14 submissions, eight accepted

This exciting program could not have been put together without the hard work of a number of dedicated and forward-thinking researchers serving as the Organizing Committee, listed on the following pages. Members of the Program Committee and the Scholarship Committee as well as publication, advertising, and local arrangements chairs worked tirelessly to put together this event. They were supported by the government sponsors, the area chairs, and the reviewers. I thank them for their efforts on behalf of the community. In addition, we gratefully acknowledge the support of our sponsors: the Office of Naval Research – N00014-15-1-2463 and N00014-16-1-2274, the National Science Foundation – IIS-1523458, and the Army Research Office – W911NF-14-1-0023. Enjoy the proceedings!

April 2016   Kathleen M. Carley

Organization

Conference Co-chairs

Kathleen M. Carley	Carnegie Mellon University, USA
Nitin Agarwal	University of Arkansas at Little Rock, USA
Jeffrey Johnson	University of Florida, USA

Program Co-chairs

Dongwon Lee	National Science Foundation, USA
Nathaniel Osgood	University of Saskatchewan, Canada
David Reitter	The Pennsylvania State University, USA
Kevin S. Xu	University of Toledo, USA

Advisory Committee

Fahmida N. Chowdhury	National Science Foundation, USA
Rebecca Goolsby	Office of Naval Research, USA
Stephen Marcus	National Institutes of Health, USA
Paul Tandy	Defense Threat Reduction Agency, USA
Edward T. Palazzolo	Army Research Office, USA

Advisory Committee Emeritus

Patricia Mabry	Indiana University, USA
John Lavery	Army Research Office, USA
Tisha Wiley	National Institutes of Health, USA

Sponsorship Committee

Huan Liu	Arizona State University, USA
Sun Ki Chai	University of Hawaii, USA
Donald Adjeroh	West Virginia University, USA
Nitin Agarwal	University of Arkansas at Little Rock, USA

Publicity Chair

Donald Adjeroh	West Virginia University, USA

Web Chair

Therese L. Williams   University of Arkansas at Little Rock, USA

Local Area Coordination

William G. Kennedy   George Mason University, USA

Student Travel Awards

Sun Ki Chai   University of Hawaii, USA

Proceedings Chair

Kevin S. Xu   University of Toledo, USA

Journal Special Issue Chair

Kathleen M. Carley   Carnegie Mellon University, USA

Tutorial Chair

Yu-Ru Lin   University of Pittsburgh, USA

Challenge Problem Committee

Kathleen M. Carley   Carnegie Mellon University, USA
Kevin S. Xu   University of Toledo, USA
Fred Morstatter   Arizona State University, USA
Kenneth Joseph   Carnegie Mellon University, USA
David Masad   George Mason University, USA

Topic Area Chairs

David Broniatowski   George Washington University, USA
Sibel Adali   Rensselaer Polytechnic Institute, USA
Christian Lebiere   Carnegie Mellon University, USA
Wen Dong   State University of New York, Buffalo, USA
Patricia Mabry   Indiana University, USA
Walter Warwick   Tier1 Performance, USA

BRiMS Society Chair

Christopher Dancy II   Bucknell University, USA

SBP Society Chair

Shanchieh (Jay) Yang   Rochester Institute of Technology, USA

BRiMS Steering Committee

Christopher Dancy II   Bucknell University, USA
William G. Kennedy   George Mason University, USA
David Reitter   The Pennsylvania State University, USA
Dan Cassenti   US Army Research Laboratory, USA

SBP Steering Committee

John Salerno   Exelis, USA
Ariel Greenberg   Johns Hopkins University/Applied Physics Laboratory,
                             USA
Shanchieh (Jay) Yang   Rochester Institute of Technology, USA
Huan Liu   Arizona State University, USA
Sun Ki Chai   University of Hawaii, USA
Nitin Agarwal   University of Arkansas at Little Rock, USA

BRiMS Executive Committee

Brad Best   Adaptive Cognitive Systems, USA
Brad Cain   Defense Research and Development, Canada
Daniel N. Cassenti   US Army Research Laboratory, USA
Bruno Emond   National Research Council, USA
Coty Gonzalez   Carnegie Mellon University, USA
Brian Gore   NASA, USA
Kristen Greene   National Institute of Standards and Technology, USA
Jeff Hansberger   US Army Research Laboratory, USA
Tiffany Jastrzembski   Air Force Research Laboratory, USA
Randolph M. Jones   SoarTech, USA
Troy Kelly   US Army Research Laboratory, USA
William G. Kennedy   George Mason University, USA
Christian Lebiere   Carnegie Mellon University, USA
Elizabeth Mezzacappa   Defence Science and Technology Laboratory, UK
   Bharat Patel
Michael Qin   Naval Submarine Medical Research Laboratory, USA
Frank E. Ritter   The Pennsylvania State University, USA
Tracy Sanders   University of Central Florida, USA
Venkat Sastry   University of Cranfield, UK
Barry Silverman   University of Pennsylvania, USA

David Stracuzzi   Sandia National Laboratories, USA
Robert E. Wray   SoarTech, USA

SBP Steering Committee Emeritus

Nathan D. Bos   Johns Hopkins University/Applied Physics Lab, USA
Claudio Cioffi-Revilla   George Mason University, USA
V.S. Subrahmanian   University of Maryland, USA
Dana Nau   University of Maryland, USA

Technical Program Committee

Mohammad Ali Abbasi   Bruno Emond
Myriam Abramson   William Ferng
Sibel Adalı   Michael Fire
Donald Adjeroh   Wai-Tat Fu
Kalin Agrawal   Liz Ginexi
Shah Jamal Alam   Ariel Greenberg
Samer Al-Khateeb   Kristen Greene
Ling Bian   Omar Guerrero
Halil Bisgin   Kyungsik Han
Lashon Booker   Walter Hill
Nathan Bos   Shen-Shyang Ho
David Bracewell   Shuyuan Mary Ho
Erica Briscoe   Xia Hu
David Broniatowski   Robert Hubal
Magdalena Bugajska   Nguyen Huy Quoc
Brad Cain   Terresa Jackson
Kathleen Carley   John Johnson
Ernesto Carrella   Randolph Jones
Daniel Cassenti   Byeong-Ho Kang
Subhadeep Chakraborty   Bill Kennedy
David Chin   Jong Kim
Kelvin Choi   Masahiro Kimura
Fahmida Chowdhury   Shamanth Kumar
Dave Clark   Kiran Lakkaraju
Gordon Cooke   Stephanie Lanza
Andrew Crooks   Othalia Larue
Rachel Cummings   Christian Lebiere
Peng Dai   Jongwuk Lee
Hasan Davulcu   Dongwon Lee
Yves-Alexandre de Montjoye   Zhuoshu Li
Jana Diesner   Jiexun Li
Wen Dong   Ee-Peng Lim
Koji Eguchi   Yu-Ru Lin
Jeffrey Ellen   Huan Liu

Lyle Long
Deryle W. Lonsdale
Andreas Luedtke
Patricia Mabry
Jonathas Magalhães
Juan F. Mancilla-Caceres
Stephen Marcus
Venkata Swamy Martha
Liza Mezzacappa
Geoffrey Morgan
Sai Moturu
Ernest Moy
Keisuke Nakao
Radoslaw Nielek
Wendy Nilsen
Kouzou Ohara
Byung Won On
Brandon Oselio
Nathaniel Osgood
Alexander Outkin
Fatih Özgül
Hemant Purohit
S.S. Ravi
David Reitter
Amit Saha
Philip Schrodt
Samira Shaikh
David Shoham
Shade Shutters

Barry Silverman
Amy Sliva
David Stracuzzi
Samarth Swarup
George Tadda
Robert Thomson
Anil Kumar Vullikanti
Melissa Walwanis
Xiaofeng Wang
Zhijian Wang
Changzhou Wang
Rik Warren
Wei Wei
Elizabeth Whitaker
Paul Whitney
Rolf T. Wigand
Tisha Wiley
Robert Wray
Kevin Xu
Laurence T. Yang
S. Jay Yang
Yong Yang
Bei Yu
Mo Yu
Serpil Yuce
Reza Zafarani
Qingpeng Zhang
Kang Zhao
Yanping Zhao

Contents

Information, Systems, and Network Sciences

Methodology

Military and Intelligence Applications

Behavioral and Social Sciences

Improving Donation Distribution
for Crowdfunding: An Agent-Based Model

Yi-Chieh Lee, Chi-Hsien Yen$^{(\boxtimes)}$, and Wai-Tat Fu

Department of Computer Science, University of Illinois Urbana-Champaign,
Champaign, USA
be341341@gmail.com, {cyen4,wfu}@illinois.edu

Abstract. Donation-based crowdfunding has the potential to democ-
ratize capital raising by soliciting donations directly from the public
through the Web and social media. These crowdfunding platforms, however, often function as unregulated open markets, in which there is minimal intervention to influence donation distribution across projects. In
fact, research on crowdfunding hints that donation distribution in most
crowdfunding platforms are suboptimal: while the overall success rates of
crowdfunding projects are often low, a significant proportion of projects
receive donations way over their targets. In this paper, we propose a new
donation distributing system that aim to (a) distribute donations more
effectively among the projects, and (b) align the allocation of donations
with the preferences of donors. An agent-based model was developed
to test the proposed system. Results showed that the proposed system
not only increased the overall success rates of projects, but also led to
more successes for projects preferred by donors. Implications to future
crowdfunding platforms are discussed.

Keywords: Crowdfunding · Fundraising · Market

1 Introduction

Crowdfunding websites have received much attention recently [1], as exemplified
by the growing number of projects and donations to sites such as Kickstarter,
DonorsChoose.org, and GiveForward [10–13]. The appeal of crowdfunding is that
everyone can raise money directly from crowdfunding websites to help accomplish
his or her design projects or various purposes, bypassing traditional sources such
as venture capitalists or financial institutions. In addition to raising capital, some
companies have also adopted crowdfunding websites for testing purposes because
activities in crowdfunding sites may predict whether potential customers and
communities will embrace their new products.

An essential characteristic of crowdfunding websites is that each project
aggregates capital from many people with a small donation. While different
crowdfunding sites have different policies for determining success, in general, a

---

Y.-C. Lee and C.-H. Yen—These authors contributed equally to this work.

K.S. Xu et al. (Eds.): SBP-BRiMS 2016, LNCS 9708, pp. 3–12, 2016.
DOI: 10.1007/978-3-319-39931-7_1

project is considered successful when the donations add up to a target amount of money specified by the project creator when the project is launched. However, crowdfunding websites do not provide direct assistance to raise capital for any specific project. Rather, project creators have to actively promote their projects within a specific period of time to increase their chance of success.

Previous studies [7,9] show that the success rates of the projects on crowdfunding websites are often low. While the majority of successful projects received donation close to their target amount, a significant proportion of projects received donations way over (>200 %) their target amount [9]. In addition, many projects received donation only in the first few days after the projects were launched and slowly lost attention of potential funders and eventually failed. These observations led us to speculate that the current unassisted process of matching of the massive amount of small donations to the projects is suboptimal, in the sense that the distribution of donations can be improved such that more high quality projects can be successful. We speculate that the current "open marketplace" of crowdfunding sites is the reason why the distribution of donations to the sets of available projects is suboptimal. This is consistent with previous studies that show that social information systems that rely on user dynamics to distribute resources may often lead to higher inequality and unpredictability [6].

Similar to many social information systems, the process of crowdfunding is highly dynamic; earlier events have a large impact on later events. In addition, the behavior of potential donors is often affected by various dynamic indicators in the projects, as well as the general policies of different crowdfunding sites. For example, Wash et al. [2] showed that the return policies of the crowdfunding websites influence the efficiency of crowdfunding. Specifically, they compared platforms that adopt an "all-or-nothing" policy, i.e., the creators get nothing unless they reach a specific donation goal (e.g., Kickstarter) to those that adopt an "incremental" policy, i.e., one can get whatever is donated even if the total is below the goal (e.g., Indiegogo). They found behavior in these platforms differed substantially, as potential donors in all-or-nothing sites likely pay more attention to projects that are perceived to have a high chance of success. Beltran et al. [4] proposed a crowdfunding system to allow donors to make conditional donations, which may motivate donors to contribute more. These studies [2,7,8] support the notion that some form of intervention by crowdfunding websites, such as policies about how donations can be made, how donations are allocated, and how donors are rewarded (or acknowledged), will have significant impact on the general matching process between donations and projects.

In this work, we develop and test one such intervention in crowdfunding. The goal is to understand the how the proposed intervention may impact the complex dynamics behind crowdfunding, to the extent that it could assist donors to make their donations more effective by enlarging the benefit to more crowdfunding projects. Specifically, the proposed intervention is inspired by an existing theory and algorithm in economic research [5], which was originally used to optimally allocate students to high schools that they preferred. Through this system, students can list the schools they want to attend in order, and the algorithm will iteratively match the students who have been rejected in the last round until most of the students match a school. Inspired by this algorithm [5], we apply a modified version of this algorithm to crowdfunding websites. First, we preliminarily extract a feature of this algorithm, matching students' preferred ranking with schools, to investigate their usefulness for crowdfunding websites. Then, we construct an agent-based model to simulate the algorithm's effect in various crowdfunding environments and analyze how well it can help match donors to projects.

Research Goals. Our goal is to explore the possibility of applying the new method of donation to crowdfunding websites. The previous method that only allows each donor to choose a single project to donate at a time. In other word, a donor has to decide how much money he/she want to donate to each project. Nevertheless, the capital of each individual is limited. When donors are interested in multiple corwdfunding projects, it is hard to find a good way distributing the limited capital into multiple projects to effectively support those projects. Therefore, our method allows donors to select multiple projects a time and decide the total amount of donation. Our system will help the donors spread out their donation among their chosen projects based on dynamics of the ongoing crowdfunding activities. Moreover, our method can reallocate donations when the prior funded project fails to achieve the donation goal in the end or has acquired sufficient funding. As previous studies mentioned that donors would like to see their donations utilized effectively.

This work proposes a new way to help donors spread out their donations. Specifically, our first research goal is (1) to use an agent-based model to explore the impact of our new method on the success rate of crowdfunding projects, (2) to investigate if the success rate of crowdfunding projects will be influenced when donors provide a list of ranked preferred projects. and (3) to test the number of projects that donors choose each time may affect the success rate of crowdfunding projects. We propose a new model and present the results of a simulation using an agent-based model.

2 Model Description

On current crowdfunding platforms, donors are allowed to choose a single project to donate to at a time and decide how much money they want to give to each project. In our system, donors have the option to select multiple projects they want to contribute to. They can specify how much money they want to donate in total and also rank their selected projects based on their preferences. Our system will automatically allocate their donations to the projects based on their preferences and thus better utilize the monetary resources. We propose to apply the new method of donation to crowdfunding websites, changing the one (donor)-to-one (project) model to one (donor)-to-multiple (projects) model. The money could be better utilized by our one-to-multiple model because each donation would have chance to be reallocated to other projects if the prior project fails to achieve the donation goal in the end, or it has already acquired sufficient funding. The aim of this algorithm is to find the best intervention to maximize the number of projects reaching their funding goals and enhance the effectiveness of each donation.

2.1   Agents (Donors)

Methods of Donation. Using the agent-based simulation model, we aim to answer the following questions: Will the success rates of crowdfunding websites increase when donors can select multiple crowdfunding projects to donate? If so, how impact does the selection improve the success rates? If donors are allowed to select and rank multiple projects to donate, how will the selection and ranking influence the success rates of projects and the overall efficiency of the crowdfunding system? To answer these questions, we developed and compared four types of crowdfunding platforms using different donation methods, and used agent-based model to simulate behavior in these platforms:

1. **Single selection (SS):** This method already exists in current crowdfunding websites donors only choose one project ($n = 1$) at a time and decide how much to donate. This method provides a baseline measurement of existing crowdfunding platforms.
2. **Multiple selection without ranking (MS-NR):** Donors select up to n projects and decide how much money they would like to donate in total.
3. **Multiple selection with ranking (MS-R):** Donors can select up to n projects and decide the total donation amount, but they are required to rank all of their selected projects according to their preference. Our system allocates their money to the higher ranked projects first.
4. **Multiple selection with mixed ranking (MS-M):** This method combines the second and the third approaches. After choosing up to n projects, the donors places their selected projects into different ranking levels according to their preferences, and each ranking level contains multiple projects that are equally preferred by the donor. This is the most general way to structure their preference toward the selected projects.

The parameter n is the maximum number of projects each donor can select, and it is adjustable in our simulation. By setting n equal to 1, all the methods are equivalent to the first baseline method. When n is greater than 1, the donors can select 1 to n projects.

2.2   Donation Distribution

The algorithm for donation distribution in our system uses the following rules when allocating money:

1. When a new donation is made, our system will allocate funds to the selected projects based on the sequence of the donors preference. The system will

allocate a donation to the highest ranked project first, but will not donate any more than is needed to reach the projects funding goal. If there is money left after donating to the first project, our system will put the rest of money towards the donors next preference sequentially until there is no money left. If multiple projects are equally preferred by the donor (when using MS-NR and MS-M methods), the algorithm will donate to the projects that will end sooner first, followed by the projects that need less money to reach their funding goal. If all of the selected projects have met their goal, the remaining money will be allocated to the highest ranked project. This rule is demonstrated in Figure X. When t = 1, the donation is assigned to project A first and then project B based on their deadline and how much they need. When t = 2, the money is assigned to project C because the highest ranked project (project A) already succeeds.

2. When a project expires and fails to achieve its goal, the donations it received will be reallocated to other projects based on the donors ranking. Along with the first rule, this ensures that the money will only be allocated to a lower ranked project when the higher ranked project has either been fully funded or has expired and failed. This rule is demonstrated in Figure X when t = 3, where project B is expired and thus its donation is reallocated to project C based on the preference.

3. When a donation is assigned to a project that is already fully funded, our system will try to move the same amount of money from the previous donations of the project to other projects if the reallocation satisfies donors preferences. Therefore, more projects will benefit from the reallocated money, while the successful project still receive more or equal funding. This rule is demonstrated in Figure X when t = 3, where $50 is assigned to project C which already succeeds. To help more project, $50 from the second donation is moved from project C to project D by our algorithm.

4. After each donation is made and allocated to one or more projects, it might be reallocated because of the above rules. We set a fixed time period (currently set to 30 days) during which each donation could be reallocated to other projects; otherwise, the donor would never know which final projects their money is allocated to. When distributing donations, our system will only consider the money that can be reallocated.

Figure 1 shows an example of our donation algorithm and process.

3   Experimental Setup

The agent-based simulation model was implemented in Python. We crawled Kickstarter and DonorsChoose.org collecting information from the years 2014 and 2015 about donation goals project status, and the amount of donation for each donor. These data provide help our simulation with realistic scenarios for crowdfunding websites.

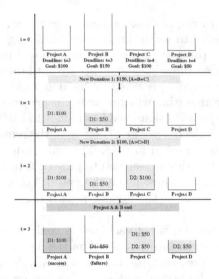

Fig. 1. An example of our donation algorithm. At the beginning of the simulation ($t = 0$), there are four projects with different donation goals and deadlines. When $t = 1$, a new donation is made with Project A, B, and C at the same ranking level; thus, the donation allocation follows the first rule. When $t = 3$, Project B fails, its donation is reallocated to Project C by rule 2, and because Project C is fully funded, our system assigns part of the donation D2 from Project C to Project D by rule 3 (Color figure online).

3.1 Donation Setting

The information released from the several crowdfunding platforms revealed that the average amount donated by an individual donor is between 50 to 100 USD. After analyzing the data from Kickstarter, we observed that the distribution of donation was closed to a logarithmic normal distribution. Consequently, our model adopted this distribution to randomly produce the amount of donation for each donor. In addition, based on the real data from crowdfunding websites, this model will randomly generate 210000 donors a month, and their preferred projects will be randomly selected from the live projects at the time the donor entered. Each donor randomly chooses 1 to n projects with a ranking that is also randomly generated from all possible combinations.

3.2 Project Setting

The variety of projects on crowdfunding platforms is diverse. According to different fundraising purposes, the donation goals and the duration of the crowd-funding projects are various. On Kickstarter, the donation goals for technology and design projects are usually higher than for other kinds of projects. Moreover, Kickstarter restricts the duration of fundraising campaigns to two months. In contrast, the projects on DonorsChoose.org are related to education. Their donation goals are often under 1500 USD, and project duration can be more than two months.

Therefore, the characteristics of every crowdfunding website are highly different, and we can only use some of them in this study. In our model, we randomly generated a donation goal between 100 USD to 10000 USD and a duration of 7 to 60 days for each project. Also, our model randomly created 3250 projects a month, which is similar to the real situation on crowdfunding websites.

In order to examine the influence of donation methods on the success rate of crowdfunding projects, we ran multiple simulations with different methods and controlled the maximum number of projects ($n = 1$ to $n = 10$) each donor could choose. The results of success rates reported in this paper are averaged from 30 simulation months (around 97500 projects) for each data point.

4   Results and Discussion

Fig. 2 shows that when each donor can only choose one project to donate, the success rate is approximately 54%, which is 8% – 10% higher than on Kickstarter. A possible explanation for this difference is that the projects selected by the donors in our simulation model were randomly generated, but donors on real crowdfunding websites are usually affected by various dynamic indicators of the projects resulting in an uneven distribution of donations across projects, resulting a lower success rate.

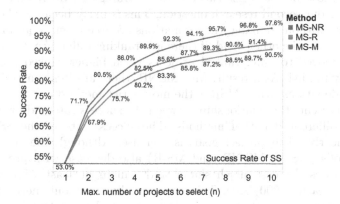

Fig. 2. This figure shows the results of our simulations. Success rate versus maximum number of projects (n). The x-axis is the maximum number of projects ($n = 1$ to 10) that each donor can choose at a time in the simulation, and the y-axis is the success rate across all crowdfunding projects. Color shows details about different methods. The data points are labeled with average success rate (Color figure online).

The success rates of the donation methods (MS-NR, MS-R, and MS-M) are all apparently greater than the baseline (SS) method. This finding answers our first research question, and suggesting that the new donation methods may distribute the donations more effectively, which could benefit more crowdfunding projects.

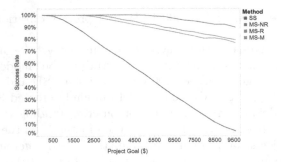

Fig. 3. The trend of success rate with respect to goal (dollars). Color shows details about methods. ($n = 10$)

Figure 2 also answers our third research question, showing that the success rate of our donation methods increases to n. That is, if donors choose more projects, their donations will have more flexibility to invest to other projects when the prior funded projects were failed. According to Fig. 2, as the maximum number of projects (n) a donor can choose increases above 5, the success rate will gradually saturate. This finding may be of interest for future research that explores how to optimize the efficiency of crowdfunding websites and fundraising strategies.

Furthermore, as shown in Fig. 2, the success rates of MS-NR were 3 % – 6 % higher than those of the other two donation strategies (MS-R and MS-RM), which answers our second research question. This is likely because when donors select projects without ranking, their donations are free to move around among those projects. On the other hand, if there is a ranking difference, the donations can only be moved to lower ranked projects when higher ranked projects are failed or fully funded. As a result, with MSNR, there are less constraints when distributing donations, while MSR is the most strict method.

Figure 3 presents the relationship between the success rate and project goals when using different donation methods. The success rate of the SS method decreases linearly when project goals is increase. Although the success rates of our methods (MS-M, MS-NR, and MS-R) also decrease with project goals, the success rates decrease much slower and remain at least 75 % when the project goals reach 10000 USD. This figure implies that our methods do not take away money from the high-goal projects to make low-goal projects succeed, but to rather distribute donations more effectively so that projects with any goal amount all have increased success rates.

Figure 4 shows the relationship between success rate and project duration when using different donation methods. The success rate of the SS method increases linearly with respect to project duration because longer projects have more time to receive donations. Our methods have better success and a more rapid increases with project duration. When the duration of a project is longer than about 22 days, the project success rate will approach 100 % in our system. This may indicate that our method can help crowdfunding projects reach their donation goals more quickly.

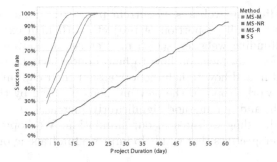

Fig. 4. Success rate versus project duration (day) when $n = 10$ (Color figure online).

In summary, our findings suggest that our system can help crowdfunding websites distribute donations more effectively by enlarging the benefit to more crowdfunding projects.

5 Limitation

Previous studies [2,3] performed an experiment in a lab setting to evaluate the performance and efficiency of their strategies for crowdfunding platforms. However, their experiment designing was far from the real crowdfunding platforms' experience. Therefore, this work constructs an agent-based model to simulate our algorithm in crowdfunding environments, that draws on data from real crowdfunding websites. However, the characteristics of the projects and the donation behavior on different crowdfunding platforms are divergent and difficult to simulate. This is the major limitation of this work, and future work will focus on improving the simulation model of crowdfunding platforms. Specifically, the goals and duration of crowdfunding projects were generated by uniform distributions further research should improve this model to reflect the real situation on each crowdfunding platform. Also, the donor's behavior in the real world will be affected by the quality, status, and other aspects of the projects, and thus would be different from uniform random selection of projects as simulated in our model. Although these limitations are present in our simulation, the comparison of the four methods was done with the same simulation settings. The trends reported in this study would therefore be comparable when adopting more realistic settings.

Besides, in order to deploy our methods to increase the efficiency of crowdfunding websites, it would be good to conduct a large-scale user study to understand the impact of the method on actual donors. How many projects would a real user select? How would they structure their preferences? Will the reallocating feature affect their motivation to donate?

6 Conclusion

Crowdfunding platforms have fundamentally changed the way we fundraise, and there are still lots of avenues to be explored in future research to improve the efficiency of crowdfunding websites. In current platforms, donations may usually be distributed to a small proportion of projects with high exposure and promotion on crowdfunding websites, which may cause an uneven distribution of donations. Therefore, we focus on how to help donors match the projects they may be interested in and how to help donors assign their capital to the projects they prefer. This work proposed the new methods inspired by the algorithm in economic research, and our method significantly increases the success rates of projects on crowdfunding websites in our simulation. We hope this work provides some new perspectives that will help improve fundraising on crowdfunding websites.

References

1. Wash, R.: The value of completing crowding projects. In: Association for the Advancement of Artificial Intelligence, AAAI (2013)
2. Wash, R., Solomon, J.: Coordinating donors on crowdfunding websites. In: Proceedings of CSCW 2014, pp. 38–48. ACM Press (2014)
3. Solomon, J., Ma, W., Wash, R.: Don't wait! How timing affects coordination of crowdfunding donations. In: Proceedings of CSCW 2015, pp. 547–555. ACM Press (2015)
4. Beltran, J.F., Siddique, A., Abouzied, A., Chen, J.: Codo: fundraising with conditional donations. In: Proceedings of UIST 2015, pp. 213–222. ACM Press (2015)
5. Abdulkadiroglu, A., Pathak, P.A., Roth, A.E.: Strategy-proofness versus efficiency in matching with indifferences: redesigning the New York City high school match (No. w14864). National Bureau of Economic Research (2009)
6. Salganik, M.J., Dodds, P.S., Watts, D.J.: Experimental study of inequality and unpredictability in an artificial cultural market. Science 311, 854–856 (2006). American Association for the Advancement of Science
7. Xu, A., Yang, X., Rao, H., Fu, W.T., Huang, S.W., Bailey, B.P.: Show me the money! an analysis of project updates during crowdfunding campaigns. In: Proceedings of the SIGCHI Conference on Human Factors in Computing Systems, CHI 2014 (2014)
8. Kim, J.G., Kong, H.K., Karahalios, K., Fu, W.T., Hong, H.: The power of collective endorsements: credibility factors in medical crowdfunding campaigns. In: Proceedings of the SIGCHI Conference on Human Factors in Computing Systems, CHI 2016 (2016)
9. Mollick, E.: The dynamics of crowdfunding: an exploratory study. J. Bus. Ventur. 29(1), 1–16 (2014)
10. Kickstarter. https://www.kickstarter.com/
11. DonorsChoose. http://www.donorschoose.org/main.html
12. Indiegogo. https://www.indiegogo.com/
13. Giveforward. http://www.giveforward.com/

Formality Identification in Social Media Dialogue

Partha Mukherjee[1]([⊠]) and Bernard J. Jansen[2]

[1] College of Information Science and Technology,
Pennsylvania State University, University Park, USA
pom5109@ist.psu.edu
[2] Qatar Computing Research Institute, HBKU, Doha, Qatar
jjansen@acm.org

Abstract. Researching second screen interactions that form a social soundtrack concerning a major broadcast media event, we perform statistical analysis on more than 800 K postings and 50 K blogs of Super Bowl XLIX on Instagram and Tumblr respectively for three categories (commercials, music and game) during three phrases (*Pre*, *During*, and *Post*) identifying the influence of different social soundtrack features of the postings on formality of contents during three phrases (*Pre*, *During*, and *Post*). For Instagram, the positive influence of URL-based postings in relative scale on formality is significant, but other features have significant negative impact in *Pre* and *Post* phases. For Tumblr, undirected broadcast pattern of conversation and number of sentences in relative scale in *Pre* and *Post* phases have a positive influence on formality. The *During* phase does not show any significant influence between any of the social soundtrack feature and formality of the postings for either Instagram or Tumblr. It is important to note that formality is significantly increased on Instagram, but it exhibits significant reduction on Tumblr. We further evaluate the effects of categories on top of the influence of social interaction features on contents of social media platforms for a fixed effects model. For Instagram's formality aspect, the fixed effects estimate of the game category significantly outperforms the other two categories in all three phases, while for Tumblr, the music category fixed effects plays the lead role in *Pre* and *Post* phases. These results assist in identifying the strength of linkage among broadcast categories, social media postings, and inherent formality, providing insights into viewer reactions to the broadcast of In-Real Life events.

Keywords: Social soundtrack · Formality · Fixed effects · Instagram · Tumblr

1 Introduction

The integration of broadcast media events, mobile technologies, and social media sites has facilitated synergic online interactions that impart feelings of togetherness, information sharing, and conversation among people in dispersed locations, leading to the creation of an online conversation for events and associated content, such as advertising. Viewers exchange information related to the event via second screen devices in terms of posting of social media comments [8]. The use of secondary screens affords the creation

© Springer International Publishing Switzerland 2016
K.S. Xu et al. (Eds.): SBP-BRiMS 2016, LNCS 9708, pp. 13–22, 2016.
DOI: 10.1007/978-3-319-39931-7_2

of what we refer to as the social soundtrack, the online interactions with others regarding broadcast programs, particularly for In-Real Life event (IRL) events such as sporting events and award shows. The effect on the role of the viewer is profound, as the nature of viewership is transiting from a passive to an active one, where the viewer can, to some degree, take action while watching and engaging in an event.

In this research, we consider Super Bowl XLIX as an IRL broadcast media event. We investigate the use of second screens in the *Pre*, *During*, and *Post* phases of Super Bowl XLIX, specifically examining if second screen formalities, or use of proper linguistic terms and syntax, from viewer interactions concerning Super Bowl commercials, game, and music categories are related to the social soundtrack features in each of event's phases. We select Instagram and Tumblr as our data collection sites. The temporal change in patterns used in social media discourse and the quantum of sentences and unique words in the social soundtrack conversation intuitively are the factors for temporal shift in observance of rules of social media etiquette. This intuition motivates our research.

The research is important as changes in language style indicate the credibility and rapport building between people in second screen conversations who do not know each other, which influences the impact of the information shared [6]. The formality of language also has significant impact on how messages are received, and it can be used to identify disparate user groups. Additionally, it is essential to understand how users engage and leverage the affordances of their technology devices for information sharing, which can have a profound effective on areas such as online advertising.

2 Related Work

There has been limited research on formality analysis on social media conversations. Understanding the social soundtrack formality of the conversations can shed light on the goals, needs, and desires of the conversation participants while viewing an event.

Sabater [12] examined Facebook messages of native and non-native English speakers to identify the stylistic variations in their online writing. The result showed that non-native speakers exhibited more formal traits in a university context. In another study on postings of two communities on Twitter, it was found that there was marked difference in formality and tone between the two user groups and the underlying differences of communication goal was cited as the reason [11]. In a separate research using WhatsApp in an university context, Alamri revealed that, over time, instructor discourse became informal [1]. Similar characteristics were found for blogs that claimed to be more informal, and it was shown that the personality of the author influenced the formality of text [9, 10]. Lee, Ham, Kim, and Kim [7] used Twitter as the social media platform to assess people's interest in Super Bowl 2012 car commercials.

None of these prior research studies assessed the temporal interaction effects of social networks and second screens from the temporal strength of linkage between social soundtrack features, patterns, and the content of second screen conversations from a formality perspective concerning live broadcast of major IRL events. Understanding the temporal aspects of formality within the social soundtrack has implications for leveraging social media within a variety of domains, including online marketing and public service communication. Also, much of the previous studies are limited to a single social media platform.

3 Research Question

In our research, we classify second screen interactions appearing in the social soundtrack concerning Super Bowl XLIX into three event categories: (a) commercials, (b) music and (c) game. There is considerable sharing of feelings in the social soundtrack on three aforementioned categories not only during but before and after the event. We label these temporal phases of the social soundtrack as: (a) *Pre* phase, (b) *During* phase and (c) *Post* phase. As we collect data related to Super Bowl XLIX from the 10th of January 2015 and continued till the 24th of February 2015 on Instagram and Tumblr, the *Pre* phase begins on 1/10/2015-00:00:00 and continue till 2/1/2015-18:29:59 (till the start of the kick off). The *During* phase is the period of the live broadcast of the event, i.e., from 2/1/2015-18:30:00 to 2/1/2015-22:30:00. The *Post* phase is the social soundtrack beginning on 2/1/2015-22:30:00 and lasting till 2/24/2015-00:00:00. We amass 811,262 Instagram media posts and 51,569 Tumblr blogs using respective APIs and secret tokens. We chose Instagram and Tumblr as they are major social media platforms with limited investigation in prior work, relative to Twitter.

In this study, we extracted the social soundtrack features in terms of count of posts corresponding to (a) pattern of viewers' conversation, (b) number of sentences in the postings, and (c) number of unique words present in the texts of the posts. The identifiers for categories of social media post patterns are listed in Table 1.

Table 1. Categories of social soundtrack conversation patterns common to the social media platforms

Category	Description
Referral (RF):	Any full length or shortened URL.
Response (RS):	Postings intentionally engaging another user by means of '@' symbol which does not meet the other requirements of containing referrals.
Broadcast (BC):	Undirected statements (i.e., does not contain any addressing) which allow for opinion, statements and random thoughts to be sent to the author's followers. Any undirected statement followed by questions '?' belongs to Broadcast (BC) category.

For Instagram and Tumblr, RF categories may contain the URLs for images and videos in addition to general full length or shortened URLs in Instagram captions and Tumblr blogs. For Instagram captions, and Tumblr blogs, we set the priority order as: RS > RF > BC. The sentences of the posts are parsed based on the punctuations such as ".", "?" and "!". The number of unique words within each posting is determined by excluding the stop words and the hashtags present in the sentences of each post.

We have an intuition that relationship between interaction features present in social media postings and the sentiments extracted from the social soundtrack conversation regarding specific categories changes in phases. The social soundtrack feature-formality linkage will also most likely change in specific phases for different categories. These feature-formality relationship of language shed light on the manner of information processing and dissemination with the social soundtrack.

Based on this intuition, we frame our research question that deals with influence of interaction features on formality of social media conversations in each phase on different categories. The research question is evaluated by linear regression on balanced panel data.

RQ1. Do the features of social soundtrack conversations on different social media platforms affect the formality of social media conversations in each phase?

4   Research Design

We classify the Instagram and Tumblr data into the three categories of second screen interaction. We identify the categories by means of the keywords collected from relevant websites related to Super Bowl commercials [2, 15], Super Bowl music [16], and Super Bowl game [13]. The list of Super Bowl commercial keywords contains the ad titles, titles of the themes / videos for the ads, hashtags associated with the spots (e.g., '#realstrength', '#likeagirl' etc.), and the first and last names of actors participating in Super Bowl commercial videos. The list of Super Bowl music keywords contains the first name and last name of the performers of the halftime and the pre-game show, terms that describes the half time show (e.g., 'shark', 'palm' etc.), and the songs (e.g., 'california girls'). The list of keywords related to Super Bowl game contains the first name and last name of the players, coaches, umpires, referees, commentators, the field positions(e.g., 'quarter-back', 'red zone' etc.), team names and other key terms related to game (e.g., 'punt', 'fumble' etc.). We assign the posts on Instagram and Tumblr to Super Bowl commercials, music, or game categories, depending on the presence of terms from the respective keyword lists. Prior to any analysis, we perform the following pre-processing steps as: (a) remove punctuations from the sentences of the posts, (b) remove the hashtags, as this does not contribute to the frequency of parts of speech (POS) tags, (c) remove the usernames addressed by "@" and "RT" within the messages, (d) remove the special characters such as "@", "RT", "via" and URLs, (e) replace all contraction of verb forms to the corresponding verbs (e.g., "'ll" to "will", "'ve" to "have", "'re" to "are") (f) replace all negations ("neither", "nor", "never", "no", "negative", "not", "n't", "won't" etc.) to "not", (g) replace a sequence of repeated characters by two characters (e.g., "coooool" to "cool", "ooooh" to "ooh" etc.), (f) lowercase the letters and expand the acronyms in the posts to its meaning extracted from relevant resources.

After pre-processing, we use the Stanford POS Tagger to identify the Part of Speech (POS) tags from the tokenized posts. We use the standard tag-set defined by Penn Treebank to identify the tags as element of POS class (i.e., noun, pronoun, verb, etc.) from the tokens by means of Stanford POS Tagger. Formality is expressed as a function of such POS class elements present in a post. We compute the F-score for formality as defined in [5], and thereafter, we calculate the aggregated F-score in five-minute intervals in phase-category space, used as the unit of analysis for testing the research hypotheses. Higher F-score of formality indicates more careful and less casual social soundtrack conversations.

We further segregate the posts regarding volume of posts, patterns of conversations exist in posts, number of sentences present in the posts and number of unique words in sentences of posts into five minutes intervals for *Pre*, *During* and *Post* phases. We have an intuition that the volume, sentence and unique words will affect formality score, as those attributes are functions of the social media texts. We transform the five minute time count data regarding volume, each of the social features and the derived formality score into a relative scale using equation: $rel\_value_j^i = Score_j^i / Max_i \{Score_{ij}^i\}$, i denotes the index of the five minute time slot within a specific phase and j is the specific attribute of the posting. Score denotes the values of attributes. *Max* function returns the highest value of the count for a specific attribute within a phase. Here, the attributes are social features (i.e., volume of posts, each of the conversation patterns, sentences, and unique words) and quantized formality.

Once the computation of relative scaling of the attributes is done for the social soundtrack conversation, we organize the categorical time count data into a balanced panel [4] data for both the social media platforms, where each of the three categories has relative values of the social soundtrack attributes across total number of five minute slots for data collection in each phase. Panel data, also known as cross-sectional time series data, can control for variables whose behavior cannot be observed (i.e., behavior of Super Bowl categories). In our study, for each phase, the balanced panel dataset can be viewed as a three dimensional space where the dimensions are (1) Super Bowl categories (commercial, music, game), (2) time stamps for each category (number of five minute time slots i.e., 6558, 49, and 6534 for *Pre, During* and *Post* phases respectively), and (3) social media platforms (Instagram and Tumblr). In each social media dataset, we have a total of 19674 (3 × 6558), 147 (3 × 49) and 19062 (3 × 6354) records each with relative values of attributes of posts for *Pre, During* and *Post* phases, respectively. Each record is the unit of analysis in evaluating the research question for each phase in two different social media platforms.

5   Methodology

We use panel data regression with fixed effects [3, 14] on balanced feature-formality panel data to evaluate the relationship between the features of social soundtrack conversations and formality concerning Super Bowl categories. In the regression model, relative formality scores data is the dependent variable, while the relative values of social soundtrack features (i.e., volume, patterns of social soundtrack conversations from Table 1, number of sentences, and number of unique words) are the cofactors. We conduct the fixed effects regression model on the panel data. The fixed effects model assumes that individual specific effect is correlated with the independent variable. We set the Super Bowl commercials as the baseline category for finding relative categorical effect in the fixed effects model. We are estimating the pure effect of social soundtrack features by controlling the unobserved heterogeneity with the addition of dummy variables for each category in the fixed effects model.

6   Result

The estimates of regression coefficients of the social soundtrack features (cofactors) identify how much second screen formality changes over time on average per category when the respective cofactor is increased by one unit. In the fixed effects regression model, we also evaluate the effect of categories (i.e. unobserved variable) on the linkage between social soundtrack features (cofactors) and the quantified formalities (response) in each phase on Instagram and Tumblr.

6.1   Instagram

From Table 2, we find that formality increases by 10.76 times in relative scale with one unit increase of URL-based captions (RF) in the *Pre* phase for Instagram; however, the unit increase of other cofactors reduces the formality significantly. In the *During* phase, the coefficients of the majority of cofactors are large and positive except sentences and unique words; however, they are not significant (p-value > 0.05) in measuring the effect of the cofactors on the formality of the Instagram captions. In the *Post* phase, formality increases 3.73 times with a unit increase of referral (RF) pattern. It is important to note that increased number of postings with captions increases formality significantly in *Pre* and *Post* phases, while in *During* phase the effect is also positive but not significant (i.e., p-value > 0.05). The variance explained (R^2) in *Pre* and *Post* phases in Instagram is lower than that explained in *During* phase.

Table 2. Fixed effects model results for Instagram

Phase	Cofactors	Coeff	p-value	R^2
Pre	**volume**	**2.834**	**0.011***	0.23
	mention	−4.173	0.057	
	referral	**10.758**	**1.4e-05***	
	broadcast	**−5.720**	**0.009***	
	sentences	**−3.349**	**7.6e-13***	
	unique words	**−2.845**	**0.021***	
During	volume	18.262	0.141	0.57
	mention	39.692	0.662	
	referral	24.200	0.785	
	broadcast	27.728	0.758	
	sentences	−13.142	0.083	
	unique words	−25.610	0.175	
Post	volume	1.967	0.091	0.25
	mention	**−8.815**	**1.4e-06***	
	referral	**3.731**	**0.010***	
	broadcast	**−8.519**	**2.8e-06***	
	sentences	**−3.890**	**2.2e-16***	
	unique words	**−8.543**	**9.9e-08***	

Figure 1 depicts the effects of music and game categories on linkage between social soundtrack features and formality on Instagram. It is seen from Fig. 1 that music and game categories have significant increased effect on feature-formality linkage relative to Super bowl commercials in all three phases. So, among three categories, the fixed effects of Super Bowl commercials is least, while the fixed effects of game category is the highest on the relationship between social soundtrack feature and formality of the content in all phases for Instagram.

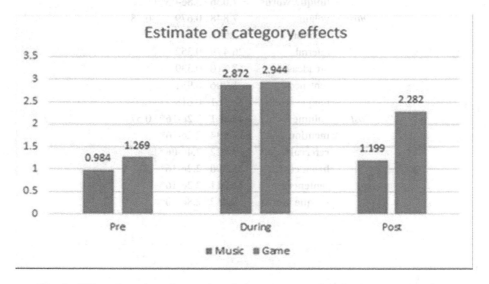

Fig. 1. Effect of music and game in relation to commercials for Instagram by phase

6.2   Tumblr

We present the results of fixed effects regression model done on the panel data for Tumblr in all three phases in Table 3. We find that formality significantly increases by 19 times and 41.19 times in relative scale for a one unit increase of undirected broadcast (BC) pattern and number of sentences respectively in the *Pre* phase for Tumblr, while for unit increase of other cofactors the formality reduces significantly. In the *During* phase, the coefficients of the cofactors are large and positive, but they are not significant (p-value > 0.05) in measuring the effect of the cofactors on the formality of the Tumblr blogs. In *Post* phase, formality increases 25.7 times and 43.21 times with unit increase of undirected broadcast (BC) pattern and number of sentences, respectively. It is important to note that the increased number of Tumblr postings reduces formality significantly in *Pre* and *Post* phases, while in *During* phase the effect is positive but not significant (i.e., p-value > 0.05). Unlike Instagram, variance explained (R^2) in Tumblr by the model in *During* phase is lower than that explained in *Pre* and *Post* phases.

Figure 2 depicts the effects of music and game categories on linkage between social soundtrack features and formality on Tumblr. It is seen from Fig. 2 that music and

Table 3. Fixed effects model results for Tumblr

Phase	Cofactors	Coeff	p-value	R^2
Pre	**volume**	**−44.977**	**2.2e-16***	0.49
	mention	**−14.988**	**1.2e-14***	
	referral	**−7.306**	**0.000***	
	broadcast	**19.034**	**2.2e-16***	
	sentences	**41.195**	**2.2e-16***	
	unique words	**−7.036**	**3.8e-08***	
During	volume	7.848	0.679	0.28
	mention	12.579	0.647	
	referral	26.479	0.352	
	broadcast	17.810	0.539	
	sentences	1.796	0.931	
	unique words	8.161	0.618	
Post	**volume**	**−43.561**	**2.2e-16***	0.57
	mention	**−19.534**	**2.2e-16***	
	referral	**−9.922**	**7.4e-06***	
	broadcast	**25.700**	**2.2e-16***	
	sentences	**43.211**	**2.2e-16***	
	unique words	**−8.462**	**2.8e-10***	

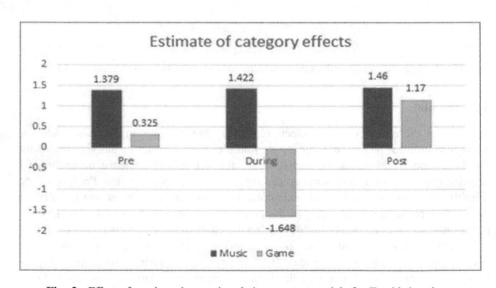

Fig. 2. Effect of music and game in relation to commercials for Tumblr by phase

game categories have significant increased effect on feature-formality linkage relative to Super Bowl commercials in the *Post* phase. In *Pre*, though relative fixed effects of music and game categories are positive, the effects of game in *Pre* phase is not significant (p-value > 0.05). In the *During* phase, there is no significant effect of music and game categories relative to commercials, though the effect of game is least in the *During* phase. The fixed effects of Super Bowl music category is most pronounced on the relationship between social soundtrack feature and formality of the content in all phases for Tumblr.

7 Discussion and Implication

The variation of conversation pattern based formality on Instagram and Tumblr facilitates to identify the demography of viewers from the change in stylometric variation for categories in different phases. It is also observed that game category has higher estimates of fixed effects on Instagram feature-formality relationship, while the impact of commercials is the lowest. This informality inherent in commercial related posts increase the personalization of the brands and marketing, allowing brands to communicate with viewers in a manner they are comfortable with via social media platforms. In *Pre* and *Post* phases for Instagram, referral or URL based recommendations (RF) pattern has the positive influence on social soundtrack formality, while the other patterns have significant negative influence (see Table 2). For Tumblr, the blogs with more undirected broadcast patterns become more formal in *Pre* and *Post* phases, while blogs containing more of other conversation patterns become less formal (see Table 3). From a feature-formality influence perspective, the relative volume of postings has positive correlation with Instagram formality, while for Tumblr it is negative (see Tables 2 and 3). The feature-formality relationship is insignificant in *During* phase, while the R^2 value is higher for Instagram. This is because of the huge difference in the sample sizes (i.e., 6500 for the *Pre-* and *Post-* phases, only 49 for the *During*). From category effect perspective, for Instagram game category outperforms other two categories in the strength of feature-formality linkage in all three phases. In Tumblr, the music category plays the lead role. This seems to infer that the media based posts that contain URLs in Tumblr are more informal relative to Instagram. It is also observed that game category has higher estimates of fixed effects on Instagram feature-formality relationship, while the impact of commercials is the lowest. This informality inherent in commercial-related posts increases the personalization of the brands, allowing brands to communicate with viewers in a manner they are comfortable with via social media platforms.

8 Conclusion

Our research provides contributions concerning understanding user behavior and interaction in terms of the shift in users' temporal formality concerning effects of categories treated as unobserved variables in the IRL event in a traditional formality computation framework. In future research, we will analyze the relationship between temporal informality of posts and different features of second screen interaction taking care of idiosyncrasies of IRL event related social media texts on more social media platforms with the estimates of fixed and random models, comparing the results with formality settings presented in this research.

References

1. Alamri, J.: (In) formality in social media discourse: The case of instructors and students in Saudi higher education. In: Proceedings of Global Learn (AACE), pp. 101–108 (2015)
2. Anonymous: 2015 Super Bowl commercials (2015). http://www.superbowl-commercials. org/2015
3. Bell, A., Jones, K.: Explaining fixed effects: Random effects modeling of time-series cross-sectional and panel data. Polit. Sci. Res. Methods **3**(01), 133–153 (2015)
4. Berrington, A., Smith, P., Sturgis, P.: An overview of methods for the analysis of panel data. NCRM Methods Review Papers. NCRM/007, pp. 1–57 (2006)
5. Heylighen, F., Dewaele, J.-M.: Formality of language: definition, measurement and behavioral determinants. Interner Bericht, Center "Leo Apostel", Vrije Universiteit Brüssel (1999)
6. Jansen, B.J., Sobel, K., Cook, G.: Classifying ecommerce information sharing behaviour by youths on social networking sites. J. Inf. Sci. **37**(2), 120–136 (2011)
7. Lee, H., Kim, Y.K., Kim, K.K., Han, Y.: Sports and social media: Twitter usage patterns during the 2013 super bowl broadcast. In: Proceedings of the International Conference on Communication, Media, Technology and Design, (ICCMTD), pp. 250–259 (2014)
8. Mukherjee, P., Jansen, B.J.: Social TV and the social soundtrack: significance of second screen interaction during television viewing. In: Kennedy, W.G., Agarwal, N., Yang, S J. (eds.) SBP 2014. LNCS, vol. 8393, pp. 317–324. Springer, Heidelberg (2014)
9. Nowson, S.: The Language of Weblogs: A study of genre and individual differences. Doctoral Thesis. University of Edinburgh (2006)
10. Nowson, S., Oberlander, J., Gill, A.J.: Weblogs, genres and individual differences. In: Proceedings of the 27th Annual Conference of the Cognitive Science Society (CogSci), pp. 1666–1671 (2005)
11. Paris, C., Thomas, P., Wan, S.: Differences in language and style between two social media communities. In: Proceedings of the AAAI International Conference on Weblogs and Social Media (ICWSM) (2012)
12. Pérez-Sabater, C.: The linguistics of social networking: A study of writing conventions on facebook. Linguistik Online **56**(6), pp. 111–130 (2013)
13. Schalter, T.: Super Bowl XLIX: Power ranking the top 25 players in this year's game. http:// bleacherreport.com/articles/2343013-super-bowl-xlix-power-ranking-the-top-25-players-in-this-years-game/page/2
14. Schmidheiny, K., Basel, U.: Panel data: fixed and random effects. Short Guides to Microeconometrics, pp. 1–16 (2011)
15. Staff, A.A.: Super Bowl XLIX ad chart: Who bought commercials in Super Bowl (2015). http://adage.com/article/special-report-super-bowl/super-bowl-xlix-ad-chart-buying-big-game-commercials/295841/
16. Wikipedia: Super Bowl XLIX halftime show (2015). http://en.wikipedia.org/wiki/Super_Bowl_XLIX_halftime_show

Modeling and Simulation of Sectarian Tensions in Split Communities

Christopher Thron[✉] and Rachel McCoy

Department of Science and Mathematics, Texas A&M University-Central Texas,
1001 Leadership Place, Killeen, TX 76549, USA
thron@tamuct.edu

Abstract. We present a simple agent-based model for the evolution of between-group attitudes (measured on a linear scale) in a community that is divided between two distinct social groups (which may be distinguished by religion, ethnicity, etc.). We derive approximate analytical equations to predict the change in mean attitudes over time given certain assumptions. The model predicts that social pressures cause each group to tend towards extremes of hostility or acceptance towards the other group. Under some conditions, groups can have stable extremist and moderate factions, but very small changes in system parameters can upset the stability. Interpersonal cohesion (the degree to which individuals influence each other's opinions) plays a significant role in controlling within-group polarization. We show that strategies to improve intergroup relations that target subsets of each group are much less effective when cohesion is low.

Keywords: Sectarian · Tension · Affinity · Hostility · Conflict · Reconciliation · Opinion dynamics · Agent-based model · Mathematical model

1 Purpose and Scope

Many local communities, especially in the Third World, are markedly divided between two or more very distinct ethnic or religious groups. In sub-Saharan Africa for instance, many villages, towns, and cities comprise distinct groups of "Christian" and "Muslim" residents. In some cases, residential neighborhoods are completely integrated; in others, districts or quarters belong to one or the other religion. These situations may or may not be accompanied by interreligious tensions. For examples and references see [1].

This paper proposes and analyzes a heuristic mathematical model of intergroup relations within a community consisting of two distinct groups clearly identified by religion or ethnicity. We shall use our model to address the following questions:

- What stable distributions of affinity/hostility are possible for different model conditions?
- Do stable distributions vary continuously with changes in conditions, or are there cases where small changes in conditions produce large differences in the distribution of affinities/animosities?

© Springer International Publishing Switzerland 2016
K.S. Xu et al. (Eds.): SBP-BRiMS 2016, LNCS 9708, pp. 23–32, 2016.
DOI: 10.1007/978-3-319-39931-7_3

- What strategies that target model conditions are sufficient to bring about desired changes in intergroup attitudes?
- What model conditions may produce deterioration in intergroup attitudes?

The remainder of the paper is structured as follows. In Sect. 2 we survey the current literature on mathematical models of intergroup violence, and we discuss examples of programs designed to mitigate tensions between divided communities. In Sect. 3 we present the model's assumptions and its mathematical specification. In Sect. 4, we provide mathematical analyses of the model under various particular conditions. In Sect. 5 we present several characteristic simulations and their implications for model behavior. In Sect. 6 we summarize our findings, draw conclusions, and list references.

2 Background

In this section we give a brief survey of relevant literature. A more extensive review is given in [1].

Numerous empirical studies exist of factional conflicts on a national scale, including civil wars. Several of these studies are summarized in [2].

Several researchers have used spatial agent-based models in an attempt to understand civil violence, beginning with Epstein [3]. In such models, agents are located on a grid, and exert influence on neighboring agents by causing them to move or change state. More sophisticated spatial agent-based models that include graph- and game-theoretic components have also been proposed, for example in [4].

Other researchers have used nonspatial models to study opinion dynamics and cluster formation within a given population [5]. In these models, agents' opinions is measured on a linear scale, and these opinions change as a result of interactions between agents (which may or may not be binary).

Several non-governmental organizations have worked to reduce tensions in divided communities through sports programs [6]. Such programs promote positive interactions between particular subgroups drawn from antagonistic groups.

Our model includes elements from several of the above sources. Like the spatial models of civil violence, we model the antagonistic interactions of distinct populations. Like the opinion dynamics models our model is nonspatial, and we measure attitudes on a linear scale. Also, taking a cue from sports programs we consider the effects of directed positive interaction between subgroups from each population.

3 Model Specification

Our mathematical model is based on assumptions concerning human interactions that are supported by sociological research. An extended list of supporting references is given in [1]. We first state these assumptions, and then provide corresponding mathematical characterizations.

3.1   Model Assumptions

The following principles are assumed to govern interpersonal interactions in a community consisting of two strongly distinct social groups.

1. Individuals within each group have varying degrees of affinity towards the other group, which can be ranked on a linear scale. On the one end are the extremists, who despise and avoid the other group and are prone to inciting violence. On the other end are the moderates, who treat individuals from the other group on an almost equal basis. This assumption conforms to the common practice of opinion dynamics models described above.
2. Constructive (cooperative) interactions between individuals in different groups tend to improve their affinities for each other, due to the mutual benefit derived from the interaction. Such contacts may occur through daily commerce, education, community development, sports, and so on.
3. Isolation and lack of contact between groups tend to increase hostility: separatism leads to rumor-mongering, mistrust, and misunderstanding. On the other hand, more contact leads to more accommodation.
4. Individuals that are more hostile towards the other group are more likely to avoid and discourage interactions with the other group.
5. In interpersonal interactions, individuals tend to influence other individuals towards their own opinion.

3.2   Mathematical Formulation

The above assumptions have been translated into mathematical form as follows:

1. The community consists of two groups of equal size, each group modeled as a set of N agents.
2. Each individual in each group has an *affinity* towards the other group, which is a number between 0 (representing extreme hostility) and 1 (meaning the individual treats both groups equally). We denote the affinity of agent A as $a(A)$.
3. Over the course of the simulation, individuals interact with each other, and their affinities change as a result of the interaction. For simplicity, we assume that individual interactions involve just two agents, and that interactions occur sequentially. Each interaction is modeled as follows:
   (a) Randomly choose a random agent A_1 who participates in the interaction.
   (b) Randomly choose a second agent A_2 as follows. With probability $1 - a(A_1)/2$, A_2 is in the same group as A_1. Otherwise, A_2 is chosen from the other group. (This rule reflects assumption (4), since it implies that the frequency of interaction with the other group is positively correlated with affinity.) Note that $a(A_1) = 1$ implies that A_2 is chosen from either group with equal probability, while $a(A_1) = 0$ means that A_2 is always from the same group as A_1.
   (c) Change the two agents' affinities based on the interaction, as follows.

If the two agents are in the same group, then:

$$a(A_1) \to a(A_1) - b_1 + c \cdot h(a(A_2) - a(A_1)) + \sigma v_1;$$
$$a(A_2) \to a(A_2) - b_1 - c \cdot h(a(A_2) - a(A_1)) + \sigma v_2.$$

If the two agents are in different groups, then:

$$a(A_1) \to a(A_1) + b_2 + c \cdot h(a(A_2) - a(A_1)) + \sigma v_1;$$
$$a(A_2) \to a(A_2) + b_2 - c \cdot h(a(A_2) - a(A_1)) + \sigma v_2.$$

In these equations, b_1 denotes the (negative) drift in affinity due to a single same-group interaction; b_2 denotes the (positive) drift in affinity due to a single inter-group interaction; c denotes the strength of "cohesion", that is the tendency of two interacting agents to influence each other towards their own opinion; σ is a noise variance; v_1, v_2 are independent, identically distributed normal random variables with mean 0 and variance 1; and h is an "influence function" which expresses the effect of the affinity difference between two individuals on the affinity adjustment resulting from interaction between those individuals. Here we take $h(x) = x$: see [1] for other cases.

Since affinities are limited to lie in the range [0,1]; the values of a generated by the above two equations are clipped to lie in this range: $a \to \max(0, \min(1, a))$.

The agent selection rule 3(b) corresponds to model assumption (4). In the equations in 3(c), The b_1 and b_2 terms reflect assumptions (2) and (3) respectively. The c terms reflect assumption (5), which implies that interactions between individuals tend to bring those individuals' opinions closer together.

4 Theoretical Analysis

We may gain considerable insight into the model behavior by considering two special cases. These cases also have practical significance, which we will elucidate below. In both cases, we are interested in changes in composition over time of each group. In particular, we consider how the percentages of extremists and moderates changes over time; and we investigate how conditions (represented by model parameters) may be changed to tilt the balance towards increased or reduced affinity.

The special cases that we will consider are as follows:

Case 1. All individuals in each group begin with the same affinity (although the two groups' affinities may differ). In this case, we are particularly interested in the eventual steady-state affinities of the groups, depending on the initial affinities.

Case 2. Both groups are initially divided into extremist (affinity = 0) and moderate (affinity = 1) factions. In this case, we are interested in the stability of configurations consisting of given proportions of extremist and moderate factions.

In this section, we provide approximate mathematical results for Cases 1 and 2. These theoretical results are compared with simulations presented in the following section. Full derivations of the following theoretical results are given in [1].

4.1   Case 1: Unanimous Starting Affinities for Each Group

In order to facilitate the analysis, we make the following simplifying assumption:

> *"Sticky assumption"*: If all agents in a particular group start out with the same affinity, then they will maintain nearly the same affinity. In other words, once the distribution of affinities for a particular group has coalesced to a single value, all the affinities move together.

The sticky assumption is never strictly accurate, but in some cases it is approximately true, and it does have the advantage that it enables an exact solution. Furthermore, we shall find in Sect. 5.1 that it gives accurate indications of the evolution of distribution medians, even when within-group affinities do in fact diverge.

Given that the sticky assumption holds, we let $x_1^{(0)}$ and $x_2^{(0)}$ be the mean starting affinities for the two different groups. Then in the case where $\sigma = 0$, the formula for the mean affinities after m steps (denoted $x_1^{(m)}$ and $x_2^{(m)}$ respectively are:

$$x_1^{(m)} = e^{2\delta(1-2k\gamma)m} \cdot \exp(-\theta[e^{(1+\beta)\delta m}-1])q_1 + e^{2(1+\beta)\delta m}q_2 + k;$$

$$x_2^{(m)} = -e^{2\delta(1-2k\gamma)m} \cdot \exp(-\theta[e^{(1+\beta)\delta m}-1])q_1 + e^{2(1+\beta)\delta m}q_2 + k,$$

where

$$\beta \equiv b_2/b_1; \gamma \equiv c/b_1; \delta \equiv 0.25b_1/N; k \equiv 2/(1+\beta);$$

$$q_1 \equiv \left(x_1^{(0)} - x_2^{(0)}\right)/2; q_2 \equiv \left(x_1^{(0)} + x_2^{(0)}\right)/2 - k.$$

4.2   Case 2: Polarized Starting Distributions

Rather than starting with unanimous groups as in Case 1, we may also consider the case where each group begins with extremist and moderate factions with affinities 0 and 1, respectively. In the following, we consider the question of when such polarized-group distributions can remain stable.

The analysis is similar to the unanimous case, except that group j ($j = 1,2$) is divided into two subgroups ja and jb which consist of extremists and moderates, respectively. Initially, these subgroups' affinities are $x_{ja} = 0$ and $x_{jb} = 1$, and contain fractions $1-f_j$ and f_j respectively of the agents in Group j. Under the sticky assumption, the following equations may be derived for the evolution of the subgroups' affinities:

$$x_{ja}^{(k+1)} \approx x_{ja}^{(k)} + (1/4N)\left(1-f_j\right)\{(b_1 + 3c)f_j + (b_2 + c)f_{3-j} - 4b_1\};$$

$$x_{jb}^{(k+1)} \approx x_{jb}^{(k)} + (1/4N)f_j\{(b_1 + 3c)f_j + (b_2 + c)f_{3-j} - 3b_1 + b_{2-4}c\}.$$

The conditions that $x_{ja}^{(k+1)} \leq x_{ja}^{(k)}$ and $x_{jb}^{(k+1)} \geq x_{jb}^{(k)}$ lead to the following conditions for stability of polarized groups:

$$c \leq (b_2 + b_1) / 4 \text{ and } b_2 \geq b_1.$$

5   Simulations

In this section, we describe computer simulations that confirm our theoretical analysis, and further elucidate the general behavior of the model. Matlab/Octave code for the simulations is available at the CoMSES Computational Model Library [7].

5.1   Unanimous-Group Starting Conditions (Case 1)

In order to verify the accuracy of the "sticky assumption", we tracked the behavior of simulations where all individuals in Group j starts with equal affinity x_{j0} ($j = 1,2$): this condition corresponds to Case 1 described in Sect. 5.1. The entire region $(x_{10}, x_{20}) \in [0,1]^2$ was sampled, and the system parameters used are summarized in Table 1. Parameters are given in terms of b_1 because the system's behavior depends on the ratios b_2/b_1, c/b_1, and σ/b_1, and is nearly independent of the value of b_1.

Table 1. Simulation parameters for Case 1 (unanimous starting affinities)

Symbol	Significance	Value(s)
b_1	Negative drift from within-group interactions	0.002
b_2	Positive drift from intergroup interactions	$3b_1$
c	Cohesion parameter	$0.25b_1$, $4b_1$
N	Number of agents in each group	250
–	Number of iterations	$500\ N$
σ	Noise parameter in individual interactions	b_1

Figure 1 shows system trajectories under different starting conditions, for the large-cohesion case ($c = 4b_1$, left) and the small-cohesion case ($c = 0.25b_1$, right). The state space is the unit square, which we have divided into quadrants with dotted lines $x_1 = x_2$ and $x_1 = 1-x_2$. The system is symmetric about the $x_1 = x_2$ line since the two groups have identical parameters. Asterisks indicate the different starting points used for different simulation scenarios: the x_j coordinate corresponds to the starting affinity value for all members of Group j ($j = 1, 2$). Solid lines show the subsequent trajectories predicted by the "sticky model" equations derived above. Dotted red lines show the trajectories followed by groups' median affinities in the agent-based simulations. Also shown for selected starting points are $(10^{th}, 10^{th})$ (blue lines with 'o' markers) and $(90^{th}, 90^{th})$ percentiles (green lines with '+' markers) of affinities for (Group 1, Group 2) in the simulations.

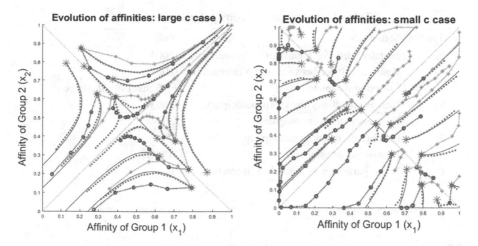

Fig. 1. Comparison of nonlinear "sticky model" predictions with simulations in which each group initially has a single characteristic affinity. See the text for detailed explanation of trajectories and markings. (Color figure online)

In both the small-c and the large-c case, the agent-based model medians follow quite closely the theoretical sticky model trajectories. In the large c case, the 10th and 90th percentile trajectories remain close to the median trajectories, which implies that the sticky assumption is accurate. In the small c case the tracking is somewhat looser, especially in cases where the two groups' initial affinities are substantially different.

In most case, the system tends either towards all extremist $(0, 0)$ or all moderate $(1, 1)$. In these cases, the eventual fate is extremist (resp. moderate) if (x_1, x_2) is above (resp. below) the line $x_1 + x_2 = 4/(1 + b_2/b_1)$. If the initial position lies exactly on this line, in the small c case is it possible for groups to split into extremist and moderate factions. Since the size of the cohesion parameter c reflects the degree to which interacting individuals influence each other, it stands to reason that when c is large, the two groups should end up with the same overall affinity.

Sports program simulations. We may simulate sports programs such as described in Sect. 2 by modifying the model to include occasional "sporting events" which involve a select group of agents, half from each group. Sporting events occur randomly at a given frequency, and each sporting effect improves the affinity of all participants by a determined amount b_{2+}. Parameters for the simulations are shown in Table 2.

Figure 2 shows the equilibrium proportion of moderates (in grayscale) as a function of the initial affinity (all individuals in both groups begin with the same affinity) and fraction of sports program participants from each group. The diagrams show the low-cohesion case ($c = b_1/3$) at left and the high-cohesion case ($c = 3b_1$) at right. When the initial unanimous affinity falls below a certain threshold the effectiveness of sports programs is limited to program participants only; but if the initial unanimous affinity is above the threshold, the influence of the program is propagated throughout both groups, and virtually 100 % of both groups attain moderate affinities. The threshold decreases gradually as the proportion of participants increases. The thresholds for the large-c case

Table 2. Parameters for sports program simulations

Symbol	Significance	Value
b_1	Negative drift for within-group interactions	0.003
b_2	Positive drift for between-group interactions	0.007
c	Cohesion parameter	$b_1/3$(small c); $3b_1$(large c)
b_{2+}	Increase in affinity for sports participants	0.003
–	Frequency of sports programs	0.01
N	Number of agents in each group	250
–	Number of iterations	2000 N
σ	Noise parameter in individual interactions	0

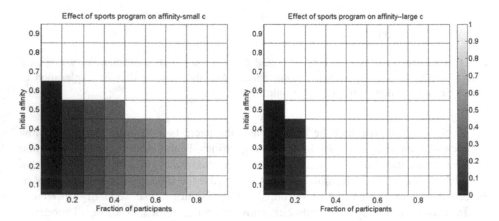

Fig. 2. Equilibrium proportions of moderates (affinity > 0.9) for communities with simulated sports programs for $c = b_1/3$ *(left)* and $c = 3b_1$ *(right)*.

are much lower than for the small-c case, and only exist when the rate of participation is under about 20 percent. In all other cases, moderation pervades the entire population.

These results point to the effectiveness of sports programs (or similar programs designed to produce positive interactions between subgroups) when the relational situation has not degenerated too seriously. The effectiveness of such programs is enormously enhanced when social cohesion is elevated.

Polarized starting conditions (Case 2). Simulations were performed for the case where both groups are initially divided between extremist (affinity = 0) and moderate (affinity = 1) fractions. Parameters used in the simulation are listed in Table 3.

Simulation results are depicted in Fig. 3 *(left)* and *(right)*, which show the equilibrium proportion of moderates (affinity > 0.9) in Groups 1 and 2 respectively, as a function of different (f_1, f_2) starting configurations. The black regions correspond to initial (f_1, f_2) configurations that eventually lead to all extremists; while the white regions correspond to initial configurations that eventually lead to all moderates. The intermediate region give stable (f_1, f_2) pairs. The region of (f_1, f_2) stability for the agent-based model is somewhat smaller than theoretical predictions derived in [1]

Table 3. Parameters for polarized-group simulations (Case 2)

Symbol	Significance	Value
b_1	Negative drift for within-group interactions	0.003
b_2	Positive drift for between-group interactions	0.007
C	Cohesion parameter	0.001
N	Number of agents in each group	250
–	Number of iterations	4000 N
σ	Noise parameter in individual interactions	0.003

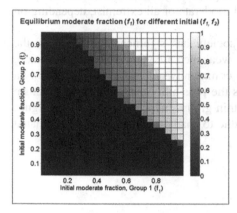 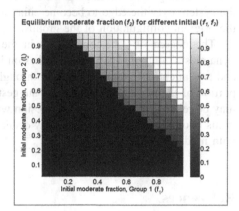

Fig. 3. Equilibrium moderate fractions for Group 1 (*left*) and Group 2 (*right*) as a function of starting (Group 1, Group 2) moderate fractions in Case 2. The equilibrium moderate fractions are indicated by the greyscales at right.

based on the sticky assumption, reflecting the fact that the agent-based dynamics produces fluctuations that undermine the stability of the system.

6 Discussion and Conclusions

The theoretical and simulation results presented above have potentially important implications for policy decisions for dealing with divided communities with the potential for intergroup violence.

Simulations show that the ratio of positive between-group drift to negative within-group drift (b_2/b_1) is critical in determining the fate of the divided community. Interventions should focus on reducing b_1 and increasing b_2. There is a minimum level of effective intervention, below which the situation will progressively deteriorate and above which the situation will eventually achieve universal moderation. Early intervention is required, because larger changes in b_1 and b_2 are required to reverse trends as populations become more extreme. In practice, positive drift may be enhanced by mutual educational, economic, cultural and/or social advantages derived from intergroup interactions.

Both theory and simulations point out the unstable nature of the system. Social pressures produce polarization, and individuals either tend increasingly towards moderation or extremism. The middle ground between the two polar extremes eventually vanishes.

Simulations indicate the effectiveness of sports programs when relationships between groups have not seriously degenerated. But if animosity is already strong, sports programs may actually worsen the situation by singling out participants and creating distance between them and others in their own community.

It is possible for a divided population to have stable moderate and extremist factions in each group. But in some cases a very slight change in overall affinity can push the system over the edge to become either all moderate or all extremist, depending on the original configuration of the groups.

The degree of cohesion present in the society has significant influence on the dynamics. Communities in which cohesion is weakened are vulnerable to polarization, and are much more difficult to reach through community outreaches that target only a portion of the population. Our study suggests the possibility that increased extremism may be related to reduced social cohesion within groups, which Robert Putnam argues in his book "Bowling Alone" [8] may occur due to changes in socialization associated with modernization.

References

1. Thron, C., McCoy, R.: Affinity and Hostility in Divided Communities: A Mathematical Model (2015). http://dx.doi.org/10.2139/ssrn.2687353
2. Dixon, J.: What causes civil wars? Integrating quantitative research findings. Int. Stud. Rev. **11**(4), 707–735 (2009)
3. Epstein, J.M.: Modeling civil violence: an agent-based computational approach. Proc. Natl. Acad. Sci. **99**(suppl. 3), 7243–7250 (2002)
4. Luo, L., Chakraborty, N., Sycara, K.: An evolutionary game-theoretic model for ethno-religious conflicts between two groups. Comput. Math. Organ. Theory **17**(4), 379–401 (2011)
5. Lorenz, J.: Continuous opinion dynamics under bounded confidence: A survey. Int. J. Mod. Phy. C **18**(12), 1819–1838 (2007)
6. Selliaas, A.: Can sport start reconciliation between peoples and states in conflict? Play the Game Conference, Reykjavik (2007). (Archived by WebCite® at http://www.webcitation.org/6cqn7foAI)
7. Thron, C.: Affinity/Hostility in Divided Communities, Version 1. CoMSES Computational Model Library (2015). https://www.openabm.org/model/4780/version/1
8. Putnam, R.D.: Bowling Alone: The Collapse and Revival of American Community. Simon & Schuster, New York (2000)

The Role of Reciprocity and Directionality of Friendship Ties in Promoting Behavioral Change

Abdullah Almaatouq[✉], Laura Radaelli, Alex Pentland, and Erez Shmueli

Massachusetts Institute of Technology, 77 Massachusetts Ave,
Cambridge, MA 02139, USA
{amaatouq,shmueli}@mit.edu

Abstract. Friendship is a fundamental characteristic of human beings and usually assumed to be reciprocal in nature. Despite this common expectation, in reality, not all friendships by default are reciprocal nor created equal. Here, we show that reciprocated friendships are more intimate and they are substantially different from those that are not. We examine the role of reciprocal ties in inducing more effective peer pressure in a cooperative arrangements setting and find that the directionality of friendship ties can significantly limit the ability to persuade others to act. Specifically, we observe a higher behavioral change and more effective peer-influence when subjects shared reciprocal ties with their peers compared to sharing unilateral ones. Moreover, through spreading process simulation, we find that although unilateral ties diffuse behaviors across communities, reciprocal ties play more important role at the early stages of the diffusion process.

Keywords: Social networks · Contagion · Adoption · Reciprocity

1 Introduction

Friendship is a fundamental characteristic of human relationships and individuals generally presume it to be reciprocal in nature. Despite this common expectation [13], in reality not all friendships are reciprocal [13,15]. The implications of friendships on an individual's behavior depend as much on the identity of his friends as on the quality of friendships [14]. Among qualities of a relationship, reciprocity can substantially differentiate a friendship from many others. It is reasonable to think that relationships that are reciprocated are substantially different from those that are not [14].

Moreover, in recent years, peer-support programs are emerging as highly effective and empowering ways to leverage peer influence to support behavioral change [10]. One specific type of peer-support programs is the "buddy system," in which individuals are paired with another person (i.e., a buddy) who has the responsibility to support their attempt to change their behavior. Such a system

© Springer International Publishing Switzerland 2016
K.S. Xu et al. (Eds.): SBP-BRiMS 2016, LNCS 9708, pp. 33–41, 2016.
DOI: 10.1007/978-3-319-39931-7_4

has been used to shape people's behavior in various domains including smoking cessation [19], weight loss [22], diabetes management or alcohol misuse [21].

Consequently, the need to understand the factors that impact the level of influence individuals exert on one another is of great practical importance. Recent studies have investigated how the effectiveness of peer influence is affected by different social and structural network properties, such as clustering of ties [5], similarity between social contacts [6], and the strength of ties [4]. However, how the effectiveness of social influence is affected by the reciprocity and direction-ality of friendship ties is still poorly understood.

When analyzing self-reported relationship surveys from several experiments, we find that only about half of the friendships are reciprocal. These findings suggest a profound inability of people to perceive friendship reciprocity, perhaps because the possibility of non-reciprocal friendship challenges one's self-image.

We further show that the asymmetry in friendship relationships has a large effect on the ability of an individual to persuade others to change their behavior. Moreover, we show that the effect of directionality is larger than the effect of the self-reported strength of a friendship tie [4] and thus of the implied 'social capital' of a relationship. Our experimental evidence comes through analysis of a fitness and physical activity intervention, in which subjects were exposed to different peer pressure mechanisms, and physical activity information was collected passively by smartphones. In this experiment, we find that effective behavioral change occurs when subjects share reciprocal ties, or when a unilateral friendship tie exists from the person applying the peer pressure to the subject receiving the pressure, but not when the friendship tie is from the subject to the person applying peer pressure.

Our findings suggest that misperception of friendships' character for the majority of people may result in misallocation of efforts when trying to pro-mote a behavioral change.

2   Results

2.1   Reciprocity and Intimacy

Despite the unique characteristics and importance of reciprocal friendships, reci-procity is implicitly assumed in very many scientific studies of friendship net-works: in their analysis they either mark two individuals as friends of each other, or as not being friends. However, not all friendships are reciprocal, as we proceed to demonstrate.

We analyze surveys that were used to determine the closeness of relationships (i.e., friendships) among participants in the Friends and Family study. Each participant in the study scored other participants on a $0-7$ scale, where a score of 0 meant that the participant was not familiar with the other, and 7 that the participant was very close to the other.

The self-reported closeness scores were then used to build the friendship net-work. Similar to [1], we considered only explicit friendship ties (*closeness* $>$ 2).

In this network, we consider a friendship tie to be "reciprocal" when both participants identify each other as friends. Alternatively, the tie is "unilateral" when only one of the participants identifies the other as a friend. Figure 1 depicts the resulting network which consists of 122 nodes and 698 edges (i.e., explicit friendships), of which 315 are reciprocal (i.e., 45 %) and 383 are unilateral (i.e., 55 %). Surprisingly, more than half of the participants' friendship ties are not reciprocated, which indicates the non-intuitive observation that people are very vulnerable to misjudging their friendship relationships and implies that people are unable to perceive reciprocity [2].

We find this result to be consistent across many self-reported friendship networks that we have analyzed: only 45 % (315 out of 698) of friendships are reciprocal in the Friends and Family dataset [1], 34 % (28 out of 82) in the Reality Mining dataset [9], 35 % (555 out of 1596) in the Social Evolution dataset [16], and 49 % (102 out of 208) in the Strongest Ties dataset [20]. The first three surveys were collected at an American university, and the fourth at a European university.

Similarly, a previous study [23] in which adolescents were asked to nominate at most 10 of their best school friends (5 male and 5 female) found that only 64 % of the reported friendships were indeed reciprocal. Our findings reinforce this finding by investigating multiple datasets from two continents, and by using complete nomination networks (in which each participant is asked about every other participant), resulting in an even more prominent lack of reciprocity.

Finally, analyzing the closeness scores associated with the two types of ties in the Friends and Family friendship network reveals that participants that share a reciprocal friendship tend to score each other higher (on average) when compared to participants that share unilateral friendship. More specifically, the average closeness score of reciprocal ties (4.7) is almost one point higher than the average score of unilateral ties (3.9) and the difference is statistically significant (two-sample T-test $p < 0.0001$).

2.2 Induced Peer Pressure

Social scientists have long suspected that reciprocal friendships are more intimate, provide higher emotional support, and form a superior resource of social capital when compared to those that are not reciprocated. This holds whether or not any party of the dyad is aware of the status of reciprocity embedded in their relationships [23]. However, we hypothesize that 'reciprocity' and 'directionality' of friendships may be critical factors in promoting peer influence, beyond the mere effect of the total tie 'strength' bound up in the relationship.

To support our hypothesis, we investigate the FunFit experiment – a fitness and physical activity experimental intervention – conducted within the Friends and Family study population during October to December of 2010. The experiment was presented to participants as a wellness game to help them increase their daily activity levels. Subjects received an 'activity app' for their mobile phone which passively collected their physical activity data and showed the participants how their activity level had changed relative to their previous activity

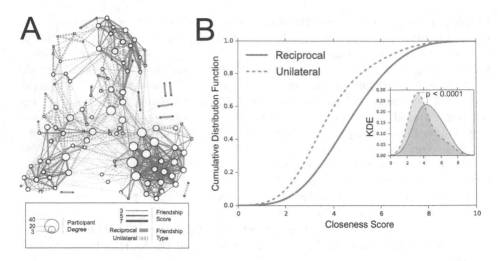

Fig. 1. Subfigure A depicts the undirected friendship nomination graph in the Friends and Family study, where nodes represent participants and edges represent friendship ties. **Subfigure B** shows the distribution of closeness scores for reciprocal and unilateral ties. ECDF and KDE of closeness scores are computed separately for unilateral ties (dashed line) and reciprocal ties (solid line) (Color figure online).

level, and the amount of money they had earned by being more active. 108 out of the 123 active Friends and Family subjects at that time elected to participate and were allocated into three experimental conditions, allowing us to isolate different incentive mechanisms varying monetary reward, the value of social information, and social pressure/influence:

- **Control:** subjects were shown their own progress and were given a monetary reward based on their own progress in increasing physical activity relative to the previous week.
- **Peer See:** subjects were shown their own progress and the progress of two "buddies" in the same experimental group, and were given a monetary reward based on their own progress in increasing physical activity relative to the previous week.
- **Peer Reward:** subjects were shown their own progress and the progress of two "buddies" in the same experimental group, but their rewards depended only on the progress of the two "buddies". This condition realizes a social mechanism based on inducing peer-to-peer interactions and peer pressure [18].

However, for the purpose of our analysis in this section, we combine the samples from the two peer pressure treatments, as we are interested in peer pressure regardless of the incentive structure, and omit the control group.

During the initial 23 days of the experiment (Oct 5 – Oct 27), denoted as P1, the baseline activity levels of the subjects were collected. The actual intervention period is denoted as P2. During the intervention period, the subjects were given feedback on their performance in the form of a monetary reward. The monetary reward was calculated as a function of the subject's activity data relative to the previous week and was divided according to the subject's experimental condition (i.e., Peer See and Peer Reward). Note that the physical activity was measured passively by logging the smartphone accelerometer (as opposed to self-reported surveys) and the game was not designed as a competition, every subject had the potential to earn the maximal reward. A previously non-active participant could gain the same reward as a highly active one, while the highly active person would need to work harder.

The results in [1] show that the two social conditions (i.e. Peer See and Peer Reward) do significantly better than the control group. Furthermore, the results suggest that there is a complex contagion effect [7], due to the reinforcement of the behavior from multiple social contacts [5,7], related to pre-existing social ties between participants. Our analysis here focuses on the role of reciprocity and directionality of friendship ties in this contagion process.

In order to investigate the role of reciprocity and directionality of friendship ties in the contagion process, we performed a regression analysis in which the dependent variable was the change in physical activity between the post-intervention phase and the pre-intervention phase (i.e., the average daily physical activity in P2 divided by the average daily physical activity in P1).

For our study, we refer to a participant whose behavior is being analyzed as "ego", and participants connected to the ego (i.e., experimental "buddies") are referred to as "alters". Because friendship nominations are directional, we studied the three possible types of friendships (from the prospective of the ego) as independent variables: an "ego perceived friend", in which an alter identifies an ego as a friend (i.e., incoming tie); an "alter perceived friend" in which an ego identifies an alter as a friend (i.e., outgoing tie); and a "reciprocal friend," in which the identification is bidirectional (i.e., reciprocal tie). Finally, we also included the tie strength (i.e., the sum of the closeness scores between an ego and his or her alters) as a control variable, which has been previously investigated as a moderator of the effect of social influence [4].

Figure 2 reports the effects found in our regression analysis (recall that the dependent variable in our model is the change in activity for the egos). We find that the reciprocity and directionality of a friendship have an effect on the amount of induced peer pressure, and these effects are much larger than the total tie strength.

The strongest effect for both treatment groups ($N = 76$) in this study was found for the reciprocal factor ($p < 0.01$) even when controlling over the strength of the tie (the tie strength is weakly significant $p = 0.07$). That is, alters in reciprocal friendships have more of an effect on the ego than alters in other types of friendships.

Interestingly, when the ego was perceived as a friend by the alters (i.e., incoming edges from the alters to the ego), the effect was also found to be positive and significant ($p < 0.05$). On the other hand, no statistically significant effect was found when the alters were perceived as friends by the ego (i.e., outgoing edges from the ego to the alters). Therefore, the amount of influence exerted by

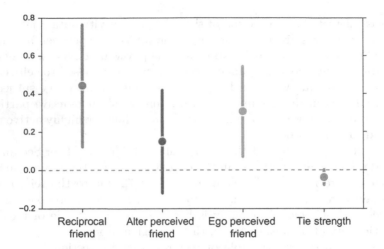

Fig. 2. Change in physical activity under experiment conditions shows that the type of friendship is relevant to the effectiveness of the induced peer pressure. The plot shows the mean effect size of the covariates (solid circles) and the 95 % confidence intervals (bars) (Color figure online).

individuals on their peers in unilateral friendship ties seems to be dependent on the direction of the friendship.

Unlike previous works on social contagion effects [8,11], which were conducted without peer-to-peer incentives, we find that influence does not flow from nominated alter to nominating ego. Surprisingly, alter's perception of ego as a friend would increase alter's ability to influence ego's behavior when ego does not reciprocate the friendship. We attribute this difference to the fact that there is a peer-to-peer incentive mechanism, and therefore there are likely to be differences in communication when the alters believe the ego to be their friend versus when they do not.

2.3   Reciprocity and Global Adoption

In order to understand the effect of reciprocal ties on global behavior adoption, we experimented with a variation of the classic epidemic spreading model, Susceptive-Infected (SI) model. We refer to this variant as the Bi-Directional Susceptive-Infected (BDSI) model. Unlike the classic SI in which behavior is transmitted along edges with a constant probability, the proposed BDSI model considers the direction in which behaviors can be transmitted with different probabilities based on the direction and type of edges – i.e. p_{rec} for reciprocal edges and p_+/p_- for the two possible directions of unilateral edges.

In order to observe the effects of reciprocal edges on diffusion, we employ an edge percolation process in which we measure the coverage (i.e., number of infected nodes), denoted by Z, and time to infect, denoted by T, when removing reciprocal and unilateral edges successively (i.e., perturbation F). That is, F is the percentage of edges removed in perturbation. We find the nature of the

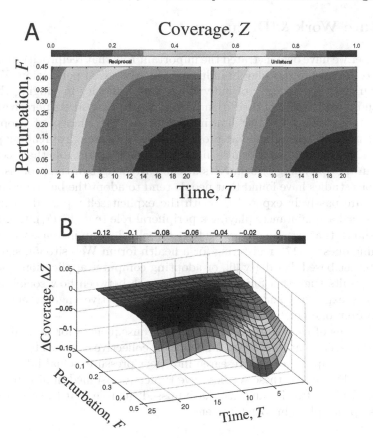

Fig. 3. Subfigure A demonstrates the effect of perturbation on the coverage and speed of adoption. **Subfigure B** illustrates how the coverage and speed decay faster when removing reciprocal edges in comparison with unilateral edges (Color figure online).

simulation results are qualitatively independent of the choice of p_{rec}, p_+ and p_- given that $p_{rec} \geq p_+ \geq p_-$.

Figure 3 shows the behavior adoption coverage when simulating the BDSI model on the self-reported friendship network from the Friends and Family dataset. As can be seen in the figure, the coverage Z decays much faster when removing reciprocal edges (left figure) compared with removing the same amount of unilateral edges (right figure). Moreover, the difference in coverage ΔZ is affected remarkably by the removal of reciprocal edges most notably in the early stages of the diffusion process (e.g., $T \in [5, 10]$). This can be attributed to the rapid diffusion within a single community through reciprocal edges, corresponding to fast increases in the number of infected users in early stages of the diffusion process, followed by plateaus, corresponding to time intervals during which no new nodes are infected the behavior escapes the community (i.e., through the strength of weak tie [12]) to the rest of the network through unilateral edges.

3 Future Work & Discussion

In this paper we have demonstrated the important role that reciprocity and directionality of friendship ties play in inducing effective social persuasion. We have also shown that the majority of individuals have difficulty in judging the reciprocity and directionality of their friendship ties (i.e., how others perceive them), and that this can be a major limiting factor for the success of cooperative arrangements such as peer-support programs. Finally, through spreading process simulation, the experimental results highlight the important role that reciprocal ties play in the spreading of behaviors at the early stages of the process.

Previous studies have found that people tend to adopt the behaviors of peers that they are passively exposed to, with the explicit self-reported friends and intimate social acquaintances playing a peripheral role (e.g., [5, 17]). Other studies have shown that passive exposures to peer behavior can increase the chances of becoming obese [8, 17], registering for a health forum Web site [5], signing up for an Internet-based diet diary [6], or adopting computer applications [3]. However, our results suggest a fundamental difference between how social learning (i.e., passive exposure) and social persuasion (i.e., active engagement) spread behaviors from one person to another.

The findings of this paper have significant consequences for designing interventions that seek to harness social influence for collective action. This paper also has significant implications for research into peer pressure, social influence, and information diffusion as these studies have typically assumed undirected (reciprocal) friendship networks, and may have missed the role that the directionality of friendship ties plays in social influence.

References

1. Aharony, N., Pan, W., Ip, C., Khayal, I., Pentland, A.: Social fMRI: investigating and shaping social mechanisms in the real world. Pervasive Mob. Comput. **7**(6), 643–659 (2011)
2. Almaatouq, A., Radaelli, L., Pentland, A., Shmueli, E.: Are you your friends' friend' poor perception of friendship ties limits the ability to promote behavioral change. PLoS ONE **11**(3), 1–13 (2016)
3. Aral, S., Walker, D.: Identifying influential and susceptible members of social networks. Science **337**(6092), 337–341 (2012)
4. Aral, S., Walker, D.: Tie strength, embeddedness and social influence: a large scale networked experiment. Manage. Sci. **60**, 1352–1370 (2014)
5. Centola, D.: The spread of behavior in an online social network experiment. Science **329**(5996), 1194–1197 (2010)
6. Centola, D.: An experimental study of homophily in the adoption of health behavior. Science **334**(6060), 1269–1272 (2011)
7. Centola, D., Macy, M.: Complex contagions and the weakness of long ties1. Am. J. Sociol. **113**(3), 702–734 (2007)
8. Christakis, N.A., Fowler, J.H.: The spread of obesity in a large social network over 32 years. New Engl. J. Med. **357**(4), 370–379 (2007)

9. Eagle, N., Pentland, A.: Reality mining: sensing complex social systems. Pers. Ubiquit. Comput. **10**(4), 255–268 (2006)
10. Ford, P., Clifford, A., Gussy, K., Gartner, C.: A systematic review of peer-support programs for smoking cessation in disadvantaged groups. Int. J. Environ. Res. Public Health **10**(11), 5507–5522 (2013)
11. Fowler, J.H., Christakis, N.A.: Dynamic spread of happiness in a large social network: longitudinal analysis of the framingham heart study social network. BMJ. Br. Med. J. **338**, 23–27 (2009)
12. Granovetter, M.S.: The strength of weak ties. Am. J. Sociol. **78**, 1360–1380 (1973)
13. Hartup, W.W.: Adolescents and their friends. New Dir. Child Adolesc. Dev. **1993**(60), 3–22 (1993)
14. Hartup, W.W.: The company they keep: friendships and their developmental significance. Child Dev. **67**(1), 1–13 (1996)
15. Laursen, B.: Close friendships in adolescence. New Dir. Child Adolesc. Dev. **1993**(60), 3–32 (1993)
16. Madan, A., Cebrian, M., Moturu, S., Farrahi, K., Pentland, A.: Sensing the "health state" of a community. IEEE Pervasive Comput. **11**(4), 36–45 (2012)
17. Madan, A., Moturu, S.T., Lazer, D., Pentland, A.S.: Social sensing: obesity, unhealthy eating and exercise in face-to-face networks. In: Wireless Health 2010, pp. 104–110. ACM (2010)
18. Mani, A., Rahwan, I., Pentland, A.: Inducing peer pressure to promote cooperation. Sci. Rep., vol. 3 (2013)
19. May, S., West, R.: Do social support interventions (buddy systems) aid smoking cessation? a review. Tob. Control **9**(4), 415–422 (2000)
20. de Montjoye, Y.A., Stopczynski, A., Shmueli, E., Pentland, A., Lehmann, S.: The strength of the strongest ties in collaborative problem solving. Sci. Rep., vol. 4 (2014)
21. Rotheram-Borus, M.J., Tomlinson, M., Gwegwe, M., Comulada, W.S., Kaufman, N., Keim, M.: Diabetes buddies peer support through a mobile phone buddy system. Diab. Educ. **38**(3), 357–365 (2012)
22. Stock, S., Miranda, C., Evans, S., Plessis, S., Ridley, J., Yeh, S., Chanoine, J.P.: Healthy buddies: a novel, peer-led health promotion program for the prevention of obesity and eating disorders in children in elementary school. Pediatrics **120**(4), e1059–e1068 (2007)
23. Vaquera, E., Kao, G.: Do you like me as much as i like you? friendship reciprocity and its effects on school outcomes among adolescents. Soc. Sci. Res. **37**(1), 55–72 (2008)

Exploratory Models of Trust
with Empirically-Inferred Decision Trees

John B. Nelson[✉], William G. Kennedy, and Frank Krueger

George Mason University, Fairfax, USA
{jnelso11,wkennedy,fkrueger}@gmu.edu

Abstract. What is the relationship between an individual's values and their propensity to trust other people? To explore this question, we built decision trees on the microdata provided by the World Value's Survey. Our findings confirm the extant literature while also hinting at cultural heterogeneity. We propose that studying nationally-specific decision trees based on survey data allows for easy-to-intuit representations of complex social problems. Moreover, for the sake of pragmatism, decision trees developed in this manner offer researchers a good tool in terms of cost-to-benefits.

1 Introduction

Social theorists have long documented the role of trust in social stability. Trust, a form of social capital, affords the opportunity for cooperation and collective action (Coleman 1990). By contrast, with distrust comes disunity, and all the maladies entailed (Putnam et al. 1994). This motivates a simple research question: *why are some people trusting, and others not?* The answer – accounting for the drivers of individual-level trust – allows the researcher to understand cultural-level problems.

At the psychological level, there is evidence of universal cognitive structures which relate identifiable personal values to one another (Schwartz 1992). But, the very recognition of cultural variation suggests group-level, sociological effects. To a computational social scientist, methodological individualism demands accounting for agents, with local experiences and information processing (Cioffi-Revilla 2013). The assumption is that the cultural variation emerges from the structural variations which pattern individual interactions. Recognizing this, we first explore the relationship between individual values – and, the effect of different values on trust – within a set of cultures. The larger goal is to develop an agent-based model of trust. Instead of reporting the larger (in-progress) findings, this paper documents how inferred decision trees allow for pragmatic preliminary analysis. That is, the tools of machine learning rapidly identify important features of the data and hint as to possible behavior, which informs subsequent theorizing and development.

2 Background

In human psychology, some needs are more important than others (Maslow 1943). According to Schwartz (1992), expressed values similarly form a hierarchy, as they

© Springer International Publishing Switzerland 2016
K.S. Xu et al. (Eds.): SBP-BRiMS 2016, LNCS 9708, pp. 42–50, 2016.
DOI: 10.1007/978-3-319-39931-7_5

are directly related to needs. Expounding, values are beliefs which motivate behavior. Contrast this to a belief that something is or is not true, but is otherwise disconnected from motive. Values drive and justify behavior whereas beliefs express general expectations.

Trust appears to be a mixture of basic values, in particular, honesty, fairness, and benevolence. Yet, there are broadly two kinds of social trust: trust in institutions, and trust in individuals. Critically, they may be antagonistic. A person who values conformity and tradition – that is, assigns this concept high importance in their value system – tends to trust institutions. Whereas, a person who values autonomy and responsibility tends to trust in individuals (Devos et al. 2002).

Scholars often rely upon surveys to characterize the relationships between expressed values. Particularly tailored to this task is the World Values Survey (WVS), which surveys individual values across time and space (World Values Survey Association 2016). At the time of writing, the six WVS waves (i.e. survey periods) span 192 countries and 34 years. These are microdata, with each record reporting the responses of a single individual. On a per-country basis, there are typically thousands of records (i.e. individuals surveyed).

Using the WVS, Tausch (2015) finds two dimensions which explain the variation in human values well. The first dimension is that of "traditional vs secular-rational" values; the second, "survival vs self-expression." Further demonstrating the prior art in modeling trust, both Jen et al. (2010) and Morselli et al. (2012) use the same data to find support for a link between trust and individual health, while also testing Schartz's theory.

3   Methodology

During preliminary investigation, there are benefits to rapid development and exploration. For reasons detailed in the later discussion, this motivated the use of decision tree learning. A variety of tree inference algorithms exist (Mitchell 1997, Ch 3). We chose the well-tested and popular scikit-learn Python package (Pedregosa et al. 2011), which implements a decision tree classifier that uses an optimized version of the CART algorithm[1].

Succinctly, CART performs a search over attributes when assembling trees. At each stage, the attribute selected as the next node is the one that best partitions the space of examples with respect to the target variable. Here, best means that, by performing the candidate split, the resulting sets are more "pure." As the purity criteria, the algorithm uses entropy, defined as:

$$H(X) = -\sum_{i=1}^{n} p(x_i) \log_b p(x_i)$$

where x_i is the observed frequency of an outcome over the discrete random variable, X. Intuitively, the more uniform the observed frequencies, the less pure

---

[1] The open-source tool built and used to generate the trees is available at https://bitbucket.org/johnnybjorn/wvstreeview. It provides an IPython notebook GUI interface to the underlying WVS data.

it is (i.e. the entropy is greater). For example, <10, 10, 10> has more entropy than <20, 5, 5>. In the context selecting an attribute, the best node is the one gives the greatest information gain (i.e. Kullback–Leibler divergence):

$$\underset{\theta}{\operatorname{argmax}} = H(X) - H(X \mid \theta)$$

Note, this is a greedy heuristic. Conceivably, something like beam search – which operates over sets of nodes at a time, rather than single ones – would perform better. However, in practice, information gain is a good and performant heuristic (Mitchell 1997, Ch. 3).

4   Results

To show the usefulness of this methodology, four countries from Wave 6 (2010–2014) of the World Value Survey serve as examples: Germany, India, Morocco, and United States[2]. Figure 1 shows the macro-level variation in trust (V24)[3]. It plots the proportion of people answering the question most "people can be trusted" affirmatively by country. The opposing response corresponds to "you need to be very careful."

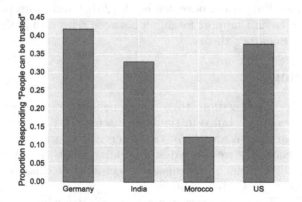

Fig. 1. Trust in others by example country

The cultural map of the world (Inglehart and Welzel 2005) inspired this selection of countries, an illustrative coverage over the reported abstract regions. Said differently, the examples draw from different regions in Tausch's (2015) two-dimensional space. The trees explained in the examples below analyze the with-incountry variance over the same question.

---

[2] These examples are not cherry-picked. India, Germany, and the US were targets of a larger effort, and thus drove our interest. Morocco was subsequently added to better cover the cultural map for this paper.

[3] WVS variable names documented in parentheses.

4.1   WVS Wave 6, Germany

Figure 2 portrays the decision tree emitted for Germany over the World Value Survey's 6th wave. Below each node, the plot describes the proportion of people answering that they trust others. Next to this number is n, the number of respondents which exist at this point in the tree traversal.

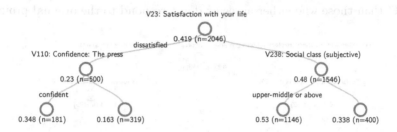

Fig. 2. Germany decision tree

This decision tree has an ROC/AUC score of 0.64. ROC/AUC is the Area Under the Curve (AUC) for the Receiver Operating Characteristic (ROC). The ROC plots the true positive rate as a function of the false positive rate. A classifier that performs no better than random would have a AUC/ROC of 0.5, illustrated by a diagonal line. More intuitively, assume you applied your classifier to one random negative and one random positive example. The AUC/ROC measures the probability that the positive example ranks higher than the negative one, according to some ranking function[4].

Here, the root node partitions the space with respect to "satisfaction with your life" (V23). Respondents answering 'dissatisfied' fall to the left; those satisfied fall to the right[5]. Those satisfied are twice as likely to answer, "most people can be trusted" (i.e. 0.48 : 0.23).

The left child of the root then splits the set over confidence in the press (V110). Of the set remaining, those who are confident in the press are about two times more likely to express trust in most people than those who are not. The right child of the root splits the set over subjective social class, with {Upper, Upper-Middle} branching on one side and {Lower, Working, Lower-Middle} on the other. Those belonging to a higher social class are approximately 50-percent more likely to be trusting than those from the lower classes.

---

[4] This assumes a function with a real-valued codomain. A decision tree produces a binary response. However, mapping the proportion of 'yes' to 'no' responses in each node allows for computation of AUC/ROC.

[5] Actually, respondents answer on a scale of 1–10, dissatisfied-satisfied. In this case, the decision tree implicitly dichotomized the space according to $x \leq 5$. This is a safe operation over the ordinal values, which are common in WVS.

4.2 WVS Wave 6, India

For India, the inferred decision tree (Fig. 3), has an ROC/AUC of 0.661. The root of the tree partitions responses by respect for immigrants. Variable V46 prompts, "when jobs are scarce, employers should give priority to people of this country over immigrants." Those who disagree (branching to the right) are more trusting. They are a bit more than twice as likely to answer "most people can be trusted" than those who either agree or don't respond to the original prompt.

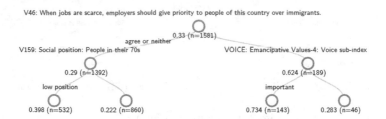

Fig. 3. India decision tree

Traversing the tree down to the left-child of the root, a prompt captures a mixture of respect for authority and traditional values in asking, what is the social position of people in their 70's (V159). Those who believe this cohort has a low position in society are a bit less than twice as likely to trust people than those who lack such respect. On the right-branching child of the root, a compound index which captures the degree to which freedom of expression matters partitions the space (VOICE). Those who value freedom of expression are approximately 2.5 times more likely to trust other people than those who do not.

4.3 WVS Wave 6, Morocco

The decision tree for Morocco (Fig. 4) achieved an ROC/AUC of 0.658. The root of the tree partitions by the prompt, "how much respect is there for individual human rights nowadays in this country" (V142). Those who believe there is a great deal of respect are about five times more likely to trust other people than everyone else.

Following the leftward path – those who believe there is a great deal of respect – the child node then decides on the basis of "people over 70 are viewed with respect" (V163). Respondents answering less affirmatively are approximately 3.5 times more likely to trust than those who answer "very likely to be viewed that way." On the right branch of the root, political party membership (V29) discriminates. Those who do not engage in politics by means of party membership are less trusting than those who at least are party members, even if they are inactive.

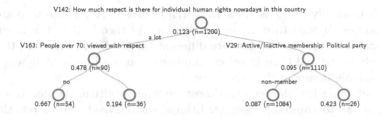

Fig. 4. Morocco decision tree

4.4   WVS Wave 6, US

The tree for the United States achieved an AUC/ROC of 0.667. The root node (Fig. 5) discriminates based on the respondents trust or distrust of secular institutions (SKEPTICISM). Those who mostly trust these institutions – the courts, the army, and the police – are twice as likely to trust individuals than those who do not.

For those that do trust secular institutions (left branch), neighborhood security (V170) is the next discriminating factor. Those who feel very secure in their neighborhoods are roughly 50 % more likely to trust individuals than everyone else. On the right child of the root, satisfaction with the household's financial situation partitions the space (V59). Those who satisfied are approximately twice as likely to trust compared to those dissatisfied.

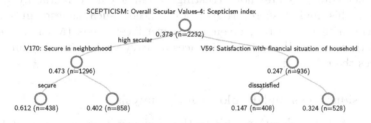

Fig. 5. US decision tree

5   Discussion

5.1   Stylized Facts

The emitted decision trees comport to the expectations of the extant literature. Categorizing the node according to Tausch (2015)'s space is subjective, but does not stretch credulity. Looking at the decision trees, many stylized facts are apparent. The following examples are a small sample of those available.

The root node for the US tree – that is, the most discriminatory node, according to information gain – splits based on whether the respondent trusts in secular institutions. This aligns well with the idea that, among Western nations,

America is markedly religious (Pew Research Center 2011). Those who are religious (traditional) tend to devalue the secular. As theorized in the literature, institutional trust and social trust are different things. But, in the US, they are more closely correlated than in other countries – so much so that it was not a selected variable in any of the other trees[6].

The root node for Morocco is almost counter-intuitive. As per the world culture map, Moroccans emphasize traditional and survival values. Yet, the tree learner selected a question that concerns respect for individual rights. However, given the type of data – microdata survey – this makes sense. If you are in the minority of people who value individual rights, such a value must especially important. Therefore, given the hierarchical structure of values, your propensity to trust is likely to be high. In the context of the decision tree, this split is likely to result in relatively pure subsets.

The case of India is particularly interesting. The root node splits those who favor their in-group (i.e. not immigrants) when economic opportunity is scarce. This may obliquely capture elements of xenophobia, but it should also reflect a high-value on traditionalism. Given the realities of a country with rich ethnic/racial heterogeneity and a high population growth rate, this seems to be a critical survey question. And, the decision tree learner uncovered the importance immediately.

Finally, the case of Germany provides evidence for frequent assertion: in general, the more satisfied you are, the greater your propensity to trust. The between-country comparisons for trust reflect the plausibility of the same assertion – richer countries are more trusting. From a classic hierarchy of needs perspective this makes sense. Higher level needs such as esteem and self-actualization rise to the top, given met lower-level needs (Maslow 1943). In service of these needs, the individual places a heavy weight on autonomy, which emphasizes the role of trust.

5.2   Decision Trees and Exploratory Analysis

The results of this exploration adds to the evidence in favor of extant theories on trust and values. Yet, the design is not especially novel. The literature is replete with studies that use the same data to arrive at similar conclusions by means of different methods. Consequently, the reported results do not add much weight to the scales of evidence. However, the use of decision tree learning as a exploratory tool proved especially pragmatic.

The algorithms for inferring decision trees are fast and robust. As a consequence, we were able to design an IPython (Pérez and Granger 2007) interface that ran analyses in real-time, given modeler input. This allowed for on-demand country/wave-specific runs. In the context of trust, this is important

---

[6] We dropped highly-correlated trust variables during the data cleaning phase. Partially, this required reading the code book and striking variables that interrogated trust with an alternate wording. But, as a backup, the software calculated correlations. We retained the SCEPTICISM variable after observing a correlation lower than expected.

for discovery. Between-country differences in trust dwarf within-country ones (Inglehart and Welzel 2010). Said differently, between-country differences explain more variance. But, these drivers are more sociological. Assuming an interest in psychological drivers, the within-country factors are more relevant.

Ignoring performant computation, the emitted trees are easy to interpret. This matters for the modeler, for whom time binds inelastically. Compare the visual portrayal of a decision tree to the tabular output of something like a logistic regression. The latter requires effort interpreting explained variance to deduce relative importance of each factor. In the decision tree, the nodes location portrays this information quickly. And, this ignores the costs of the more expensive operation – pruning the factors included in the regression to combat the problem of collinearity. The decision tree is self-pruning by the nature of the inference algorithm. Given collinearity, the selected node makes the collinear one unimportant from an information gain perspective. It simply drops out of the analysis, without any numerical issues.

Of course, an experienced researcher may find cause for concern in this methodology. It is positivist to the point of being a-theoretical. And, given the opportunity of point-and-click analysis, it is easy to find results that are seem novel but are truly arbitrary. This is true, and it warrants caution. But, as a tool for preliminary analysis, it strikes a good balance. In any early-stage research project, there is a period of pure exploration. The researcher becomes more familiar with the data by forming expectations and testing them. When done in a perfunctory manner, this process is prone to error. Consequently, cursory tests are less decisive than they could be, so clarity emerges slowly. Given the robustness and obviousness of decision trees, such tests over expectations are more robust. And, assuming commensurate effort or time, it should leave the researcher with more a refined understanding of the target system.

6   Conclusion

Research is messy; fraught with dead ends; and, riddled with bad paths. Decision trees allow for pragmatic exploration. As familiarity with the problem domain increases, such a tool may cease to be useful. But, during the early stages, it allows for parsimonious explanations while favoring large effects. That is stable ground upon which to build.

We used a decision tree learner to explore the question of how individuals' values affect their propensity to trust. From this exploration, we validated the extant literature. That is, using a different methodology, we generated evidence in agreement with the prior theory. Moreover, the trees made country-specific factors immediately salient. Perhaps, these factors would not surprise a cultural anthropologist with domain expertise on the specific country. But, they are easy to miss when generalizing over all countries at once, in deference to methodological convenience. With decision trees, the data acts as a tour guide, allowing for a more nuanced perspective, even in absence of domain expertise. Consequently, they grant the modeler a more intimate understanding of the problem.

References

Cioffi-Revilla, C.: Introduction to Computational Social Science: Principles and Applications. Springer Science & Business Media, New York (2013)

Coleman, J.S.: Foundations of Social Theory. Cambridge, Belknap (1990)

Devos, T., Spini, D., Schwartz, S.H.: Conflicts among human values, trust in institutions. Br. J. Soc. Psychol. **41**(4), 481–494 (2002). Wiley Online Library

Inglehart, R., Welzel, C.: Modernization, Cultural Change,and Democracy: The Human Development Sequence. Cambridge University Press, New York (2005)

Inglehart, R., Christian, W.: Changing mass priorities: the link between modernization and democracy. Perspect. Polit. **8**(02), 551–567 (2010). Cambridge Univ Press

Jen, M.H., Sund, E.R., Johnston, R., Jones, K.: Trustful societies, trustful individuals, and health: an analysis of self-rated health and social trust using the world value survey. Health Place **16**(5), 1022–1029 (2010). Elsevier

Maslow, A.H.: A theory of human motivation. Psychol. Rev. **50**(4), 370 (1943). American Psychological Association

Mitchell, T.M.: Machine Learning. McGraw Hill, Burr Ridge (1997)

Morselli, D., Spini, D., Devos, T.: Human values and trust in institutions across countries: a multilevel test of Schwartz's hypothesis of structural equivalence. Surv. Res. Methods **6**(1), 49–60 (2012)

Pedregosa, F.G., et al.: Scikit-learn: machine learning in Python. J. Mach. Learn. Res. **12**, 2825–2830 (2011)

Pew Research Center, The American-Western European Values Gap. Report, Pew Global Attitudes Project (2011)

Pérez, F., Granger, B.E.: IPython: a system for interactive scientific computing. Comput. Sci. Eng. **9**(3), 21–29 (2007). IEEE Computer Society

Putnam, R.D., Leonardi, R., Nanetti, R.Y.: Making Democracy Work: Civic Traditions in Modern Italy. Princeton University Press, Princeton (1994)

Schwartz, S.H.: Universals in the content and structure of values: theoretical advances and empirical tests in 20 countries. Adv. Exp. Soc. Psychol. **25**(1), 1–65 (1992)

Tausch, A.: Towards new maps of global human values, based on world values survey (6) data. Based on World Values Survey (6) Data (March 31, 2015) (2015). Available at SSRN: http://ssrn.com/abstract=2587626

World Values Survey Association. WORLD VALUES SURVEY Wave 6 2010-2014 OFFICIAL AGGREGATE V.20150418. Aggregate File Producer: Asep/JDS, Madrid Spain (2016)

Predicting Privacy Attitudes
Using Phone Metadata

Isha Ghosh and Vivek K. Singh[✉]

School of Communication and Information,
Rutgers University, New Brunswick, USA
{isha.ghosh,v.singh}@rutgers.edu

Abstract. With the increasing usage of smartphones, there is a corresponding increase in the phone metadata generated by individuals using these devices. Managing the privacy of personal information on these devices can be a complex task. Recent research has suggested the use of social and behavioral data for automatically recommending privacy settings. This paper is the first effort to connect users' phone use metadata with their privacy attitudes. Based on a 10-week long field study involving phone metadata collection via an app, and a survey on privacy attitudes, we report that an analysis of cell phone metadata may reveal vital clues to a person's privacy attitudes. Specifically, a predictive model based on phone usage metadata significantly outperforms a comparable personality features-based model in predicting individual privacy attitudes. The results motivate a newer direction of automatically inferring a user's privacy attitudes by looking at their phone usage characteristics.

Keywords: Privacy attitudes · Social signals · Phone metadata · Call logs

1 Introduction

Recent results have pointed to a significant awareness of the dangers of sharing information and pictures on social networking sites (SNSs), and young adults today are careful about constructing identity and online information disclosure [6, 7]. However, the fact that a number of interactions happen over smartphones which constantly receive and send out data signals containing personal information often slips under the radar [17, 21, 22]. Enormous amounts of personal data are being captured from user's smartphones to personalize user requirements however, there is little work done to understand how this mobile metadata (particularly call logs,) can be used to build personalized privacy settings for users. In this paper, we propose pivoting the use of phone metadata (particularly call logs) to allow users to utilize their own data to obtain personalized privacy recommendations.

As a first step toward this goal, we test the ability of an individuals' phone metadata to predict user's privacy attitudes. Once established, such interconnections could be used to automatically define user's privacy settings without the need for manual surveys or weaving through complicated choices.

This paper makes two important contributions.

© Springer International Publishing Switzerland 2016
K.S. Xu et al. (Eds.): SBP-BRiMS 2016, LNCS 9708, pp. 51–60, 2016.
DOI: 10.1007/978-3-319-39931-7_6

1. Motivates and grounds the use of phone metadata as a method of assessing user privacy needs.
2. Lays the groundwork for building automated models that define user's privacy attitudes, without the need for explicit surveys.

2 Related Work

We focus the presentation of related work on research projects that discuss information sharing within and outside communities and their impact on privacy attitudes. As more and more aspects of human life get mediated by mobile phones, a quick and easy way to identify privacy attitudes for users may be useful to suggest default settings and configurations in a variety of applications and scenarios faced by the user. However, in order to effectively suggest such settings and configurations, it is important to gain an understanding of an individual's privacy attitudes.

There have been several attempts to define privacy attitudes. In a systematic discussion of the different notions of privacy Introna and Poloudi (1999) developed a framework of principles that explored the interrelations of interests and values for various stakeholders where privacy concerns have risen. The central idea around an individual's privacy attitude is the desire to keep personal information out of the hands of others, along with the ability to connect with others without interference. In this context, concern for privacy is a subjective measure—one that varies from individual to individual based on that person's own perceptions and values. In other words, different people have different levels of concern about their own privacy [18].

However, a concern for privacy does not translate into similar behavior. Previous research [1, 2, 27] has explored the dichotomy that exists between concerns about privacy and actual behaviors exhibited by individuals. The focus of this study is to understand the attitude and concerns of individuals towards privacy and how these concerns influence (or are influenced by) their real-world social behavior.

While there exist a number of studies to measure privacy attitudes and behaviors in online interactions, [1, 27, 30], we wanted to get a sense of privacy attitudes in both online and offline behaviors. Therefore we use the Westin's [32–38] studies to gain a holistic understanding of privacy concerns exhibited by individuals'. While there have been multiple recent efforts toward building newer measures of privacy, [1, 2, 7] Westin's work remains one of the most comprehensive approaches towards obtaining a well-rounded understanding of privacy attitudes exhibited by an individual. This paper uses the Privacy Segmentation Index [38] that categorizes individuals' into one of three categories based on their levels of privacy concerns:

- **Fundamentalists:** who feel very strongly about privacy and grant it an especially high value
- **Pragmatists:** who also have strong feelings about privacy but can also see the benefits from surrendering privacy in situations where they believe care is taken to prevent the misuse of this information; and
- **Unconcerned:** those who have no real concerns about privacy or about how other people and organizations are using information about them.

Previous research has explored various methods for improving the understanding of complex privacy settings [14, 23, 31] in SNSs. A recent study automatically generates privacy settings for any images uploaded by the user [25]; another study describes a privacy wizard for SNSs that describes a particular user's privacy preferences based on a limited amount of user input [15]. While this research work can help in preserving privacy in online SNSs, there is a gap in the literature around using cell phone metadata to generate automated privacy settings for individuals. Our research focuses on using this cellphone metadata as a predictor of an individual's privacy attitudes.

3 Study Undertaken

We adopt the Privacy Segmentation Index (PSI) to gain an understanding of privacy attitudes displayed by individuals. This survey consisted of statements designed to measure levels of concern about personal information disclosed by individuals to companies or businesses and their concerns about whether their information was being protected or not. Each of these questions required responses on a 4-point scale ranging from strongly disagree to strongly agree and individuals were classified as Fundamentalists, Pragmatists or Unconcerned based on their responses.

Participants for this study were recruited from Rutgers University, New Brunswick. During the study, participants were invited to the study site to read and sign the consent form and fill out an online survey. The survey consisted of the Privacy Segmentation Index [38], and demographic questions (e.g. gender, age). While the participants completed the survey, they were asked to install the study client on their phones. The study client collected call logs, sms logs, and location logs, over a 10 week study period. Cellphone metadata collected by the study client and the participant answers to the privacy attitude and behavior surveys were analyzed to test multiple hypotheses. A total of 53 participants completed the study i.e. installed the study client app and completed the privacy survey. Of these 31 (59 %) were men and 18 (34 %) were women (demographic data was unavailable for three participants). The majority of participants were undergraduates between the ages of 18–21 years.

Based on the information collected from this app we defined a set of features that will allow us to explore relationships between mobile phone interactions and privacy attitudes. Below are the variables and their definitions:

Privacy of user data was of utmost priority throughout this project. All data were secured and protected at standards applied to medical metadata. All metadata captured by the study clients was required to be no more detailed than those employed by typically installed apps like the Gmail and Instagram. The participants had the option to opt out of the studies at any time. The eventual goal is to design privacy apps that 'learn' user preferences over a short period of time (e.g. weeks) and can recommend privacy settings even after they stop receiving the data. Also, though outside the scope of the current work, the eventual privacy app coming out of this research project will be extended to run under the OpenPDS framework [17, 21]. OpenPDS (an open source personal data source) keeps personal data in the cloud under the purview of an individual user rather than the third parties like Google or Facebook.

Given that this is the first study to connect mobile phone metadata with privacy attitudes, we have adopted a multi-stage approach. We first tested multiple hypotheses based on existing literature connecting social behavior and privacy attitudes. We found multiple hypotheses to hold and significant associations between phone signals and privacy attitudes. This motivated the second stage of analysis where multiple features were combined into unified prediction models. The validation at each stage, yielded better confidence and interpretability for the next phase.

4 Hypotheses Testing

Existing studies [2, 13, 26] show maintaining privacy is a strong factor in determining how users present themselves and hence exerts an influence on their social interactions. For our study, we used the number of calls made by individuals over the period of study to determine their level of interaction. Past research has analyzed privacy and interaction on SNSs and has explored the relationship between privacy concerns and actual behavior on SNSs. For example, Acquisti and Gross (2006) found that one's privacy concerns were a weak predictor of the use of social network sites [1]. Based on these studies, we attempt to understand the relationships between greater phone interactions and privacy attitudes. We expect our study to show a negative relationship between number of calls and privacy attitudes.

H1: Higher call count is associated with a lower concern for privacy
H2: Longer time spent on calls is associated with a lower concern for privacy

Previous research [4, 8, 10] has shown the importance of maintaining social ties in today's world. Online SNSs support both the maintenance of existing social ties and the formation of new connections. While online social network sites offer an attractive means for interaction and communication, they also raise privacy and security concerns. Researchers [2, 12, 29] agree that maintaining these social networks require disclosure of personal information and encourage the sharing of information. Along with online social networks, phone calls are also a popular way to stay in touch. As the number of mobile phone users' increase, more and more information will be shared over phone calls. Individuals receive a number of phone calls outside of their network or known "friends and families" in a given day. These calls could be from marketing agencies, credit card sellers or even phishing, where criminals try to gain sensitive personal information using fraudulent means. In such a scenario, we believe that people who are highly concerned about their privacy would only respond to calls from within their network. Therefore, we hypothesize that a higher call response rate indicates a lower concern for privacy.

H3: Higher call response rate is associated with a lower concern for privacy.
H4: Higher missed call rate is associated with a higher concern for privacy.

The creation and maintenance of relationships is one of the chief motivations for an individuals' use of SNSs [1, 6]. Studies of the first popular social networking site, Friendster, [5, 11] describe how members create their profile with the intention of communicating news about themselves to others. The structure of these online social networks allows for the same information to be shared between close friends, strangers or acquaintances [1]. As newer contacts are included within an existing social network, there is more personal information shared within the network. Therefore, we hypothesize that people who have high privacy concerns will not readily initiate newer contacts. That is, a higher number of new contacts in a network indicates a lower concern for privacy.

H5: Higher number of new contacts initiated is associated with a lower concern for privacy.

Based on the data captured in this study, we tested the above-mentioned hypotheses, operationalized each based on the features defined in Table 1 and undertook Pearson's correlation analysis. As shown in Table 2, we found hypothesis H3, H4 and H5 to be statistically significant in the expected direction and H1, and H2 to be non-significant.

Table 1. Variables used in the study

Variable Name	Definition
Privacy Concern (Output Variable)	Score as determined by responses to Westin's Privacy Segmentation Index. Scores for each question on a Likert Scale of 1–4 were added together to get a combined Privacy Concern Score.
Call Count	*n(Calls)* Total number of calls received or made by participants in the duration of the study
Call Duration	$\sum$*(time spent on calls)* Sum of time spent on all calls received or made in the duration of this study
Missed Call Rate	*(Number of missed calls / Call Count) * 100* This is percentage of calls missed (not answered) by the participant.
Call Response Rate	*(Number of responded missed calls/ Number of missed calls) * 100* Where "responded missed calls" are those which were returned within 1 h of the missed call.
Number of New Contacts in Outgoing calls	This variable is defined as the number of calls made to new contacts; i.e. contacts that were not seen in the initial four weeks of the study but calls were initiated after the first four weeks.

Demographic analysis of the data also provided some interesting insights. Descriptive analyses of data from this study shows that while both males and females are concerned with data sharing, females tended to have a slightly higher concern for privacy with 60 % females receiving a "very concerned" privacy score on Westin's Index while only 54 % males received the same score. We had a diverse sample in terms of race and income groups with representations from Asian Americans, African Americans, Latino, and White populations. There were no significant variations in privacy attitudes across race and income groups.

Table 2. Results of hypotheses testing based on data in Rutgers Wellbeing study for n = 53

	Hypothesis Testing	Expected Direction	p-value	Pearson's Correlation Coefficient
H1	Not Significant	–	0.532	–0.088
H2	Not Significant	–	0.490	–0.097
H3	Significant	Yes	0.043	–0.279*
H4	Significant	Yes	0.017	0.425**
H5	Significant	Yes	0.025	–0.313*

Based on existing studies [1, 7, 13] analyzing information sharing and disclosure, we hypothesized that a higher call count and longer time spent on calls would be indicative of a low concern for privacy. However, our results showed that this was not the case. We found a significant correlation between the frequency of answering and not accepting calls and privacy attitudes. This implies that the calls individuals choose to answer on their smartphones may be determined by their attitude towards privacy and information sharing. Similarly, calls individuals choose to ignore or "miss" can be a reflection of their privacy concerns. Previous studies have analyzed information sharing and disclosure in online social networks, [12, 16] however, our results show that information sharing over smartphones maybe a significant predictor of privacy attitudes. There is a significant relationship between the number of new contacts included in an individual's network and their privacy attitudes. This implies that as an individual's network grows the amount of information disclosed increases. While similar results were found for online social networks, our study shows that networks formed over phone calls, can also be an important variable in determining an individual's information sharing patterns.

These preliminary results indicate that the interconnections between privacy needs and phone metadata are not yet fully understood but could yield interesting findings when analyzed systematically. This also implies that concern over privacy may not be apparent by an examination of the simplest features, but an analysis of more nuanced features, like how individuals react to phone calls in terms of the number of calls they actually respond to, may reveal a more interesting pattern.

A little over half (57 %) the surveyed participants exhibited moderate (27 %) to high (31 %) concerns for privacy, and 43 % participants had a low score or were "unconcerned" with sharing their data. This is very different from the results of the original survey conducted in 2003. Westin reported only 10 % of the population was classified as "Unconcerned," with the majority of individuals displaying moderate (64 %) to high concerns (26 %) regarding the (ab)use of their personal information [38]. While our sample includes mostly single undergraduate students and is not representative of a larger population, such a vast difference in results, begs a question about the differences in information sharing attitudes of individuals in the early or mid 2000's to the present day.

5 Towards a Predictive Model of Personal Privacy Attitudes

Multiple significant hypotheses suggest predictive potential of nuanced phone usage metadata towards privacy attitudes. Hence we used the three features found to be significant in the analysis above to build a combined predictive model for privacy attitudes. We consider two different classifications for the privacy attitudes. First, is the conventional three category classification as suggested by Westin and second is a two-class categorization based on the median value split.

In the first scenario the classes were defined based on the criteria recommended by Westin as already described in Sect. 2. This resulted in a split as follows: Privacy Fundamentalists: 5, Privacy Pragmatists: 31, Privacy Unconcerned: 17, Total: 53. Given the multiple (> 2) classes present we decided to use the MultiClass Classifier as implemented in Weka 3.6, with J48 decision tree as its underlying method. Further considering the relatively modest sample size, we decided to use Leave-One-Out cross-validation to tradeoff between the learning ability and the generalizability of the results. We also compare the proposed phone-features based approach with two other approaches. One is a baseline 'Zero-R' approach, which simply classifies all data into the largest category. The second approach is based on using Big-Five [20] personality variables, which have been shown by multiple efforts to be related with privacy attitudes [19, 26]. The same classification method was applied to the different approaches. Lastly, given the unequal size of the classes, we also report the ROC (Receiver Operating Characteristic – Area Under the Curve) statistic along with the accuracy scores. Multiple prior efforts have suggested ROC as a more interpretable metric for classification when dealing with unequal classes [9].

As shown in Table 3, the Phone-features based model performed better than both the compared approaches. Focusing on the ROC metric, the model yielded 36 % better prediction than the baseline model. Contrary to the expectations, the results also suggest that personality based metrics may not capture the right kind of signals to have predictive ability on privacy attitudes.

Table 3. Classification results using different approaches for (a) three-way classification as per Westin's taxonomy and (b) two-way classification (High vs. Low Privacy Concern).

	Three-Way Classification		Two-Way Classification	
	Accuracy	ROC	Accuracy	Accuracy
Baseline (Zero-R)	0.58	0.50	0.56	0.50
Personality Features	0.53	0.40	0.43	0.39
Phone-Usage Features	**0.66**	**0.68**	**0.74**	**0.69**

To ameliorate some of the complexities associated with multi-class (> 2) classification, we also consider a two-way classification problem, where the classes were based on a median split. Multiple participants fell at the median score (8 out of 12) and this resulted in two roughly equal classes of sizes of 30 (below or equal to median) and 23 respectively. We ran the classification in Weka 3.6 using J48 decision tree algorithm with Leave-One-Out cross validation. As shown in Table 3, this resulted in accuracy of 74 % at a two class classification task and an ROC metric of 0.69, which indicates a 38 % improvement over the baseline. Again, the relatively poor performance of personality based features suggests that traditional personality type measures may not be suited to predict privacy attitudes. Further investigation with larger samples is needed to confirm this initial evidence.

6 Discussion

Limitations of this study include that our sample is from only one university and not from a nationally representative sample. The sample population was also not very diverse in terms of age as they were mostly undergraduate students between 21–23 years. Also, we used a self-report survey on privacy rather than observing and recording the participants' behavioral patterns in terms with respect to data sharing.

This study was carried out as an exploratory field study to understand how cell phone metadata can be used to build an individuals' personal privacy signature. While, we respect completely individuals' rights to their data, we posit that the current privacy debate is heavily biased towards the sharing and protection of socio-mobile data from third parties. Comparatively, little attention has been paid towards re-pivoting the same data to suggest privacy settings to users themselves for different applications. While similar studies have been conducted using online social network data, this study is the first to motivate and ground the use of phone metadata towards identifying the privacy attitudes and needs of individuals.

While the current work has focused on relatively simple set of features and tested a small number of hypotheses, the significant jump obtained in prediction ability points to the value in exploring this direction further. In particular the direction of using more nuanced behavioral features, over a correspondingly larger sample size and degrees of freedom is part of our future work. With appropriate refinements and advancements, the proposed methodology could allow for automatic privacy attitude understanding for billions of mobile phone users.

Acknowledgements. We would like to thank Cecilia Gal, Padampriya Subramnian, Ariana Blake, Suril Dalal, Sneha Dasari, and Christin Jose, for help with conducting the study and processing the data.

References

1. Acquisti, A., Gross, R.: Imagined communities: awareness, information sharing, and privacy on the facebook. In: Danezis, G., Golle, P. (eds.) PET 2006. LNCS, vol. 4258, pp. 36–58. Springer, Heidelberg (2006)
2. Acquisti, A., Grossklags, J.: Privacy attitudes and privacy behavior. In: Camp, L.J., Lewis, S. (eds.) Economics of Information Security. Advances in Information Security, vol. 12, pp. 165–178. Springer, Heidelberg (2004)

3. Beale, R.: Supporting social interaction with smart phones. IEEE Pervasive Comput. **4**(2), 35–41 (2005)
4. Bourdieu, P., Wacquant, L.: Classification struggles and the dialectic of social and mental structures. In: An Invitation to Reflexive Sociology, p. 14. University of Chicago Press, Chicago (1992)
5. Boyd, D.: Friends, friendsters, and myspace top 8: Writing community into being on social network sites (2006)
6. Boyd, D., Ellison, N.: Social network sites: definition, history, and scholarship. J. Comput.-Mediated Commun. **13**(1), 210–230 (2007)
7. Buchanan, T., Paine, C., Joinson, A.N.: Internet Privacy Scales. Journal of the American Society for Information Science and Technology (2007)
8. Burke, M., Kraut, R., Marlow, C.: Social capital on Facebook: Differentiating uses and users. In: Conference on Human Factors in Computing Systems, CHI 2011. ACM (2011)
9. Chawla, N.V.: Data mining for imbalanced datasets: An overview. In: Maimon, O., Rokach, L. (eds.) Data Mining and Knowledge Discovery Handbook, pp. 853–867. Springer, Heidelberg (2005)
10. Coleman, J.: Social capital in the creation of human capital. In: Knowledge and Social Capital, vol. 94, pp. 17–41 (1988)
11. Donath, J., Boyd, D.: Public displays of connection. BT Technol. J. **22**(4), 71–82 (2004)
12. Dwyer, C., Hiltz, S.R., Passerini, K.: Trust and privacy concern within social networking sites: a comparison of Facebook and MySpace. In: Proceedings of the Americas Conference on Information Systems AIS 2007, Keystone (2007)
13. Ellison, N., Steinfield, C., Lampe, C.: The benefits of Facebook "friends": exploring the relationship between college students' use of online social networks and social capital. J. Comput. Mediated Commun. **12**, 1143–1168 (2007)
14. Egelman, S., Oates, A., Krishnamurthi, S.: Oops, I did it again: mitigating repeated access control errors on facebook. In: Proceedings of the SIGCHI conference on Human Factors in Computing Systems, pp. 2295–2304. ACM (2011)
15. Fang, L., LeFevre, K.: Privacy wizards for social networking sites. In: Proceedings of the 19th International Conference on World Wide Web, pp. 351–360 (2010)
16. Fogel, J., Nehmad, E.: Internet social network communities: Risk taking, trust, and privacy concerns. Comput. Hum. Behav. **25**(1), 153–160 (2009). doi:10.1016/j.chb.2008.08.006
17. Hang, A., Von Zezschwitz, E., De Luca, A., Hussmann, H.: Too much information! User attitudes towards smartphone sharing (2012)
18. Introna, L., Pouloudi, A.: Privacy in the information age: Stakeholders, interests and values. J. Bus. Ethics **22**(1), 27–38 (1999)
19. Junglas, I.A., Johnson, N.A., Spitzmüller, C.: Personality traits and concern for privacy: an empirical study in the context of location-based services. Eur. J. Inf. Syst. **17**(4), 387–402 (2008)
20. John, O.P., Naumann, L.P., Soto, C.J.: Paradigm shift to the integrative big five trait taxonomy. In: Handbook of Personality: Theory and Research, 3rd edn., pp. 114–158 (2008)
21. Karlson, A., Brush, A., Schcchter, S.: Can I borrow your phone?: understanding concerns when sharing mobile phones. In: CHI, pp. 1647–1650 (2009). doi:10.1145/1518701. 1518953
22. Nordichi 2012: Making sense through design. In: Proceedings Of The 7th Nordic Conference On Human-Computer Interaction, pp. 284–287. doi:10.1145/2399016.2399061
23. Reeder, R.W., Karat, C.-M., Karat, J., Brodie, C.: Usability challenges in security and privacy policy-authoring interfaces. In: Baranauskas, C., Abascal, J., Barbosa, S.D.J. (eds.) INTERACT 2007. LNCS, vol. 4663, pp. 141–155. Springer, Heidelberg (2007)

24. Schlegel, R., Kapadia, A., Lee, A.: Eyeing your exposure: Quantifying and controlling information sharing for improved privacy. In: Proceedings of the 7th Symposium on Usable Privacy and Security, SOUPS 2011 (2011). doi:10.1145/2078827.2078846

25. Squicciarini, A.C., Sundareswaran, S., Lin, D., Wede, J.: A3P: adaptive policy prediction for shared images over popular content sharing sites. In: Proceedings of the 22nd ACM Conference on Hypertext and Hypermedia, pp. 261–270. ACM (2011)

26. Steinfield, C., Ellison, N.B., Lampe, C.: Social capital, self-esteem, and use of online social network sites: A longitudinal analysis. J. Appl. Dev. Psychol. 29(2008), 434–445 (2008)

27. Stutzman, F.: An evaluation of identity-sharing behavior in social network communities. Int. Digit. Media Arts J. 3(1), 10–18 (2006)

28. Rosenbaum, B.L.: Attitude toward invasion of privacy in the personnel selection process and job applicant demographic and personality correlates. J. Appl. Psychol. 58(3), 333–338 (1973)

29. Tufekci, Z.: Can you see me now? Audience and disclosure regulation in online social network sites. Bull. Sci. Technol. Stud. 11, 544–564 (2008)

30. Wang, Z., Liu, Y.: Identifying key factors affecting information disclosure intention in online shopping. Int. J. Smart Home 8(4), 47–58 (2014)

31. Watson, J., Besmer, A., Lipford, H.R.:+ Your circles: sharing behavior on Google+ . In: Proceedings of the Eighth Symposium on Usable Privacy and Security, p. 12. ACM, July 2012

32. Westin, A.F.: Privacy and freedom. Wash. Lee Law Rev. 25(1), 166 (1968)

33. Westin, A., Harris Louis & Associates: Harris-Equifax Consumer Privacy Survey (1991)

34. Westin, A., Harris Louis & Associates: Equifax-Harris Consumer Privacy Survey (1996)

35. Westin, A., Harris Louis & Associates: E-Commerce & Privacy: What Net Users Want 1998

36. Westin, A.: Freebies and Privacy: What Net Users Think (1999) http://www.pandab.org/sr990714.html

37. Westin, A., Harris Interactive: IBM-Harris Multi-National Consumer Privacy Survey (1999)

38. Westin, A.: Consumer, Privacy and Survey Research (2003). http://www.harrisinteractive.com/advantages/pubs/DNC_AlanWestinConsumersPrivacyandSurveyResearch.pdf

A Preliminary Study of Mobility Patterns in Urban Subway

Nuo Yong[1,2], Shunjiang Ni[1,2(✉)], and Shifei Shen[1,2]

[1] Institute of Public Safety Research,
Tsinghua University, Beijing 100084, China
sjni@tsinghua.edu.cn
[2] Department of Engineering Physics, Tsinghua University,
Beijing 100084, China

Abstract. Understanding human mobility patterns is of great importance to traffic forecasting, urban planning, epidemic spread and many other socioeconomic dynamics covering spatiality and human travel. Based on the records of Beijing subway, we presented a preliminary study of human mobility patterns at urban scale, including return ratio and trip distance. Especially, both linear distance and actual route distance are considered. We found that for a single mode of transportation, the displacement distribution not only decays exponentially, but also has a peak, which represents the characteristics of travel radius (CTR). The CTR of actual route distance is significantly greater than that of linear distance, which indicates that quite of the passengers make detours relative to the linear path when traveling by subway.

Keywords: Human mobility · Urban subway · Linear and route distance · CTR

1 Introduction

Understanding human mobility patterns is of great importance to traffic forecasting [1], urban planning [2], epidemic spread [3, 4] and many other socioeconomic dynamics covering spatiality and human travel [5, 6]. In recent years, the study of human mobility patterns has been paid much attention. For example, Brockmann [7] reported a quantitative assessment of human travelling statistics by analyzing the circulation of bank notes in the United States. Gonzalez [8] studied the trajectory of mobile phone users for a six-month time period to analyze human mobility patterns. Xin [9] analyzed the movements of 1.9 million mobile phone users during Haiti earthquake to explore mobility patterns especially in disaster. Jungkeun [10] presented a framework that can produce Wi-Fi users' movement patterns by wireless traces. Besides that, GPS data [11] and location-based social services [6] were also used to analyze human mobility patterns.

However, data such as trajectory of bank notes, mobile phones or GPS measurements could only obtain a very rough estimate of human mobility distribution. In reality, the exact path of different individuals, even with the same origins and destinations, varies widely for different purpose. In 2013, Yan [12] researched direct travel

© Springer International Publishing Switzerland 2016
K.S. Xu et al. (Eds.): SBP-BRiMS 2016, LNCS 9708, pp. 61–70, 2016.
DOI: 10.1007/978-3-319-39931-7_7

diaries of hundreds of volunteers and analyzed mobility patterns at the individual level, but the dataset is relatively small. In addition, as is often the case, an individual cannot reach the destination through the straight-line distance path no matter what he/she takes to travel. Because once an individual travels by means of transportation, he/she has to choose the closest path from the existing lines of transportation. The difference between straight-line distance and actual travel distance was ignored in previous studies.

In this paper, based on the records of Beijing urban subway, we presented a preliminary study of human mobility patterns at urban scale. Different from previous research, our big data could capture the exact path with explicit origins and destinations to some extent, which improved the data accuracy in the study of human mobility. Furthermore, when dealing with the trip distance of subway passengers, both linear distance and actual route distance were considered. We found that for a single mode of transportation, the displacement distribution not only decays exponentially, but also has a peak, which represents the characteristics of travel radius (CTR). The CTR of actual route distance is significantly greater than that of linear distance, which indicates that quite of the passengers make detours relative to the linear path when traveling by subway. Further discussion implies that for subway passengers in Beijing, the longer the trip is, the more detours they make.

2   Method

We take records from the Auto Fare Collection (AFC) system of Beijing subway. The total records cover a population of three million passengers for a single weekday, and two million for a single weekend. In order to facilitate the subsequent research, we obtained the following information from the records. The card ID, which is typically corresponding with a passenger. The entry and exit station number, which means the codes of entry and exit subway stations that the passenger gets through. The exact time that the passenger swipes the card when getting in and out of the subway station. In this paper, we analyze the records of fourteen continuous days in October 2014 (namely 20141013–20141026), and we use C++ programming to mine and count the general passenger flow information and the corresponding hidden information.

Especially, when dealing with the trip distance of passengers, both linear distance and actual route distance of the same trip are taken into consideration. As is known to all, when travelling by subway, passengers cannot reach the destination through the shortest path (linear distance) in most cases. They have to choose the closest path from the existing subway lines. In fact, in actual travel behavior, people, more or less, need to make a balance between the ideal path and the actual one. Therefore, studying the difference between linear distance and actual route distance is of general significance.

In order to get the linear distance, we obtained the longitude and latitude of all the subway stations from Baidu Map [13], and calculated the linear distance between any two of them. In order to get the actual route distance, based on our previous study [14], we divided the path feature into two categories. One is the path with obvious shortest route, the other is the path with competitive route. For the former, actual route distance can be obtained directly based on the topological relationship of subway maps. For the latter, we take use of the record information, namely the actual transit time, to invert the actual route distance. Specifically, for any pair of origin-destination, we take the A * heuristic algorithm and Dijkstra algorithm to get two or three alternate competitive paths to connect them. Then we calculated the transit time of each alternative competitive path, including running time, walking time and the transfer time. Finally, by comparing the transit time of each alternative competitive path with the actual transit time, we can infer the actual path and calculate the actual route distance.

3 Result

In order to depict the typical mobility patterns of urban passengers, we presented the spatial distribution of passengers' originations and destinations at first, and then we concerned about the spatiotemporal periodic and regularity of their travel behavior, such as return ratios and trip distance distribution.

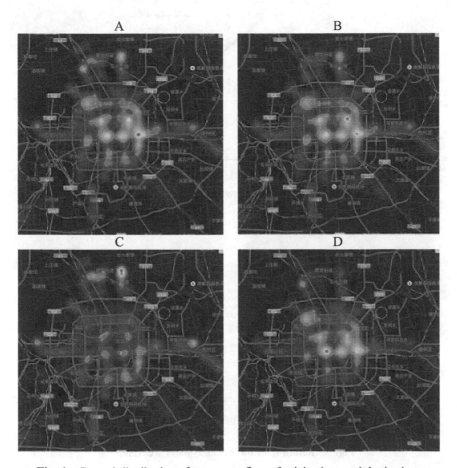

Fig. 1. General distribution of passenger flow of originations and destinations

3.1   General Distribution of Passenger Flow

Figure 1A and B presents the general distribution of passenger originations and destinations. It is obvious that the hot spots of originations and destinations have a high coincidence ratio. It is probably because of the high return ratio, which will be demonstrated in Fig. 3. The spatial distribution of the hot spots also presents a characteristics of clustering. Further analysis of different time period within a day, Fig. 1C presents the distribution of passenger source in the morning, and Fig. 1D shows the same in the afternoon. As is shown in the picture, the living quarters mainly cluster in the north of the city, and the work areas mainly locate in the center of the city. All of the statistics are the average of fourteen continuous days in October 2014.

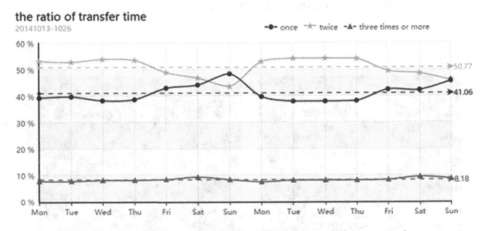

Fig. 2. Transfer time ratio for subway passengers during two weeks

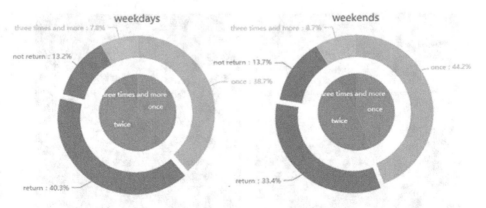

Fig. 3. Return ratio for weekdays and weekends

3.2   Transfer Time and Return Ratio

Figure 2 presents the transfer time ratio of subway passengers during the two weeks. From Monday to Thursday, the ratios of once, twice and three times or more a day are relatively stable. Twice a day has the highest proportion of 53.5 % on average. Next is once a day of 38.7 % on average. Three times or more a day has the lowest proportion of 7.8 % on average. From Friday to Sunday, the ratio of twice a day drops significantly compared with weekdays. On the contrary, the ratio of once a day increases significantly. At the same time, there is also a small rise for three times or more a day.

Further analysis of passengers who take the subway twice a day, we found that their return ratios reach up to 75.2 % on weekdays (especially from Monday to Thursday). Even on weekends (especially from Friday to Sunday), their return ratios also reach to 70.9 %. Here, return ratio means the percentage of those who have a round trip behavior. In general, for all the subway passengers, the conservative estimate of return ratio is 40.3 % for weekdays and 33.4 % for weekends. It may be the reason for the high coincidence ratio in general distribution of passenger originations and destinations.

3.3   General Distribution of Subway Trip Distance

Based on the two weeks records, we measured the probability $p(r)$ of travel distance r for all the subway stations. In order to improve the fitting precision, we investigated 14 continuous days and added them up to calculate the probability distribution. Figure 4 presents the displacement distribution $p(r)$ of linear and actual route distance data. We use LM (Levenberg-Marquardt) and UGO (Universal Global Optimization) method to fit the data and get the probability density function.

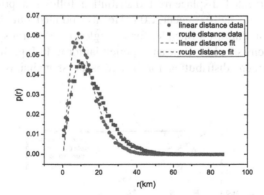

Fig. 4. Displacement distribution of linear and actual route distance

$$p(r) = \alpha \cdot \exp(\beta r) \cdot r^{\gamma}. \tag{1}$$

For linear distance distribution, $\alpha = 0.010$, $\beta = -0.198$, $\gamma = 1.619$. For actual route distribution, $\alpha = 0.006$, $\beta = -0.149$, $\gamma = 1.503$. As is shown in Fig. 4, the typical feature of trip distance distribution can be divided into two parts. For the front part,

$p(r)$ increases with r, which is controlled by power function. For the remaining parts of the image, $p(r)$ decays exponentially.

After calculating the derivative of function (1), we got the peak position $r_0 = |\gamma/\beta|$. On the one hand, the peak position can be regarded as a feature distance of subway. Within the feature distance, the number of subway passengers increases with the increase of distance. Outside the feature distance, the number of subway passengers decreases with the increase of distance. On the other hand, the peak position represents the most likely trip distance. We call it characteristics of travel radius (abbreviated as CTR). Same statistics analyses have been done to every single station. Results indicate that there is slight difference between different stations, but the average CTR is about 10 km.

4 Discussion

4.1 Trip Distance Related Work

In 2006, Brockmann [7] reported on a solid and quantitative assessment of human travelling statistics by analyzing the circulation of bank notes in the United States. He found that the distribution of travelling distances decays as a power law, with exponent 1.59. In 2008, Gonzalez [8] studied the trajectory of mobile phone users for a six-month period and found that the distribution of displacements over all users is well approximated by a truncated power-law, where the scaling exponent is 1.75. Both of them analyzed travel behavior through data unrelated to transportation, therefore, their results may not be affected by the mode of transportation.

In 2013, Yan [12] researched direct travel diaries of hundreds of volunteers and found that the aggregated displacement distribution follows a power law with an exponential cutoff, with the exponent 1.05. He also pointed out that from the perspective of maximum entropy principle, for a single mode of transportation, the displacement distributions should follow an exponential form. In Fig. 4(c) of ref [12], the cumulative displacement distribution for bus trips in Shanghai is similar to Fig. 5

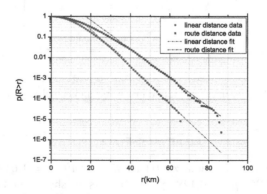

Fig. 5. Cumulative displacement distribution of linear and actual route distance

in this paper. It suggests that public transportation like bus and subway may capture the same fundamental mechanism driving human travel patterns.

Furthermore, different from ref [12], we believe that for a single mode of transportation, the displacement distribution not only decays exponentially, but also has a peak value through the data analysis above. Based on the physical meaning of peak position, it is speculated that different modes of transportations have different CTR. For example, CTR of walking must be less than taking the subway, whereas CTR of taking the airline flight must be greater than taking the subway.

4.2   Relationship Between Linear and Actual Route Distance

As can be seen in Figs. 4 and 5, the distributions of linear distance and actual route distance are not the same. And CTR of actual route distance is significantly greater than that of linear distance. It indicated that quite of the passengers make detours relative to the linear path when traveling by subway. But what is the relationship between linear and route distance reflected by the statistical results? To simplify the problem, we assume that statistically, for any path, the difference between linear distance (denoted by r_{il}) and actual route distance (denoted by r_{ir}) is denoted by Δr_i.

$$r_{ir} = r_{il} + \Delta r_i. \tag{2}$$

The difference Δr_i is affected by many factors. For example, Δr_i may come from the restriction of subway network topology or the diversity of travel preference. So the function relation of Δr_i is extremely complex. From the perspective of qualitative analysis, we only consider three typical cases: Δr_i is a constant value, or has positive correlation with linear distance, or has negative correlation with linear distance. We analyzed all the three possible hypotheses by means of numerical simulation. The method is shown below. As already described above, the probability density distribution of subway trip distance is as function (1). Firstly, ten thousand random numbers that follow the distribution above were generated to simulate the linear travel distance of ten thousand subway passengers. Secondly, Δr_i of three different forms were added to generate the actual route distance of the passengers. Figure 6 presents the linear and actual route distance distributions of three kinds of simulations.

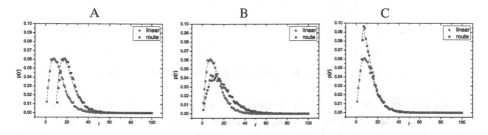

Fig. 6. Numerical simulation of three possible relationships

As is shown in Fig. 6A, when $\Delta r_i = const$, the distribution of actual route distance is only a translation transformation on the horizontal axis of linear distance. It indicates that if, for any path, the difference is a constant value, then CTR also increases the constant value, but the shape feature of their distributions stay the same. When $\Delta r_i \sim r_{il}$, the distribution of actual route distance is smoother than that of linear distance and CTR has increased proportionally. As is shown in Fig. 6B, the shape feature of both linear and route distance distribution is most similar to real statistical result in Fig. 4. It is speculated that for subway passengers in Beijing, the longer the trip is, the more detours they make. When $\Delta r_i \sim exp(-r_{il})$, as is shown in Fig. 6C, the distribution of actual route distance is steeper compared with that of linear distance, and CTR increased not that significantly.

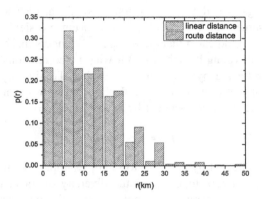

Fig. 7. Displacement distribution of linear and actual route distance in GUOMAO

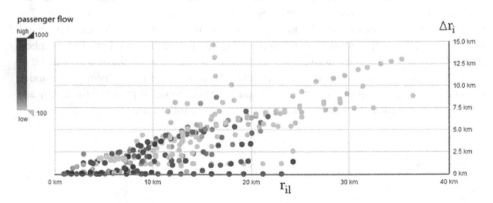

Fig. 8. Relationship between r_{il} and Δr_i of GUOMAO station

Take the hot spot, both in originations and destinations, GUOMAO station for example. Select the records of 20141015 (Wednesday) with a total of 74589 passengers starting from that station. As is shown in Fig. 7, the frequency statistics distribution of all the passengers' actual route distance is smoother than that of linear distance. CTR increased, but not that obviously. The distribution is similar to Fig. 4 to a certain degree. Figure 8 shows the scatter-plot of all of the 74589 passengers' r_{il} and Δr_i. The color of scatters represents the corresponding passenger flow. To some extent, with the increase of r_{il}, there exists a growing trend of Δr_i, but the actual relationships is much more complicated than we imagine.

5 Conclusion

In conclusion, based on the big data of subway records, we analyzed the general distribution of passenger flow and studied the trip characteristics of passengers, including return ratio and trip distance. Especially, when dealing with the trip distance of passengers, both linear distance and actual route distance of the same trip are considered.

We found that for a single mode of transportation, the displacement distribution not only decays exponentially, but also has a peak value through the data statistics. The peak position also represents the characteristics of travel radius (CTR). For subway, CTR is about 10 km. What's more, the distributions of linear distance and actual route distance are not the same. And CTR of actual route distance is significantly greater than that of linear distance, which indicates that quite of the passengers make detours relative to the linear path when traveling by subway. Further discussion implies that for subway passengers in Beijing, the longer the trip is, the more detours they make.

This paper presents a more precise estimate of human mobility distribution at urban scale, which can be applied to city planning or traffic simulation. Besides that, analyzing the difference between ideal and actual travel distance proposes a new perspective of the research on human mobility and may provide support on other socioeconomic dynamics covering spatiality and human travel. Further study could include the analysis of other transportation modes, such as bus or passenger car, to establish the urban transportation system framework.

Acknowledgments. The authors deeply appreciate support for this paper by the National Natural Science Foundation of China (Grant No. 91546111 and 71573154), the Research on the development strategy of national public safety science and technology (Grant No. 2014-ZD-02) and the Collaborative Innovation Center of Public Safety.

References

1. Horner, M.W., O'Kelly, M.E.: Embedding economies of scale concepts for hub network design. J. Transp. Geogr. **9**, 255–265 (2001)
2. Um, J., Son, S.W., Lee, S.L., Jeong, H., Kim, B.J., Stanley, H.E.: Scaling laws between population and facility densities. Proc. Natl. Acad. Sci. **106**, 14236–14240 (2009)
3. Balcan, D., Vespignani, A.: Phase transitions in contagion processes mediated by recurrent mobility patterns. Nat. Phys. **7**, 581–586 (2011)
4. Shunjiang, N., Wenguo, W.: Impact of travel patterns on epidemic dynamics in heterogeneous spatial metapopulation networks. Phys. Rev. E **79**, 016111 (2009)

5. Zheng, V.W., Zheng, Y., Xie, X., Yang, Q.: Collaborative Location and activity recommendations with GPS history data. In: International Conference on World Wide Web, pp.1029–1038 (2010)
6. Scellato, S., Noulas, A., Mascolo, C.: Exploiting place features in link prediction on location-based social networks. In: Proceedings of the 17th ACM SIGKDD International Conference on Knowledge Discovery and Data Mining, pp.1046–1054. ACM (2011)
7. Brockmann, D., Hufnagel, L., Geisel, T.: The scaling laws of human travel. Nature **439**, 462–465 (2006)
8. González, M.C., Hidalgo, C.A., Albert-László, B.: Understanding individual human mobility patterns. Nature **453**, 779–782 (2008)
9. Xin, L., Linus, B., Petter, H.: Predictability of population displacement after the 2010 haiti earthquake. Proc. Natl. Acad. Sci. **109**, 11576–11581 (2012)
10. Yoon, J., Noble, B.D., Liu, M., Kim, M.: Building realistic mobility models from coarse-grained traces. In: Proceedings of the 4th International Conference on Mobile Systems, Applications and Services, pp.177–190. ACM (2006)
11. Jiang, B., Yin, J., Zhao, S.: Characterizing the human mobility pattern in a large street network. Phys. Rev. E – Stat. Nonlinear, Soft Matter Phys. **80**, 1711–1715 (2009)
12. Yan, X.Y., Han, X.P., Wang, B.H., Zhou, T.: Diversity of individual mobility patterns and emergence of aggregated scaling laws. Sci. Rep. **3**, 454 (2013)
13. Baidu Map. http://api.map.baidu.com/lbsapi/getpoint/index.html
14. Zhao, B., Ni, S., Yong, N., Ma, X., Shen, S., Ji, X.: A preliminary study on spatial spread risk of epidemics by analyzing the urban subway mobility data. J. Biosci. Med. **3**, 15 (2015)

An Agent-Based Simulation of Heterogeneous Games and Social Systems in Politics, Fertility and Economic Development

Zining Yang[1,2](✉)

[1] Claremont Graduate University, 170 E. Tenth St., Claremont, CA 91711, USA
zining.yang@cgu.edu
[2] La Sierra University, 4500 Riverwalk Pkwy, Riverside, CA 92505, USA

Abstract. This paper studies both the macro and micro level, as well as the linkage between the two, to answer the question of how economic, political, and demographic factors impact a country's development trajectory. Combining system dynamics, agent-based modeling, and evolutionary games in a complex adaptive system, I formalize a simulation framework of Politics of Fertility and Economic Development (POFED) to understand the relationship between those factors over time. I validate the original system dynamics model with updated data and measure, fuse the endogenous attributes with non-cooperative game theory in an agent-based framework, and simulate the heterogeneous interactions between individuals. This paper demonstrates the linkage between macro environment and micro behavior. Simulations of real world scenarios show network emergence under different environments. The results suggest policy implications for societies at different stages of development.

1 Introduction

Rooted in international political economy, POFED is a quantitative, trans-disciplinary approach to understanding growth and development through the lens of interdependent economic, demographic, social and political forces at multiple scales, from individuals to institutions and society as a whole. In each country's development path, macro structure provides political, social and economic environment that constrains or incentivizes micro level human behavior, while micro level human agency can act, react and interact, thus shapes macro environment.

Previous literature in this field mostly focuses at macro level. Countries are used as unit of analysis or specific cases. Empirical research uses macro structural, society level variables, like GDP, fertility rate, and literacy rate among others to test different theories. Each one of these indicators is the sum of millions of human choices, sampled at arbitrary annual frequencies from an imperfect data and population distribution. However, the micro level is very poorly studied, and the linkage between macro constraints and micro level choices remains undiscovered.

Therefore, this paper studies income level, fertility decision, and education at micro level of human agency, to better understand how individuals behave under different environments. Additionally, I investigate individual choice feedback mechanism on

© Springer International Publishing Switzerland 2016
K.S. Xu et al. (Eds.): SBP-BRiMS 2016, LNCS 9708, pp. 71–82, 2016.
DOI: 10.1007/978-3-319-39931-7_8

macro societal trends and conditions. The results confirm POFED theory, demonstrate individual strategy choice matters significantly to development, and offer policy implications for different societies.

2 POFED in Complex Adaptive Systems

Scholars in international political economy have done considerable research on development, focusing on the interrelationship between income, fertility, human capital, and political development, measured as political stability and political capacity [4, 8, 9, 12, 13]. Feng et al. [13] presents a formal model that characterizes the two trajectories of development – a poverty trap with persistent economic stagnation, and industrialization and rising incomes, and establishes that the interaction between politics and economics determines which path a nation travels. In more recent POFED literature, Feng et al. [12] presents a dynamic general equilibrium model that formalizes the political mechanisms that prompt demographic change and augment economic development. Abdollahian et al. [1] emphasizes the dynamic interrelationships between income, fertility, human capital, political effectiveness, and social stability. They show that fertility rates b depend on income level y; and that income depends on past income and political conditions. There is generational feedback on the creation of human capital h, as increased education would increase political capacity x and income, thus reduces political instability s. Instability also has a temporal feedback and impacts fertility decision. Their system of equations describes how the five main components work at society level, but is not empirically tested.

However, one major flaw associated with dynamic general equilibrium models is the assumption of a perfect world, which does not hold in reality [11]. Another general critique of formal or empirical macro level, structural analysis across most of social science is that aggregate structures often help explain or predict necessary, but not sufficient conditions of political, economic and social phenomena, as the emergent behavior is not captured [7, 11]. Policy makers often face information about complex systems different from what is assumed by traditional analysis [17]. Many individual level explanations, spanning positive political theory, microeconomics and game theoretic behavior might provide insights into human agency and thus offer the promise of theory sufficiency.

As macroscopic structures emerge from microscopic events lead to entrainment and modification of both, co-evolutionary processes are created over time. I posit a new approach where agency matters: individual game interactions, strategy decisions and outcome histories determine an individual's experience. These decisions are constrained or incentivized by the changing macroeconomic, demographic pattern, social and political environment via POFED theory, conditioned on individual attributes at any particular time. Emergent behavior results from individuals' current feasible choice set, conditioned upon macro environment. Conversely, progress on economic development, the level of internal instability, and population structure emerge from individuals' behavior interactions.

In order to create a simulation to capture the complexity of development, I extend Abdollahian et al.'s system dynamic representation of POFED theory towards integrated macro-micro scales in an agent-based framework. I first instantiate two systems of difference equations that are empirically validated using updated data from World Bank [26] and new measure from Fisunoglu [14] and Kugler and Tammen [16] using Three Stage Least Square estimation. This technique permits correlations of the unobserved disturbances across several equations, as well as restrictions among coefficients of different equations, and improves upon the efficiency of equation-by-equation estimation by taking into account such correlations across equations. Since understanding the interactive effects of macro-socio dynamics and individual agency in intra-societal transactions are key elements of a complex adaptive systems approach, I then fuse the system dynamic component to agent attribute changes with a non-cooperative Prisoner's Dilemma game following Axelrod [5, 6] and Nowak and Sigmund [20, 21]. This design allows the simulation of intro-societal economic transactions, which, in return, shapes the macro system where interactions take place. I finally explore the parameter space, conduct sensitivity test and simulate societies at different stages of development.

3 An Agent-Based Model

I propose an agent-based model in a complex adaptive system framework that captures both macro level changes and micro level behavior by incorporating system dynamics component and game theory component. Following the work by Abdollahian et al. [2, 3], my agent-based model has both the interactive effects and feedbacks between individual human agency as well as the macro constraints and opportunities that change over time for any given society. Individual decisions are affected by other individuals, social context, and system states. These elements have first and second order effects, given any particular system state or individual attributes. Such an approach attempts to increase both theoretical and empirical verisimilitude for some key elements of complexity processes, emergence, connectivity, interdependence and feedback found throughout several disciplines across all scales of modernization and human development. Figure 1 depicts the high level process and multi-module architecture. There are three modules in the agent-based model: micro agent process, macro society process, and heterogeneous evolutionary game process.

In micro agent process, individual agents behave as a system in terms of updating income, fertility and education decisions. Each individual agent carries all three variables that are randomized from the society's distribution. I maintain individual agent variable relationships and changes following the latest POFED literature [1]. Here I use empirically validated parameter values from Three Stage Least Square estimation as a good first approximation. This method has been widely used by many scholars [2, 3] to simulate the dynamic process at individual level. These endogenously derived individual agent variables impact how economic transaction games occur, based on society variables either increasing or decreasing individual wealth and ultimately societal productivity [6].

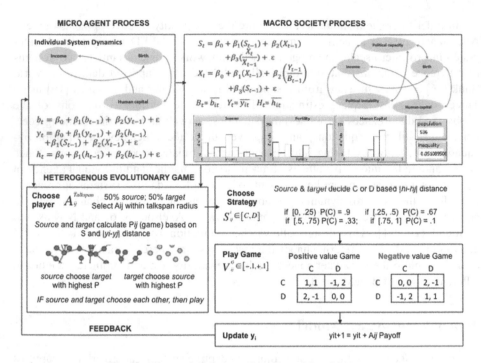

Fig. 1. Three-module agent-based model architecture.

In macro society process, instead of taking each individual agent as a system, this module takes the entire society as the system, with political instability, political capacity, economic condition, human capital, and fertility rate as main attributes. This module is critical as it connects micro individual level and macro society level. Society economic condition is aggregated from individual wealth by taking the mean. Human capital is aggregated from individual level of education, and fertility rate is also aggregated from individual level in the same way. The feedback loop is completed in the way that initial individual variables are randomized from the society distribution, get updated in micro agent process and evolutionary game process, then get aggregated at society level and interact with other society variables, while society variables also impact the evolutionary game process. I also use empirically validated parameter values from Three Stage Least Square estimation in this module. The updated instability is brought into the evolutionary game process to affect the probability that agents interact with each other. This feedback loop allows the study on how individual behavior changes macro environment, and how environment in turn impacts individual behavior.

In heterogeneous evolutionary game process, I choose to focus on non-cooperative game in the macro political stability environment. Prisoner Dilemma game is chosen because it allows agents to choice between maximizing individual benefit and mutual benefit. Evolutionary game theory provides insights into understanding individual, repeated societal transactions in heterogeneous populations (Sigmund, 1993). Social co-evolutionary systems allow each individual to either influence or be influenced by all other individuals as well as macro society [24, 27], perhaps eventually becoming coupled and quasi-path interdependent. To model communications and technology diffusion for frequency and social tie formation [18], I have agent i evaluate the likelihood of conducting a simple socio-economic transaction with agent j based on similarity of income level $|y_i - y_j|$, stability of the environment, and physical distance, which reflects level of technology the society obtains. Each agent will choose the one with the highest likelihood to conduct the socio-economic transaction game, and the interaction will take place only when both agents choose each other. This approach reflects recent work on the importance of both dynamic strategies and updating rules based on agent attributes affecting co-evolution [2, 3, 15, 19].

Once agents decide to play, they choose strategies based on $|h_i - h_j|$, since messages close to a receiver's position has little effect, while those far from a receiver's position are likely to be rejected [22]. Small difference of human capital leads to high probability of cooperation, while large difference results in high probability of defective strategy. Following Abdollahian et al. [2, 3], I specifically model socio-economic transaction games as producing either positive or negative values as I want to capture behavioral outcomes from games with both upside gains or downside losses. Subsequently, A^{ij} games' V^{ij} outcomes condition agent y^i_{t-1} values, modeling realized costs or benefits from any particular interaction. The relative payoff for each agent is calculated based on simple PD, non-cooperative game theory [10, 20, 21, 23] where T > R>P > S. The updated $y^i_{t-1} = y^i_t + A^{ij}$ game payoff for each agent then gets added to the individual's variables for the next iteration. I then repeat individual endogenous processing, aggregated up to society as a whole and repeat the game processes for $t + n$ iterations, where n is the last iterate. In this module, Ai strategies are adaptive, which affect A^{ij} pairs locally within an approximate radius as first order effects. Other agents, within the society but outside the reaching radius, are impacted through cascading higher orders.

4 Results

I implement the agent-based model in NetLogo [25]. The baseline initial population is 500 to represent a sample of any given population. The state variables for this model are fertility decision, education, and income. Global variables are level of instability and relative political capacity, which are setup at society level. Since society variables do not change on a daily basis, I approximate one time step as one month given data calibration for a simulated time span. This design allows me to study the dynamics of the key variables with reasonable frequency, and the 20-year period is also proper for a cycle in the study of political economy. In order to make generalizable model inferences, I conducted a quasi-global sensitivity with parameter for fertility, income, human capital, political capacity, political instability, and technology at low, medium, and high levels, resulting in over 17,000 runs across 240 time steps.

I compare the pooled OLS results from baseline model that only includes macro level variables and models that include both macro and micro level variables. With aggregated income as dependent variable, macro level variables only explain 20 % of the variance. With the number of micro level interactions added, the new model adds explanatory power by 2.6 %. The dramatic increase in model fit comes from individual choice. When using number of cooperation as the additional independent variable, adjusted R^2 more than doubles to 54.3 %; while counting both cooperative strategies and defective strategies gives the best model fit of 55.3 %. The sensitivity test also confirms the original POFED theory that negative value of instability significantly speeds the pace of economic development, while technology has a positive impact, as increasing individual agents' ability to reach other like-minded agents spurs cooperation. More importantly, the results demonstrate that micro level behavior helps explain macro level dynamics. While individuals communicate and make deals with each other, more products and services become available while the cost of which goes down. Benefits are derived from specialization of products and services, which outweighs the economic and social costs by achieving higher efficiency. Cooperation pays higher dividends, while defective strategies reduce social wealth. In other words, this model captures the micro level behavior that can better explain macro level phenomena.

After confirming that agent-based model more effectively captures the relationship between economic, political and demographic factors than traditional econometric models, I conduct simulation for a few societies to understand the growth path under different levels of development. In this process, I adjust parameters for population density, technology level, political capacity and instability, as well as the distribution on income, fertility, and education for the population. Cases presented below include China 1960–1980, Japan 1970–1990, and Afghanistan 1980–2000.

To explore more details of individual agents interaction, I also show below the network in the society at the beginning (tick 1, top left), one third of the time span simulated (tick 80, top right), two thirds of the simulated period (tick 160, bottom left), and final stage (tick 240, bottom right) on the right part of Fig. 2. Size of the agents indicates their wealth level, and color shows education level, with high in blue and low in red. The high population density is well presented in the graph, even from tick 1. However, at the beginning of the simulation, there are not many interactions among agents, because the level of instability is relatively high so individuals do not have the incentive to conduct socio economic transaction games, so only a few links show up. As time goes, the network density increases. Although individuals cannot reach others who are physically far away from them, high population density ensures the quantity of interactions. As instability weakens and individuals' income gap decreases, there are more and more people involved in the socio economic transaction games, as can be seen in the bottom two graphs. The size of individual agents also slightly increase in general, as a result of increased wealth due to cooperative strategies. This perfectly reflects the reality that Chinese people start to switch focus from political conflict to economic development during the period from 1960 to 1980, especially after the country opened up in 1978. Individual emergent behavior at micro level feeds back to the macro level, impacting the society's growth path.

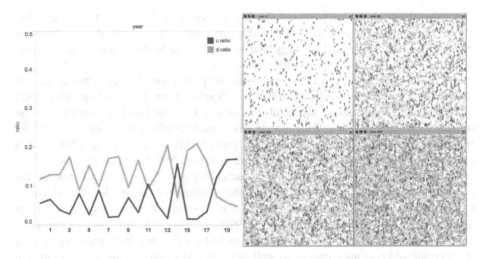

Fig. 2. Cooperation vs. defection and network formation in China's growth path.

After looking at the details of the slowly growing society, the next case I want to focus is a well-developed society, Japan, from 1970 to 1990. There is income convergence, accompanied with human capital dynamics. Due to the development of higher education and increased enrollment of women, the level of human capital increases, though at minimal level as it almost already reached the "celling" of this indicator. With more female receiving higher education and participating in workforce, people tend to marry at an older age, thus time left for giving birth to children shrinks. The slow but steady economic development also increases the opportunity cost of raising children, further reducing fertility rate across the entire period. Government capacity was relatively stable, due to the stable development of other factors. There was only a small increase, due to more control over physical resources. Then I also explore the micro level agent interactions shown in Fig. 3. The left side shows two interesting patterns that are very different from what can be seen in the previous case. Firstly, the percentage of agents interacting with each other is large. Due to high level of technology development, individuals are able to reach out to others who are far from them with telephone, cellphone, Internet and so on. The incentive of playing transaction game with each other is also high because of high level of social stability. It is consistent with the literature [13] that stable environment provides people with more incentive to invest, thus increases the potential of economic development. Across the entire simulated timespan, there are more than half of the agents interacting with each other and playing the transaction game and among them, more than half choose cooperative strategies, as the second important pattern. The proportion of individual agents playing cooperation is relatively stable over time, fluctuating between 30 % and 40 %. However the proportion of individual agents choosing to defect keeps declining over time, from 30 % to less than 5 %. The reduction of defective strategy matches with the aggregated income growth dynamics in Japan. This match also confirms the theory that more cooperative strategy and less defection leads to economic development and this is again revealed in the simulation.

From the right side of Fig. 3, one can see the population density is not as high as China's, however the networks are still dense. Starting from the very beginning, there are more interactions among agents in this society than the previous mainly because individuals are able to reach further due to advanced technology and social stability. As time goes, the density of network increases, though still not as much as China's. It is not only because the population density here is lower, but also because it is actually more difficult for individuals to match with each other once the feasible set strategy and less defective strategy accompanied with, slightly higher individual's level of education as compared to tick 1, which helps enhance the economy of the society. This pattern also coincides with real world Japan from 1970 to 1990. Although just one particular simulation, what is critical is that co-evolutionary behavior results in path dependence of economic and education change as well as being a key determinant for development outcomes. Moreover, changes towards cooperation leads to increasing wealth over time (Fig. 4).

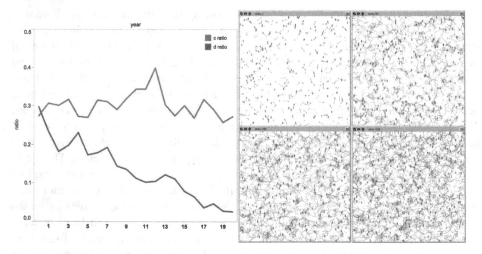

Fig. 3. Cooperation vs. defection and network formation in Japan's growth path.

Finally I run the model to simulate an underdeveloped society, Afghanistan, from 1980 to 2000. High level of political instability greatly hinders economic development, and keeps fertility rate at a very high level. Individuals choose to have more children because the cost of raising children is relatively low and the chances the children survive are also low. With a large number of young people but very little resources in the society, the level of human capital is very limited. Although increasing at a steady

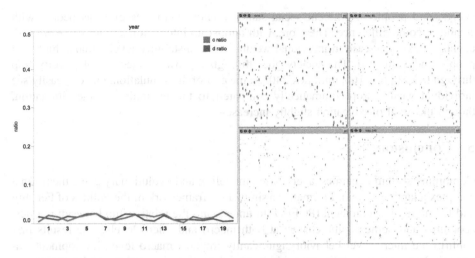

Fig. 4. Cooperation vs. defection and network formation in Afghanistan's growth path.

rate, it takes time for the society to accumulate a certain level of human capital that can enhance economic growth. Low level of economic development and human capital also constraints political capacity. The government can only extract very limited tax resources and human resources, which in turn limits its ability to facilitate education or economic growth. Turning to micro level, one can see both the red line and the green line stay at the bottom of the graph, indicating at each iteration, there are only very few individuals who are involved in the socio-economic transaction game. The main reasons are, firstly, high political instability limits individual's incentive to interact with each other; and secondly, low level of technology constraints individual's ability to reach each other. The majority of people in that society can only use telephone or mail to contact each other, which allows much less communications and transactions than what Internet and cellphone can facilitate. The other interesting pattern is the levels of cooperation and defection are very similar over time. The relatively high level of defection is also another reason why economy does not grow. This defective pattern coincides with the trend of economic stagnation and low level of political capacity that we see in the previous panel of graphs.

I also look at the network dynamics of Afghanistan between 1980 and 2000. Initially, I see low developed society with high income inequality and polarization on the education continuum. Compared to the society of China and Japan, Afghanistan a much higher proportion of undereducated population. There are almost no interactions between individuals because of three reasons: first, the society is highly instable, which reduces people's incentive to conduct socio-economic transaction games. Second, the level of technology in that society is so low that individuals have very limited capacity to reach others; so give that relatively low population density, very few interactions can be expected. Finally, there is high income inequality among individual agents, and people at different social status do not interact as often as people with smaller income difference. Therefore, with almost no interaction and no value created, agents quickly decrease in individual wealth as the instability level remains high and people undereducated. The society goes through a growth trajectory of poverty trap through tick 80, tick 160, until tick 240. At the end of the simulation, one can easily see individual wealth drops dramatically compared to the beginning of the simulation, though the level of education slightly increases.

5 Conclusion

Combining system dynamics, agent-based modeling and evolutionary game theory in a complex adaptive system, I formalize a simulation framework of the Politics of Fertility and Economic Development (POFED) to understand the relationship between politics, economic and demographic change at both macro and micro levels. The results first confirm the micro level behavior significantly impacts macro level development trajectory. The more people interact with each other, the more value can be potentially created. Besides, what matters most is individual agent's strategic choice when they play socio-economics transaction games. Cooperation pays higher social dividends on average. Individual's mutual cooperative behavior creates trust among each other, which enhances both political stability and economic growth. On the other hand, defection reduces social wealth, in addition to its negative impact to the instability in the society, which comes from second order effect. There is also feedback from macro level variables to individual agents, who update their attributes and change the pace and tempo of socio-economic transactions, which reinforce macro level development. In other words, this approach that combines both levels captures the micro level behavior that can better explain macro level phenomena.

This simulation model creates a baseline for current policy efforts, showing when poverty or growth is likely to occur. Policy implications can be inferred from three simulation cases. For developing societies like China, besides stabilizing the macro environment with increasing technology, it is essential to maintain high level of interactions among individuals, while promoting cooperative strategy and eliminating defective strategy to create more value. For well-developed societies like Japan, with stable environment and high level of technology, it is critical to maintain low level of defective strategy, while increasing the number of interactions that use cooperative strategy. For underdeveloped societies like Afghanistan, with high level of instability and low level of human capital, the effective policy of getting out of poverty trap is to increase the number of interactions among individuals while ensuring cooperative strategy and eliminating defective strategy.

References

1. Abdollahian, M., Kugler, J., Nicholson, B., Oh, H.: Politics and power. In: Kott, A., Citrebaum, G. (eds.) Estimating Impact: A Handbook of Computational Methods and Models for Anticipating Economic, Social, Political and Security Effects, pp. 43–90. Springer, New York (2010)

2. Abdollahian, M., Yang, Z., Coan, T., Yesilada, B.: Human development dynamics: an agent-based simulation of macro social systems and individual heterogeneous evolutionary games. Complex Adapt. Syst. Model. **1**(1), 1–17 (2013)
3. Abdollahian, M., Yang, Z., Neal, P.: Human development dynamics: an agent based simulation of adaptive heterogeneous games and social systems. In: Kennedy, W.G., Agarwal, N., Yang, S.J. (eds.) SBP 2014. LNCS, vol. 8393, pp. 3–10. Springer, Heidelberg (2014)
4. Arbetman, M., Kugler, J.: Political Capacity And Economic Behavior. Westview, Boulder (1997)
5. Axelrod, R.: The Evolution Of Strategies In The Iterated Prisoner's Dilemma. In: Davis, L. (ed.) Genetic Algorithms and Simulated Annealing, pp. 32–41. Morgan Kaufman, Los Altos (1987)
6. Axelrod, R.: The Complexity of Cooperation: Agent-Based Models of Competition and Collaboration. Princeton University Press, Princeton (1997)
7. Buchanan, M.: Economics: meltdown modeling. Nature **460**(7256), 680–682 (2009)
8. Chen, B., Feng, Y.: Some political determinants of economic growth: theory and empirical implications. Euro. J. Politi. Econ. **12**(4), 609–627 (1996)
9. Chen, B., Feng, Y.: Determinants Of economic growth in China: private enterprise, education and Openness. China Econ. Rev. **11**(1), 1–15 (2000)
10. Dixit, A., Skeath, S.: Games of Strategies. Norton & Company, New York (2009)
11. Farmer, J.D., Foley, D.: The economy needs agent-based modeling. Nature **460**(7256), 685–686 (2009)
12. Feng, Y., Kugler, J., Swaminathan, S., Zak, P.J.: Path to prosperity: the dynamics of freedom and economic development. Inter. Interact. **34**(4), 423–441 (2007)
13. Feng, Y., Kugler, J., Zak, P.J.: The politics of fertility and economic development. Inter. Stud. Q. **44**(4), 667–693 (2000)
14. Fisunoglu, F.A.: Beyond the Phoenix Factor: Consequences of Major Wars and Determinants of Postwar Recovery. Claremont Graduate University (2014)
15. Kauffman, S.A.: The Origins of Order: Self-Organization and Selection in Evolution. Oxford University Press, Oxford (1993)
16. Kugler, J., Tammen, R.L.: The Performance of Nations. Rowman & Littlefield, New York (2012)
17. Lempert, R.J.: A new decision sciences for complex systems. Proc. Natl. Acad. Sci. **99**, 7309–7313 (2002). suppl.3
18. McPherson, M., Smith-Lovin, L., Cook, M.: Birds of a feather: homophily in social networks. Ann. Rev. Sociol. **27**, 415–444 (2001)
19. Moyano, L.G., Sanchez, A.: Spatial Prisoner's Dilemma with Heterogeneous Agents. Elsevier Manuscript, New York (2013)
20. Nowak, M.A., Sigmund, K.A.: Strategy of win-stay, lose-shift that outperforms tit-for-tat in the prisoner's dilemma game. Nature **364**, 56–58 (1993)
21. Nowak, M.A., Sigmund, K.A.: Evolution of indirect reciprocity by image scoring. Nature **393**, 573–577 (1998)
22. Siero, F.W., Doosje, B.J.: Attitude change following persuasive communication: integrating social judgment theory and the elaboration likelihood model. J. Soc. Psychol. **23**, 541–554 (1993)
23. Sigmund, K.: Games of Life. Oxford University Press, Oxford (1993)
24. Snijders, T.A., Steglich, C.E., Schweinberger, M.: Modeling the co-evolution of networks and behavior. In: Longitudinal Models the Behavioral and Related Sciences, pp. 41–71 (2007)

25. Wilensky, U.: NetLogo, Center for Connected Learning and Computer-Based Modeling, Northwestern University, Evanston (1999). http://ccl.northwestern.edu/netlogo/
26. World Bank: The World Development Indicators. Washington, DC (2014)
27. Zheleva, E., Sharara, H., Getoor, L.: Co-evolution of social and affiliation networks. In: Proceedings of the 15th ACM SIGKDD International Conference on Knowledge Discovery and Data Mining. ACM, New York (2009)

On Discrimination Discovery Using Causal Networks

Lu Zhang[✉], Yongkai Wu, and Xintao Wu

University of Arkansas, Fayetteville, USA
{lz006,yw009,xintaowu}@uark.edu

Abstract. Discrimination discovery is an increasingly important task in the data mining field. The purpose of discrimination discovery is to unveil discriminatory practices on the protective attribute (e.g., gender) by analyzing the dataset of historical decision records. Different types of discrimination have been proposed in the literature. We aim to develop a framework that is able to deal with all types of discrimination. We make use of the causal networks, which effectively captures the existence of discrimination patterns and can provide quantitative evidence of discrimination in decision making. In this paper, we first propose a categorization for various discrimination. Then, we present our preliminary results on four types of discrimination, namely system-level direct discrimination, the system-level indirect discrimination, group-level discrimination, and individual level discrimination. We have conducted empirical assessments on real datasets. The results show great efficacy of our approach.

1 Introduction

Discrimination discovery has been an active research area recently [18,20]. Discrimination generally refers to an unjustified distinction of individuals based on their membership, or perceived membership, in a certain group, and often occurs when the group is treated less favorably than others. Laws and regulations disallow discrimination on several grounds, such as gender, age, marital status, sexual orientation, race, religion or belief, membership in a national minority, disability or illness, denoted as *protected attributes*. Various business models have been built around the collection and use of individual data including the above protected attributes to make important decisions like employment, credit, and insurance. Consumers have a right to learn why a decision was made against them and what information was used to make it, and whether he was fairly treated during the decision making process. Therefore, the historic data and the predictive algorithms must be carefully examined and monitored for potential discriminatory outcomes for disadvantaged groups.

Our society has endeavored to discover discrimination, however, we face several challenges. First, discrimination claims legally require plaintiffs to demonstrate a causal relationship between the challenged decision and a protected status characteristics. However, randomized experiments, which are gold-standard for causal relationship inferring in statistics, are not possible or not cost-effective

© Springer International Publishing Switzerland 2016
K.S. Xu et al. (Eds.): SBP-BRiMS 2016, LNCS 9708, pp. 83–93, 2016.
DOI: 10.1007/978-3-319-39931-7_9

in the context of discrimination analysis. In most cases, the causal relationship needs to be derived from the observational data not controlled experiments. Second, algorithmic decisions, which may not be directly based on protected attribute values, could still incur discrimination against the vulnerable classes of our society.

The state of the art of discrimination discovery [18,20] has developed different approaches for discovering discrimination. These approaches classify discrimination into different types including group discrimination, individual discrimination, direct and indirect discrimination. However, these work are mainly based on correlation or association-based measures which cannot be used to estimate the causal effect of the protected attributes on the decision. In addition, each of them targets one or two types of discrimination only. In real situations, several types of discrimination may present at the same time in a dataset. Thus, a single framework that is able to deal with all types of discrimination is a necessity.

We propose to investigate all types of discrimination in our research. We categorize various discrimination based on whether discrimination is across the whole system, occur in one subsystem, or happen to one individual, and whether discrimination is a direct effect or indirect effect on the decision. Then, we propose to develop a single unifying framework to capture and measure different discrimination types. We make use of the causal networks [21], which effectively captures the existence of discrimination patterns and can provide quantitative evidence of discrimination in decision making. Based on the causal networks, we present our preliminary results on system-level direct and indirect discrimination, group and individual-level discrimination. Empirical assessments for the system-level direct discrimination on two real datasets have been conducted. The results show great efficacy of our approach.

The rest of this paper is organized as follows. Section 2 presents the discrimination categorization. System-level direct and indirect discrimination is discussed in Sects. 3 and 4. Section 5 deals with group and individual-level discrimination. The experimental setup and results for system-level direct discrimination are shown in Sect. 6. Section 7 summarizes the related work. Finally, Sect. 8 concludes the paper.

2   Discrimination Categorization

We assume the historical dataset $\mathcal{D}$ contains a subset of explicitly specified protected-by-law attributes, some decision attributes, and other non-protected attributes. For ease of representation, we assume that there is only one protected attribute and one decision. We denote the protected attribute by C, associated with domain values of the protected group c^- (e.g., female) and the non-protected group c^+ (e.g., male); and denote the decision by E, associated with domain values of positive decision e^+ and negative decision e^-. Our formulation and analysis can be generalized to situations which involve multiple protected attributes and decisions.

Several types of discrimination have been proposed in the literature. In [18], discrimination is classified as group discrimination, individual discrimination, direct and indirect discrimination. Accordingly, different types of discrimination discovery techniques have been developed, e.g., association rules for group discrimination discovery [16,17], situation testing for individual discrimination discovery [11], correlation analysis considering explanatory attributes for direct discrimination discovery [6,25], rule inference for indirect discrimination discovery [5,16], and fair classification for group/individual fairness [3].

All the above approaches are mainly based on correlation or association. In discrimination discovery, it is critical to derive causal relationship, and not merely association relationship. We need to determine what factors truly cause discrimination and not just which factors might predict discrimination. Besides, we need a unifying framework and a systematic approach for determining all types of discrimination rather than using different types of techniques for some specific types of discrimination. To this end, we first categorize various discrimination types with the following two dimensions:

- Discrimination Level. The decision making process can be modeled as a stochastic system where discrimination may happen. Discrimination can exist across the whole system, occur in one particular subsystem, or happen to one particular individual. We call them system-level discrimination, group-level discrimination, and individual-level discrimination, respectively.
- Discrimination Manner. Discrimination can be either the direct causal effect of C on E or indirect causal effect which passes the effect of C on E via some intermediate attributes. We call the former as direct discrimination and the latter indirect discrimination.

It is worth pointing out that a discrimination can combine two features mentioned above. As an example, there can be a direct discrimination at the system-level, thus forming a system-level direct discrimination.

For a quantitative measurement of discrimination, a general legal principle is to compare the proportion of positive decisions between the protected group and non-protected group [18]. The comparison can be measured by differences or rates of these proportions. In the proposed research, we will use *risk difference*, i.e., the difference in the the proportion of positive decisions between the protected group and non-protected group, as our discrimination measure. The results can be easily applied to other measures such as risk ratio, odds ratio, etc. In general, risk difference can be performed within a subpop-

Table 1. University admission: row 1 is the number of applicants and row 2 is the acceptance rate

(a) Case I				(b) Case II				(c) Case III			
Math		Biology		Math		Biology		Math		Biology	
Female	Male	Female	Male	Female	Male	Female	Male	Female	Male	Female	Male
800	200	200	800	200	800	800	200	800	200	200	800
22 %	20 %	42 %	40 %	15 %	26 %	35 %	44 %	26 %	15 %	35 %	44 %

ulation under a partition using a subset of attributes. Formally, given a sub-population $\mathbf{b}$ produced by a partition $\mathbf{B}$, risk difference can be denoted as $\Delta P|_{\mathbf{b}} = P(e^+|c^+, \mathbf{b}) - P(e^+|c^-, \mathbf{b})$. We say that C negatively affects E within subpopulation $\mathbf{b}$ if $\Delta P|_{\mathbf{b}} \geq \tau$, $(\tau > 0)$, where τ is a threshold for discrimination depending on the law. For instance, the 1975 British legislation for sex discrimination sets $\tau = 0.05$, namely a 5 % difference. In the following, we use an example to illustrate why it is imperative to develop this categorization framework and determine correctly the type of discrimination under investigation.

Illustrative Example. Suppose that in a university admission system, we have three attributes, the applicants' *gender, major applied,* and *admission decision,* and assume there are two majors, *math* and *biology.* They have different acceptance standards: the competition for *math* is more challenging than that for *biology.* Meanwhile, the choice of the major depends on the gender of an applicant, as males are more likely to apply for one major whereas females prefer the other. Table 1 shows three cases. In Case I, we see the overall admission rate is 36 % for males, but only 26 % for females. However, the claimed discrimination against the whole university may be groundless. This is because, when examining each major, no major is biased against females but to some extent in favor of females, i.e., $\Delta P|_{\{math\}} = -0.02$ and $\Delta P|_{\{biology\}} = -0.02$. In Case II, the overall admission rate of females is 31 %, which is slightly higher than that of males 30 %, showing that females and males have approximately equal chances to be accepted. However, there is clear discrimination against female, as in each major the admission rate of females is significantly lower than that of males ($\Delta P|_{\{math\}} = 0.11$ and $\Delta P|_{\{biology\}} = 0.09$). In Case III, the biases in the admission rates in the two majors are opposite, i.e., $\Delta P|_{\{math\}} = -0.11$ and $\Delta P|_{\{biology\}} = 0.09$. Hence there would be insufficient evidence to litigate the university for discrimination, since university-wide discrimination, probably because of a universal prejudice against females among admission officers, or a biased admission procedure commonly adopted by all majors, should be presented in each major of the university. As we can see, solely examining either the overall admission rates or the admission rates in any one major would lead to incorrect conclusions. On the other hand, if admission decision is made at the major level and a biased admission procedure could only be adopted by a particular major, the biology major could be litigated for discrimination against females.

The phenomenon shown in Case I and II is known as the Simpson's paradox [14], which indicates we need to consider other attributes correctly when determining and measuring the discrimination. The phenomenon shown in case III implies that system-level/group-level discrimination is a negative effect persisting in all subpopulations, given partition $\mathbf{B}$. This makes discrimination different from general causalities in that it is a persistent effect. In social and psychological sciences, three sources of discrimination are generally identified: prejudice, statistical thinking, and unintentionality [18]. All these factors can be considered as persistent across the system (for system-level discrimination) or the components within a subsystem (for group-level discrimination) and hardly change. Thus, discrimination should be considered as persistent and does not reverse or disappear under situations where the sources of discrimination are supposed to exist.

While direct discrimination is about the direct causal effect of C on E, indirect discrimination concerns about the indirect causal effects that may also be considered as discrimination. In this case, C does not necessarily have direct effect on E. Instead, it affects E via some apparently neutral attributes which are correlated with the protected attribute, hence eventually results in an unfair treatment of the protected group. Our proposed causal network based discrimination discovery framework attempts to capture and measure all types of discrimination shown in the above categorization. In this paper, we present our preliminary results for system-level direct discrimination and system-level indirect discrimination.

3 System-Level Direct Discrimination

System-level direct discrimination deals with the direct causal effect of C on E across the whole system. The direct causal effect of C on E is captured by the direct arc from C to E, i.e., $C \rightarrow E$ in the causal network. Thus, not all causal paths but only the *direct* arc may represent discrimination. As discussed above, discrimination cannot be inferred directly from the presence of the direct arc due to the intrinsic differences between discrimination and general causalities. In addition to the presence of the direct arc, we need to measure the exact causal effect carried by the arc under a correct partition **B**. In order the do this, we need to suppress all other influences, some of which are spurious, some of which, although causal, can be explained by other attributes and hence are not regarded as discrimination. In other words, partition **B** must suppress influences by all other attributes. Otherwise, it cannot generate a correct and meaningful partition.

We employ the "path blocking" technique [15] to suppress all other influences. A path can be blocked by conditioning on a set of nodes not containing the two end-nodes. Upon blocked, the effect originally transmitted through the path is suppressed in each subpopulation under the partition defined by the set of nodes. If all paths other than arc $C \rightarrow E$ are blocked, all undesired influences are suppressed. We refer to the set of nodes using which we can measure the exact causal effect carried by $C \rightarrow E$ as the *block set*. As defined in [15], a node set **S** not containing C or E blocks a path p between nodes C and E if either (1) p contains at least one noncollider X in set **S**, or (2) p contains at least one collider X, and X and all its descendants are outside set **S**. A block set should block all paths from C to E. In addition, as we cannot measure the exact causal effect of $C \rightarrow E$ if we have the knowledge about the consequences caused by E, no E's descendant should be contained in the block set.

We consider the system-level direct discrimination a persistent effect across the system (e.g., university-wide discrimination should cause bias in each major of the university). Thus, given a block set **B**, the discriminatory effect presents if $\Delta P|_\mathbf{b} \geq \tau$ holds for each subpopulation **b**. If there are multiple block sets in a causal network, we observe that inconsistent conclusions can be drawn according to different block sets. If discrimination does exist, the discriminatory effect must present under each correct partition. Thus, a discrimination claim is not convincing if inconsistent conclusions drawn under different partitions. Therefore, to make a discrimination claim, we need to examine all block sets and they must reach a consistent conclusion.

Based on the above analysis, we propose our discrimination criterion. Formally, we use $\mathbf{B}$ to denote a block set as well as its defined partition, and use $\mathbf{b}$ to denote each subpopulation.

Definition 1. *Discrimination is considered to present if inequality $\Delta P|_{\mathbf{b}} > \tau$ holds for each instance $\mathbf{b}$ associated with each block set $\mathbf{B}$.*

In real situations, $\Delta P|_{\mathbf{b}}$s may vary from one subpopulation to another due to randomness in the decision making process and sampling. The $\Delta P|_{\mathbf{b}}$ values of a few instances $\mathbf{b}$ could be less than τ or even negative although the majority of $\Delta P|_{\mathbf{b}}$ values are significantly greater than τ. To better captures discrimination under the context of randomness and sampling, we propose a relaxed (τ, α)-discrimination criterion, which examines whether the likelihood requirement $P(\Delta P|_{\mathbf{B}} \geq \tau) \geq \alpha$ holds. In addition, we also propose an efficient way to test the requirement in Definition 1: instead of examining each block set $\mathbf{B}$, examining one node set $\mathbf{Q}$, which is the set of E's all parents except C, i.e., $\mathbf{Q} = \mathrm{Par}(E) \backslash \{C\}$, is sufficient for guaranteeing the requirement. Please refer to our technical report [23] for details.

4   System-Level Indirect Discrimination

While direct discrimination is about the direct causal effect through $C \rightarrow E$, indirect discrimination concerns about the indirect causal effects that are transmitted through intermediate attributes along the causal paths from C to E other than the direct arc $C \rightarrow E$. These intermediate attributes are correlated with C, hence the indirect effect eventually results in an unfair treatment of the protected group. A well-known example of indirect discrimination is redlining, where the residential Zip code of the individual is used for making decisions such as granting a loan. Although the Zip code is apparently a neutral attribute, it correlates with race due to the racial makeups of certain areas. Thus, the use of the Zip code can indirectly lead to racial discrimination.

Not all indirect causal effects of C on E should be considered as indirect discrimination. From a legal perspective, the absence of indirect discrimination can be proved if the defendant can provide an objective and reasonable justification on the using of the attributes correlated with the protected attribute. Consider a loan application dataset which contains three attributes, gender C, loan status E, and income X. The causal structure $C \rightarrow X \rightarrow E$ shows being female is the actual cause of the low income and is the indirect cause of loan denial through low income. The use of attribute X can be legally justified because of an actual legitimate causal relationship of X and E, i.e., a loan is denied if the applicant has low income. The high correlation between income X and gender C may be due to the fact that the women in the dataset tend to be underpaid.

In this case, the causal effect of gender on loan denial should not be considered as discrimination.

We refer to the attributes on the causal paths whose usage cannot be legally justified as the *redlining* attributes. Formally, the causal path $C \rightarrow \cdots \rightarrow X \rightarrow \cdots \rightarrow E$, denoted as p, corresponds to an indirect discrimination if X is a redlining attribute. We propose to identify the redlining attributes by examining the relationship represented by each arc along each causal path from C to E. If any relationship cannot be legally justified, the node that emanates the arc representing an unjustified relationship is identified as the redlining attribute. We propose to measure the indirect causal effect through the paths that each contains at least one redlining attribute. Similarly, indirect discrimination can be claimed if persistent negative effects are measured.

To measure the indirect causal effect through a set of paths **p**, we propose a simple three-step approach. First, we measure the direct causal effect through $C \rightarrow E$ by blocking all paths from C to E other than arc $C \rightarrow E$. Second, we measure the combined effect by blocking all paths from C to E other than **p** and arc $C \rightarrow E$. At last, the indirect causal effect through p can be identified by the difference of the above two measurements.

5   Group and Individual-Level Discrimination

Group-level discrimination occurs in a particular subsystem other than across the whole system. The group G can be specified by analysts to denote a subsystem. It is determined by a subset of profiling attributes. For example, when we determine whether there exists group-level discrimination in a particular major (e.g., CS) in university admission, G contains all applicants in CS. When adapting discrimination discovery techniques for system-level discrimination to group-level discrimination, we should note that the determination of block set **B** needs to be adjusted based on the given group G to form a partition within the given group. For instance, when focusing on group-level discrimination in CS major, **B** may contain test scores. Then, for each test score **b**, group-level discrimination can be claimed after we examine $\Delta P|_\mathbf{b}$ across all test scores among CS applicants.

Individual-level discrimination requires to identify discrimination for a specific individual, i.e., an entire record in the dataset. It can be considered as a special case of group-level discrimination, in which the values of all profiling attributes are given. To deal with individual-level discrimination, we propose to find two neighborhood groups that contain similar individuals from the protected group and the non-protected group. The individual is considered as discriminated if significant difference is observed between the decisions from the two groups. We propose to use the causal networks as the guideline of finding the neighborhood group. The causal structure of the system and the causal effect of each attribute on the decision can be used to facilitate the similarity measurement. Please refer to our technical report [24] for the details of our work on individual-level discrimination.

6   Experiments

We present our preliminary results for system-level direct discrimination discovery using two real data sets: the Adult dataset [10] and the Dutch Census of 2001 [13], which are widely used in discrimination discovery literature. The causal networks are constructed and presented by utilizing an open-source software TETRAD [4] which is a platform for causal modeling. We employ the original PC algorithm [21] and the significance level $\alpha = 0.01$ for network learning. The threshold τ is set as 0.05.

The Adult dataset consists of 48842 tuples with 11 attributes. Each tuple corresponds to an individual and describes the individual's personal information such as age, eduation, sex, occupation, income, etc. Since the computational complexity of the PC algorithm is an exponential function of the number of attributes and their domain sizes, for computational feasibility we binarize each attribute's domain values into two classes to reduce the domain sizes. For numerical attribute such as age or income, the domain values are binarized into low and high classes based on the median. For categorical attribute such as eduation or occupation, we select the domain value with the largest number of tuples as one class, and other domain values are combined as another class.

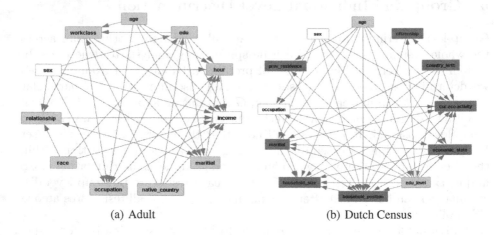

(a) Adult   (b) Dutch Census

Fig. 1. Causal networks

We treat sex (female and male) as the protective attribute and income (low_income and high_income) as the decision. The causal network is shown in Fig. 1a. We observe an arc pointing from sex to income, indicating that the two attributes are causally related and further examination is required for discovering discrimination. We find all the block sets **B**. Some subpopulations contain zero tuple from the dataset. On ignoring these subpopulations, the value of $\Delta P|_{\mathbf{b}}$ ranges from -0.614 to 0.524 across all the other subpopulations. Based on our criterion, we consider there is no discrimination against females in the Adult dataset.

Another dataset Dutch Census consists of 60421 tuples each of which is described by 12 attributes. Similarly, we binarize the domain values of attribute age due to its large domain size. We treat sex (female and male) as the protective attribute and occupation (occupation_w_low_income, occupation_w_high_income) as the decision. The causal network is shown in Fig. 1b. An arc from sex to occupation is observed in the network. We find all the block sets **B**. The value of $\Delta P|_b$ ranges from 0.062 to 0.435 across all the subpopulations, implying that females are discriminated in obtaining occupations with high income. Therefore, discrimination against females is detected in the Dutch dataset.

7 Related Work

A number of data mining techniques have been proposed to discover and measure discrimination in the literature. Pedreschi et al. [16,17] proposed to extract from the dataset classification rules, each of which consists of the protective attribute, the decision, and a set of context attributes. If the presence of the protective attribute increases the confidence of a classification rule, this classification rule is regarded as a discriminatory decision pattern in the data set. Then, discrimination can be unveiled by searching all discriminatory decision patterns. Based on that, the authors in [12] further proposed to use the Bayesian network to compute the confidence of the classification rules for detecting discrimination. Bonchi et al. in [1] proposed a random walk method based on the Suppes-Bayes causal network. Differently, conditional discrimination, where part of discrimination may be explained by other legally grounded attributes, was studied in [25]. In [22], the authors proposed the use of loglinear modeling to capture and measure discrimination and developed a method for discrimination prevention by modifying significant coefficients from the fitted loglinear model.

For individual discrimination, Luong et al. in [11] exploited the idea of situation testing. For each member of protected group with a negative decision outcome, testers with similar characteristics are searched for in a dataset. When there are significantly different decision outcomes between the testers of the protected group and the testers of the unprotected group, the negative decision can be considered as discrimination. For indirect discrimination, the authors in [5,16] studied the data mining task of discovering the attributes values that can act as a proxy to the protected groups and lead to discriminatory decisions indirectly. The authors in [19] adopted an approach based on rule inference to deal with the indirect discovery. The authors in [3] addressed the problem of fair classification that achieves both group fairness, i.e., the proportion of members in a protected group receiving positive classification is identical to the proportion in the population as a whole, and individual fairness, i.e., similar individuals should be treated similarly.

Another issue related to anti-discrimination is discrimination prevention, which aims to build non-discriminatory predictive models when the historical data contains discrimination [2,7–9]. Proposed methods focus on either modifying the historic data to remove discrimination, or tweaking the predictive model to make it discrimination free. In all the methods, discrimination needs to be identified and measured first before it can be removed. Our work complements discrimination prevention in that we provide a formal criterion and measure for discrimination, which advances theoretical understanding related to both discrimination discovery and prevention.

8   Conclusions and Future Work

We categorize different discrimination types based on discrimination level and discrimination manner. We investigated the problem of discrimination discovery for system-level direct discrimination and indirect discrimination. We establish a discrimination models based on the causal networks. In the future work, we plan the extend the results to other types of discrimination. Using our discrimination criteria, we will also study the problem of discrimination prevention, which aim to remove discrimination by modifying the based data before conducing predictive analysis.

Acknowledgment. This work was supported in part by U.S. National Institute of Health (1R01GM103309).

References

1. Bonchi, F., Hajian, S., Mishra, B., Ramazzotti, D.: Exposing the probabilistic causal structure of discrimination. arXiv preprint (2015). arXiv:1510.00552
2. Calders, T., Verwer, S.: Three naive bayes approaches for discrimination-free classification. Data Min. Knowl. Discov. **21**(2), 277–292 (2010)
3. Dwork, C., Hardt, M., Pitassi, T., Reingold, O., Zemel, R.: Fairness through awareness. In: Proceedings of the 3rd Innovations in Theoretical Computer Science Conference, pp. 214–226. ACM (2012)
4. Glymour, C., et al.: The TETRAD project (2004). http://www.phil.cmu.edu/tetrad
5. Hajian, S., Domingo-Ferrer, J.: A methodology for direct and indirect discrimination prevention in data mining. TKDE **25**(7), 1445–1459 (2013)
6. Hajian, S., Domingo-Ferrer, J., Monreale, A., Pedreschi, D., Giannotti, F.: Discrimination-and privacy-aware patterns. Data Min. Knowl. Discov. **29**(6), 1–50 (2014)
7. Kamiran, F., Calders, T.: Classifying without discriminating. In: International Conference on Computer, Control and Communication, pp. 1–6 (2009)
8. Kamiran, F., Calders, T.: Data preprocessing techniques for classification without discrimination. Knowl. Inf. Syst. **33**(1), 1–33 (2012)
9. Kamishima, T., Akaho, S., Sakuma, J.: Fairness-aware learning through regularization approach. In: ICDMW, pp. 643–650. IEEE (2011)
10. Lichman, M.: UCI machine learning repository (2013). http://archive.ics.uci.edu/ml
11. Luong, B.T., Ruggieri, S., Turini, F.: K-nn as an implementation of situation testing for discrimination discovery and prevention. In: KDD, pp. 502–510. ACM (2011)

12. Mancuhan, K., Clifton, C.: Combating discrimination using bayesian networks. Artif. Intell. Law **22**(2), 211–238 (2014)
13. Statistics Netherlands. Volkstelling (2001). https://sites.google.com/site/faisalkamiran/
14. Pearl, J.: Comment: understanding simpsons paradox. Am. Stat. **68**(1), 8–13 (2014)
15. Pearl, J., et al.: Causal inference in statistics: an overview. Stat. Surv. **3**, 96–146 (2009)
16. Pedreschi, D., Ruggieri, S., Turini, F.: Measuring discrimination in socially-sensitive decision records. In: SIAM SDM (2009)
17. Pedreshi, D., Ruggieri, S., Turini, F.: Discrimination-aware data mining. In: KDD, pp. 560–568. ACM (2008)
18. Romei, A., Ruggieri, S.: A multidisciplinary survey on discrimination analysis. Knowl. Eng. Rev. **29**(05), 582–638 (2014)
19. Ruggieri, S., Hajian, S., Kamiran, F., Zhang, X.: Anti-discrimination analysis using privacy attack strategies. In: Calders, T., Esposito, F., Hüllermeier, E., Meo, R. (eds.) ECML PKDD 2014, Part II. LNCS, vol. 8725, pp. 694–710. Springer, Heidelberg (2014)
20. Ruggieri, S., Pedreschi, D., Turini, F.: Data mining for discrimination discovery. ACM TKDD **4**(2), 9 (2010)
21. Spirtes, P., Glymour, C.N., Scheines, R.: Causation, Prediction, and Search, vol. 81. MIT press, Cambridge (2000)
22. Wu, Y., Wu, X.: Using loglinear model for discrimination discovery and prevention. Technical report, DPL-2015-002, University of Arkansas (2015)
23. Zhang, L., Wu, Y., Wu, X.: Causal Bayeisan network-based discrimination discovery and removal. Technical report, DPL-2016-001, University of Arkansas (2016)
24. Zhang, L., Yongkai, W., Xintao, W., Discovery, situation testing-based discrimination : a causal inference approach. Technical report, DPL-2016-002, University of Arkansas (2016)
25. Zliobaite, I., Kamiran, F., Calders, T.: Handling conditional discrimination. In: ICDM 2011, pp. 992–1001. IEEE (2011)

Health Sciences

Social Position Predicting Physical Activity Level in Youth: An Application of Hidden Markov Modeling on Network Statistics

Teague Henry[1]([☒]), Sabina B. Gesell[2], and Edward Ip[2]

[1] University of North Carolina at Chapel Hill, Chapel Hill, USA
trhenry@email.unc.edu
[2] Wake Forest Baptist Medical Center, Winston-Salem, USA
{sgesell,eip}@wakehealth.edu

Abstract. Social positioning has been shown to have impacts on physical activity in youth. In this study Hidden Markov Modeling is used to infer latent social positions from a set of computed network statistics in two network of youth over time. The association between physical activity and social position is analyzed. Youth in less centrally located social roles tended to have less physical activity than youth with more centrally located social positions.

Keywords: Youth · Friendship network · Social position · Physical activity · Centrality · Hidden Markov Modeling

1 Introduction

In recent years, there has been increased interest in the relation between social networks and obesity related behaviors. There are clear associations between an individual's social network and their obesity related behavior in both adults and youth [1–4] The majority of this research has focused on the coupled processes of peer influence, in which physically active or inactive individuals influence their peer's behavior, and peer selection, in which one's physical activity influences which peers one selects. For example, perceived social support from active children tends to increase activity in inactive youth [2]. This work has revealed that peer influence and peer selection processes play an important role in determining physical activity level, however little work has been done on examining the effect of an individual's position in their social network on physical activity.

S.B. Gesell—This work was supported by Award Number K23HD064700 (Gesell) from the Eunice Kennedy Shriver National Institute of Child Health and Development. The content is solely the responsibility of the authors and does not necessarily represent the official views of the funding agency.

E. Ip—This work was supported by the following grant awards (PI: Ip) NIH 1U01HL101066-01, and NSF SES-1424875.

K.S. Xu et al. (Eds.): SBP-BRiMS 2016, LNCS 9708, pp. 97–106, 2016.
DOI: 10.1007/978-3-319-39931-7_10

Individuals are embedded within a social network but their position within the network is not wholly defined by what direct peers they have. The connections among peers influences the individual's position within the network. Individuals can be more or less centrally located, their peer groups can be more or less closely knit, and individuals can act as major connectors between social groups, or might exist on the periphery of a social network. *Social position* can be thought of as a complex nonlinear interaction between various characteristics of an individual's location in a social network, defining the role that that person plays in the network structure. This definition echoes both *structural equivalence* [5] and *structural isomorphism* [6], while being less strict than either. Instead of requiring individuals to have the same peers, as in structural equivalence, or to be in the same type of relation with their peers, as in structural isomorphism, in this study we say that if individuals have similar values on a set of selected network statistics (such as betweenness, indegree, outdegree) they have the same social position. This definition renders *social position* as a latent variable that is indicated by the set of selected network statistics.

In light of the results on peer influence and social support's effect on physical activity in youth [7], it is likely that social position will have have an impact on physical activity. Youth with more ties relative to peers have more social power and with that more control over the diffusion of information and/or behavior [8]. As such, it is likely that youth with high centrality will have greater consistency over time in their physical activity. If a youth is tightly integrated into their local network – i.e., one that has a large clustering coefficient, prior research suggests that they will tend to adopt new behaviors more quickly than those who are not integrated into their local networks [8–10]. This suggests that tightly integrated youth will change their physical activity levels over time, as a reflection of the behaviors of their peers. Finally, if a youth is not integrated into their network, and doesn't have many friends, it is likely that their physical activity level will decrease over time due to lack of social support, or remain consistent due to not being influenced by peers. These effects would be independent of the physical activity level of peers, and have important consequences for intervention design.

In this paper, we study the association between the position an youth has in his or her social network and their physical activity level. Due to the use of longitudinal network data, this analysis poses important methodological challenges. First, the social position that an individual has in a network is not directly observed, rather it is indicated by set of network statistics that individual has. Secondly, the social position of an individual might change over time as the structure of a network changes. Finally, the social position must be linked to the individual's behavior, in this case their activity level.

In order to capture this latent social positioning, and to allow the social position to change over time, we propose a two step modeling procedure. The first step uses Hidden Markov Modeling [11–13] on a set of network statistics computed for each individual in the network at each time point. Then, once the states or social positions are estimated, we use those as predictors of physical activity at the next time point.

There are several benefits using the HMM approach in analyzing longitudinal social network data. First, unlike existing cross-sectional network models such as the exponential random graph [14], or advanced longitudinal network models such as the stochastic actor-based models for network dynamics [15] the HMM approach does not consider micro-mechanism and therefore avoids the direct modeling of ties between individuals. Thus the HMM approach is more flexible and allows multiple relevant global network summary statistics to be included for characterizing the social position of an individual. Second, unlikely stochastic block modeling [16] or other community detection methods, HMM does not require the presence of a community structure to determine social position. Finally, without resorting to studying individual ties, the behavioral change of individuals over time can be modeled such that the result reveals a broader picture of the evolution of social position and its relation to outcomes such as activity level, a capability that most community detection methods lack.

2 Data and Measures

The sample consisted of 81 children, averaging 7.96 years (SD=1.74). These children lived in low SES neighborhoods, attended public schools, and attended one of two afterschool programs. Both programs enrolled children from the school in which the school-based program was located, as well as children from other schools in the area. One program was new and located in a community center with the first wave of collection happening 1 week after the beginning of the program, and the other was established, organized by the local YMCA and located in a public school, and had been in progress for several years. The friendship networks from these two programs are respectively referred to as the Community and the YMCA network. Data was collected at the beginning of the programs, 6 weeks, and 12 weeks into the programs, for a total of three waves of data collection. This data set is described elsewhere [7].

2.1 Physical Activity

Physical activity levels were captured through ActiGraph GT1M accelerometers (ActiGraph LLC, Pensacola, FL). Details of the physical activity measurements are outlined elsewhere [7]. Validated thresholds were used to code each childs time spent at sedentary, light, moderate, and vigorous activity levels [17,18]. In this study minutes of time spend in moderate or vigorous activity is the outcome of interest.

2.2 Friendship Networks

For each afterschool care program, complete social network data was collected at each time point. Participants were asked "Please tell me the names of the friends you hang around with and talk to and do things with the most here in this after-school program and were allowed to nominate any number of friends".

This nomination procedure is used in other youth network studies [3]. This procedure results in directed binary networks, where each edge represents individual i nominating individual j as a friend. Thus a friendship network can be visualized through the use of a graph in which nodes are used to represent individuals and bidirectional arcs and used to represent incoming and outgoing ties between individuals. For the current study, the structure of the network evolves over time, implying that the latent social positions of the individuals in the network could change over time.

Network Statistics. In order to use HMM to infer social position and the change in social position, a set of network statistics must be used. This framework for analyzing changing social positions can be used with any number of network statistics, so the rational for inclusion of a particular network statistic will be application specific. In the case of this study, we focused on statistics that could map onto measures of popularity, expansiveness, friendship clusters, social status/power, or social support.

Indegree. Indegree is defined as the number of incoming edges to a node, or in this case, the number of received friendship nominations. This can act as a measure of popularity in social networks.

Outdegree. Outdegree is defined as the number of outgoing edges from a node. It can act as a measure of expansiveness in social networks.

Betweeness Centrality. The directed betweeness centrality for individual k is defined as the sum over every $i \neq j \neq k$ of the proportion of geodesics linking individual i to individual j that contain individual k [19,20]. Broadly speaking, this is a measure of how important an individual is for communication between two distal parts of a network. Should an individual's betweenness be high, that means that most communication along edges must pass through that individual. This (and the other centrality measures below) can act as a measure of social power or social support.

Closeness Centrality. The closeness centrality of an individual k is defined as the inverse of the sum of the lengths of the shortest path from individual k to all other individuals in the network [19]. Higher closeness suggests that the individual is more centrally located in the network, while lower closeness suggests that the individual is more isolated.

Clustering Coefficient. The clustering coefficient for an individual k can be broadly defined as the proportion of edges existing within the individual's neighborhood (the individuals he has nominated or was nominated by) divided by the number of edges there could possibly be in the neighborhood [21]. This measures how tightly connected an individuals group of friends is.

3 Model

The HMM methodology was used as a tool to delineate heterogeneity among the dynamic of network change. From a statistical perspective, the HMM approach employs multivariate longitudinal analysis to (1) identify different groups of individuals occupying different social positions in the network, and (2) determine the transition dynamic between different social positions. Operationally, the HMM proceeded in several stages. First, network statistics identified in Sect. 2.2 were used to create a multi-variable profile for describing social position. In this application hereafter we call the set of network features (statistics) the profile for social position within the peer network.

The second stage of HMM involved the determination of the appropriate number of distinct hidden states for sufficiently explaining the variation of the profile of network measures. We used the Bayesian Information Criterion (BIC), a commonly used goodness-of-fit index, for this purpose. Analogous to the determination of the number of classes in latent class analysis (LCA) [22], the HMM assumes that a finite number (usually small) of distinct states of social position can capture the heterogeneity in social positions. However, unlike LCA, which is a static analysis, the HMM allows individuals to transition to a different state of social position over time.

In the third stage of HMM analysis, parameters for the hidden states and were estimated using network statistic data. Three sets of parameter were made available from the HMM analysis (1) the conditional distributions of the variables defined by the social position profile given a specific hidden state, or $P(Y_{ijt}|Z_{it})$, where Y_{ijt} denotes the jth network feature for individual i at time t, and Z_{it} denotes the hidden state of social position of individual i at time t, $i = 1, \ldots, n$, $j = 1, \ldots, J$, and $t = 1, \ldots, T$, (2) the prevalence of the states of social position at the first time point, or $P(z_{i1})$, and (3) the transition probabilities between the states of social position, or $P(Z_{it}|Z_{it-1})$, $t = 2, \ldots, T$. The parameters in (1) allowed the interpretation of the nature of the different states of social position, and the prevalence of the states over time can be derived from (2) and the Markov assumption, which asserted that at a given time t, the state of social position at t was only dependent on the state at $(t-1)$ and not on previous time points. Additionally, the HMM assumed that the states of social position remained unchanged over time, even though it was possible that some states evolved and existed after some times. It was also possible that some states disappeared after some time. Because the HMM only probabilistically estimated how likely an individual occupied a specific state of social position at each time point, we applied the maximal rule and selected the state that had highest probability $P(Z_{it}|Y_i)$ as the individuals state. Here Y_i represents the entire vector of the observations of individual i. As a result, the HMM produced, for each individual, a trajectory of states of social position.

The trajectories of social position were further analyzed in predicting the proximal outcome of physical activity. Longitudinal analysis, specifically a lag 1 auto-regressive model with covariates was applied to the data for determining if social position predicted level of physical activity. The analysis was conducted using R version 3.2.1 [23].

4   Results

Five states were selected based on the BIC criteria. The breakdown of the state to network statistic relation is presented in Fig. 1. Individuals in State 1 have a slightly below average number of friends, while having above average clustering. This suggests that children in state 1 are part of moderately sized, moderately tight knit friendship groups. State 1 can be called the Average Cluster state. Children in State 2 have far fewer friends on average but have high clustering, suggesting that they are part of tight knit friendship groups. State 2 can be called the Tight Cluster state. State 3 has a higher number of friends, but is average in every other network characteristic. State 3 can be called the Average state. Individuals in State 4 have few friends, and very little clustering, suggesting that they are not very well attached to the rest of the social network. State 4 can be called the Loosely Connected state. Finally, children in State 5 have a large number of friends, high betweenness and low clustering, suggesting that these individuals act as connectors between different parts of the network. State 5 can be called the Connector state. Interestingly, the Connector state show relatively low levels of closeness and clustering, suggesting that children in this state attained a degree of independence and individuality.

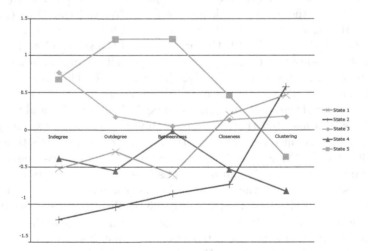

Fig. 1. 5-state model of social position. State 1= Average Cluster, State 2 = Tight Cluster, State 3 = Average, State 4 = Loosely Connected, State 5 = Connector.

Table 1 reports the transition probabilities between states of social position. It can be seen that State 1 (Average Cluster state), State 3 (Average state), and State 5 (Connector state) are most stable, respectively with high probabilities of

Table 1. Transition probabilities between states.

	State 1	State 2	State 3	State 4	State 5
State 1	0.74	0.10	0.16	0.00	0.00
State 2	0.45	0.33	0.13	0.09	0.00
State 3	0.00	0.00	0.87	0.00	0.13
State 4	0.17	0.00	0.38	0.28	0.17
State 5	0.00	0.00	0.08	0.00	0.92

74 %, 87 %, and 92 % of staying within the same state over two consecutive time points. On the other hand, State 2 (Tight Cluster state) and State 4 (Loosely Connected state) are unstable, respectively with only 33 % and 28 % of staying in the same state. For State 2, the most likely state that it would transition into is State 1 with probability 45 %. The result suggests that children that had few friends, even when they were clustered, tended to assimilate with the larger group over time and showed incremental gain in in- and out-degrees (see profiles of States 1 and 2 in Fig. 1). For State 4, the transition that is most likely is to State 3, or the Average State (38 %), suggesting that some loosely connected children over time tended to transition to mean behavior.

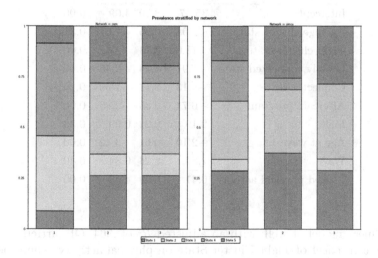

Fig. 2. Prevalence of State 1 to 5, from bottom to top in order, over time for the community network (left panel) and YMCA (right panel). (Color figure online)

Figure 2 shows the prevalence of the states over time for the two networks - the Community Network and the YMCA Network. There appears to be some differences in the way the two networks evolve. For example, State 2 (Tight Cluster) comprises a relatively high percentage of children in the Community

network but is much smaller in the YMCA network. An explanation of the
phenomenon is that the Community network is a new program and children
that entered the program were likely to start engaged only with their previous
friends going to the same program, whereas the YMCA network was more well
established and thus showed fewer children in the two transient states of Tight
Cluster (state 2) and Loosely Connected (state 4). The distribution of the states
from both programs seem to converge at the last wave of data. Note that some
states did not exist at certain time points - e.g., state 4 (Loosely Connected) for
the YMCA network at Wave 3.

4.1   Social Position Predicting Physical Activity

To link social position to physical activity, we fit a linear model predicting physical activity with lag 1 social position. To control for autoregressive effects, lag
1 physical activity was also included. Gender, age and which program the individual were in were controlled for. The Average State is used as the reference
category. Results are shown in Table 2:

Table 2. Social position at lag 1 predicting physical activity in minutes

	β	SE	t-value	p-value
Intercept	35.57	8.70	4.09	0.00
Average cluster state	−2.98	2.95	−1.01	0.31
Tight cluster state	−11.66	3.74	−3.12	0.00
Loosely connected state	−7.91	3.23	−2.45	0.02
Connector state	−4.13	3.17	−1.30	0.20
Aftercare program	−6.71	2.35	−2.86	0.00
Male	2.19	2.22	0.99	0.33
Age at wave 1	−2.10	0.61	−3.42	0.00
Wave	1.89	2.26	0.84	0.40
Lagged physical activity	0.41	0.09	4.53	0.00

The linear model fit well, with an adjusted R^2 of .364. Of particular note is
the significant effect of Tight Cluster State on physical activity, where being in
Tight Cluster State at time t predicts 11.66 fewer minutes of physical activity
when compared to the Average state time $t + 1$ (p <.05). Additionally, there is
a significant effect of Loosely Connected State of a 7.91 min decrease in physical activity (p <.05). As would be expected, there is an autoregressive effect
of physical activity. Additionally, there is a significant program effect which is
expected, as one program was held outdoors while the other was held inside,
with participation in the indoors program predicting an on average reduction in
physical activity of 6.71 min (p <.05).

5   Conclusion

There are two contributions that this paper makes. The first is the new approach of modeling global features within a network to infer latent social position. Current network methodology restricts analysts to inferring latent communities, or groups of individuals who cluster together. The approach introduced in this paper proposes a different latent classification of individuals into latent social positions, a classification that echos both structural and isomorphic equivalence, but is less restrictive than either. Furthermore, using HMM, this approach allows for the social position of an individual to change over time. Finally, this approach is extremely flexible, allowing for any individual level network statistic to be used to define social position.

The second contribution is in relating social position to physical activity in children. The analysis presented here divides up the sample into five interpretable social roles, and shows decreases in physical activity over time for two of the roles. The cause however of these decreases is not entirely clear. There are several reasons why children in a Tight Cluster State could have less physical. One explanation is that due to the smaller number of friends that children in this state have, they have fewer opportunities to participate in physical activity. Alternatively, as clustering can accelerate diffusion of behavior or information within a group but will constrain it between groups [8], norms of lower physical activity might have infiltrated the clusters. Children in the Loosely Connected State also show a significantly less physical activity over time, though this cannot be as strongly attributed to norm infiltration as the Loosely Connected State has a lower clustering coefficient. However, the Loosely Connected State also has a low number of peers, leading that to be a potential common cause for the decrease in physical activity over time.

There are several limitations to this study. The first is the small sample size of the networks and the small number of time points collected. This limits the interpretations that can be made regarding these results. Additionally, the choice of the network statistics by which to define the latent social positions may not have captured the social positions that would effect physical activity in youth. Finally, the transition matrix between the states was not discussed here, due to the focus on the association between social position and physical activity. Further research should focus on the qualities of the transition matrix between these social states.

References

1. Bahr, D.B., Browning, R.C., Wyatt, H.R., Hill, J.O.: Exploiting social networks to mitigate the obesity epidemic. Obesity **17**(4), 723–728 (2009). (Silver Spring, Md.)
2. Sallis, J.F., Prochaska, J.J., Taylor, W.C.: A review of correlates of physical activity of children and adolescents. Med. Sci. Sports Exerc. **32**(5), 963 (2000)
3. De La Haye, K., Robins, G., Mohr, P., Wilson, C.: Homophily and contagion as explanations for weight similarities among adolescent friends. J. Adolesc. Health **49**(4), 421–427 (2011). Official Publication of the Society for Adolescent Medicine

4. De La Haye, K., Robins, G., Mohr, P., Wilson, C.: How physical activity shapes, and is shaped by, adolescent friendships. Soc. Sci. Med. **73**(5), 719–728 (2011)
5. Lorrain, F., White, H.C.: Structural equivalence of individuals in social networks. J. Math. Sociol. **1**(2013), 49–80 (1971)
6. Borgatti, S.P., Everett, M.G.: Notions of position in social network analysis. Sociol. Methodol. **22**, 1–35 (1992)
7. Gesell, S.B., Tesdahl, E., Ruchman, E.: The distribution of physical activity in an after-school friendship network. Pediatrics **129**(6), 1064–1071 (2012)
8. Valente, T.W.: Social Networks and Health. Oxford University Press, New York (2010)
9. Coleman, J.S., Katz, E., Menzel, H.: Medical Innovation: A Diffusion Study. Bobs Merrill, New York (1966)
10. Rogers, E.M., Kincaid, D.L.: Communication Networks: Toward a New Paradigm for Research. Free Press, New York (1981)
11. MacDonald, I.L., Zucchini, W.: Hidden Markov and Other Models for Discrete-Valued Time Series. CRC Press, New York (1997)
12. Ip, E.H., Jones, A.S., Heckert, D.A., Zhang, Q., Gondolf, E.D.: Latent Markov model for analyzing temporal configuration for violence profiles and trajectories in a sample of batterers. Sociol. Meth. Res. **39**(2), 222–255 (2010)
13. Ip, E.H., Zhang, Q., Rejeski, W.J., Harris, T.B., Kritchevsky, S.: Partially ordered mixed hidden Markov model for the disablement process of older adults. J. Am. Stat. Assoc. **108**(502), 370–380 (2013)
14. Robins, G., Pattison, P., Kalish, Y., Lusher, D.: An introduction to exponential random graph (p*) models for social networks. Soc. Netw. **29**(2), 173–191 (2007)
15. Snijders, T.A., van de Bunt, G.G., Steglich, C.E.: Introduction to stochastic actor-based models for network dynamics. Soc. Netw. **32**(1), 44–60 (2010)
16. Snijders, T.A.B., Nowicki, K.: Estimation and prediction for stochastic blockmodels for graphs with latent block structure. J. Classif. **14**(1), 75–100 (1997)
17. Pate, R.R., Almeida, M.J., McIver, K.L., Pfeiffer, K.A., Dowda, M.: Validation and calibration of an accelerometer in preschool children. Obesity **14**(11), 2000–2006 (2006). (Silver Spring, Md.)
18. Puyau, M.R., Adolph, A.L., Vohra, F.A., Butte, N.F.: Validation and calibration of physical activity monitors in children. Obes. Res. **10**(3), 150–157 (2002)
19. Freeman, L.C.: Centrality in social networks conceptual clarification. Soc. Netw. **1**(3), 215–239 (1979)
20. White, D.R., Borgatti, S.P.: Betweenness centrality measures for directed graphs. Soc. Netw. **16**(4), 335–346 (1994)
21. Wasserman, S., Faust, K.: Social Network Analysis: Methods and Applications. Cambridge University Press, Cambridge (1994)
22. McCutcheon, A.L.: Latent Class Analysis. SAGE Publications, Thousand Oaks (1987)
23. Core, R., Team, R.: A language and environment for statistical computing. R Foundation for Statistical Computing, Vienna, Austria (2015)

Incorporating Disgust as Disease-Avoidant Behavior in an Agent-Based Epidemic Model

Christopher R. Williams[✉] and Armin R. Mikler

University of North Texas, Denton, TX 76203, USA
christopherwilliams3@my.unt.edu, mikler@unt.edu

Abstract. Kiesecker et al. demonstrated disgust behavior in nature in 1999, and further research has shown that humans also exhibit disgust as part of the "behavioral immune system" [5,8]. We present preliminary results from an agent-based model incorporating disgust as disease-avoidant behavior, the SLIPR model (susceptible, latent, infectious, presenting, removed), a modification and extension of the traditional SEIR model (susceptible, exposed, infectious, removed). The SLIPR model restructures the compartments of the SEIR model to allow for a distinct period of infectiousness occurring prior to visible disease presentation and extends it by simulating disgust as disease-avoidant behavior. SLIPR suggests that, for specific values of parameters such as disgust magnitude and population density, this disease-avoidant behavior significantly affects the spread of disease.

Keywords: Agent-based modeling · Computational epidemiology · Behavior modeling · Emotion

1 Introduction

Disgust, in the context of disease-avoidant behavior, may bring to mind an image of a shopper avoiding an aisle containing another coughing patron (what Schaller would call 'discriminatory sociality' [8]), but disgust is a more primal behavior that can be observed in animals as distantly related to humans as tadpoles [5]. The existence of such behavior across the animal kingdom suggests that it has been evolutionarily successful in inhibiting the spread of disease. Humans have been found to display various disease-avoidant behaviors, as motivated, for example, by the emotion of disgust. The psychological scientist Mark Schaller has coined the term "behavioral immune system" to describe the disease-avoidant behavior documented in such individuals [8]. Despite this, individual disease-avoidant behavior has been largely ignored in epidemic models.

The two broadest categories of epidemic models that disgust can be incorporated into can be described as deterministic and stochastic [7]. Deterministic epidemic models most often take the form of compartmental mathematical models. In compartmental models, the population is modeled as two or more fluid compartments that represent different disease states. The most well known of such

© Springer International Publishing Switzerland 2016
K.S. Xu et al. (Eds.): SBP-BRiMS 2016, LNCS 9708, pp. 107–116, 2016.
DOI: 10.1007/978-3-319-39931-7_11

epidemic models is the SIR model, first proposed by Kermack and McKendrick, with the respective compartments being susceptible, infectious, and recovered (and later sometimes expanded to include compartments for exposed and/or a final return to susceptible) [4].

We present preliminary results from an agent-based model incorporating disgust as disease-avoidant behavior, the SLIPR model (susceptible, latent, infectious, presenting, removed), a modification and extension of the traditional SEIR model (susceptible, exposed, infectious, removed).

2   Related Work

Piero Poletti et al. have proposed a mathematical model based on evolutionary game theory that accounts for spontaneous behavioral response to perceived risk of infection [6]. While Poletti et al. make no explicit reference to disgust or the behavioral immune system, their model is in concordance with Schaller's description of the behavioral immune system as being comprised of "detection and response mechanisms" which guide "decision-making strategies [...] that minimize the infection risk" [8]. Their model is composed of two parts; an extension of the SIR model for modeling disease transmission, defined as a system of differential equations accounting for asymptomatic and symptomatic infections, and a model of behavioral changes implemented using imitation dynamics from evolutionary game theory. Behavior in this context is a change in the number of contacts based on perceived risk, which they have chosen to represent using the perceived prevalence of the disease. The perceived prevalence is a calculated using, and corresponding exactly to, the number of symptomatic cases. The parameters affecting behavior are the threshold past which perceived risk causes behavioral change, the contact reduction factor, q, and the speed of behavioral change. They noted three interesting aspects of the model's predicted "effectiveness of human self-protection":

1. "A small reduction in the number of contagious contacts enacted by the population can remarkably alter the spread of the epidemic;
2. "For small values of q, multiple epidemic waves can occur.
3. "There exists a threshold for q such that smaller values do not determine a larger impact of behavioral changes on the final epidemic size, the daily peak prevalence and the peak day" [6].

As a mathematical model, however, the perceived risk of infection is dependent on population information, in particular, the number of symptomatic individuals. An agent-based model would allow for behavior modeling at an individual level, with susceptible agents potentially displaying disease-avoidant behavior in response to symptomatic agents regardless of disease prevalence.

3   The SLIPR Model: An Agent-Based Model Incorporating Disgust as Disease-Avoidant Behavior

3.1   Compartmental Models and the SLIPR Model

The SIR model has been extended for various purposes, such as the SEIR model, which adds an 'exposed' compartment, E, in order to accommodate diseases which have significant latent periods during which they are not infectious, and is likewise implemented using a system of differential equations [1]. For the purposes of modeling disease-avoidant behavior, however, we must consider the possibility of the individual becoming infectious before they are symptomatic, which would temporarily preclude disgust. Such presymptomatic transmission has been suggested to occur in diseases such as human influenza, where viral shedding has been observed up to six days before clinical onset [2]. It is therefore necessary to partition the 'exposed', E, and 'infectious', I, compartments into three new compartments, L, I, and P. These three compartments represent the latent period, the infectious period until the point the disease presents with visible symptoms, and the period in which the disease is both infectious and presenting with visible symptoms, respectively.

3.2   Agent-Based Models

In order to model behavior that is displayed on an individual level, SLIPR is implemented as an agent-based compartmental model. Agents in SLIPR have a directionality, d, which is an integer 0–8, representing the eight degrees of directional freedom in two-dimensional space within the Moore neighborhood, as it is understood in the context of movement metrics [9]. The agents begin at random coordinates within the field, are randomly assigned an initial directionality, and proceed to move at each step in the simulation according to a correlated random walk, which has a bias, probability b, toward its previous directionality, with a chance of making a turn of size t, in accordance with Eq. 1.

$$f(d_{next}) = \begin{cases} (d_{prev} + / - t) \mod 8, & \text{if } \mathbf{rand}() < b \\ d_{prev}, & \text{otherwise} \end{cases} \qquad (1)$$

The space occupied by the agents is a square two-dimensional field which wraps around on both axes to create a torus. Agent density, ρ, is calculated in agents per pixels^2 as described by Eq. 2, where x is the length of one edge of the field, and n is the number of agents.

$$\rho = \frac{1}{\sqrt{(x^2)/n}} \qquad (2)$$

This leads to agents that have generally persistent direction, occasionally making adjustments to their course [3]. At each step of the simulation, the susceptible agents are stepped through in order to determine if they are within the disgust radius of a presenting agent, or if transmission from an infectious agent has occurred.

3.3 Modeling Disgust

We have implemented disease-avoidant behavior by creating a disgust radius, r_d, within which susceptible agents recognize other infected agents who are presenting disease symptoms and attempt to perform disease avoidance. Depending on the presentation of the disease, the detection capability of the host, and 'disgustingness' of the disease presentation, it may be desirable to set this radius to be greater or lesser than the transmission radius, r_t, within which the disease has a probability of being transmitted.

When a presenting agent enters the disgust radius of a susceptible agent, the susceptible agent reverses its direction in an attempt to avoid the presenting agent, according to Eq. 3. Note that this method of avoidance will do nothing to prevent the susceptible agent from contact with non-presenting infectious agents, as disgust is dependent on perceptibility of disease presentation. Also, the susceptible agent may not be able to completely avoid presenting agents in the case that there are two or more presenting agents surrounding the susceptible agent.

$$d_{next} = (d_{prev} + 4) \bmod 8 \tag{3}$$

3.4 Visualization

A visualization was created to view the simulation while running, as seen in Fig. 1, which is useful for debugging and investigating unexpected results.

Fig. 1. A simulation running with a small field for demonstration purposes. The agents are colored by compartment: green for susceptible, yellow for latent, red for infectious, purple for presenting, and blue for removed (Color figure online)

4 Simulation

To investigate the impact of the disgust behavior in our model, we have simulated an epidemic on a synthetic population and environment with baseline parameters describing a hypothetical infectious disease that meets the necessary conditions for creating an epidemic. The field size is chosen such that there is an agent density of 1/30, 1/40, and 1/50 agents per pixels^2 at a population of 400 agents. The hypothetical disease has a transmission radius of 20 pixels and is simulated using three different disgust radii, 0 (no disgust), a 21-pixel disgust radius, and a 25-pixel disgust radius. Due to the stochastic nature of the model 50 trials are run and averaged for each configuration. We use 100 steps for each of the latent, infectious, and presenting periods. The transmission probability for each step that a susceptible agent is within the infection radius of an infectious or presenting agent is set to 0.03. The parameters can be seen in Table 1.

Table 1. Simulation parameters

Parameter	Investigated values	Baseline value
Disgust radius r_d (in units)	21, 25	0
Transmission radius r_t (in pixels)	-	20
Number of agents	-	400
Agent density ρ (in agents per pixels^2)	1/30, 1/40, 1/50	-
Latent period (in steps)	-	100
Infectious period (in steps)	-	100
Presenting period (in steps)	-	100
Transmission probability (per step)	-	0.03
Correlated walk persistence (probability b)	-	0.7

4.1   Results

Agent density proved to have a significant effect on the efficacy of disgust as disease-avoidant behavior. At a density of 1/30 agents per pixels^2 there was not a significant difference in total infections, peak infected agents, or the step at which the peak occurred. However, as the agent density decreased, the efficacy of disgust increased compared to the baseline (Figs. 2, 3, 4). In a simulated epidemic with agent density 1/50 agents per pixels^2, a disgust radius of 25 pixels reduced the total infected agents by 37.60 % as compared to the baseline, the number of

Table 2. Simulation results, as averaged over 50 trials

Parameter	No disgust			21px disgust radius			25px disgust radius		
Agent Density (in app^2)	1/30	1/40	1/50	1/30	1/40	1/50	1/30	1/40	1/50
Total infected agents	398.78	366.52	301.14	390.04	359.78	224.56	398.46	334.98	187.92
Infected agents at peak	103.5	53.96	26.1	100.96	55.32	18.46	102.3	46.46	15.34
Step at peak infection	708	1068	1332	751	1026	1532	743	1148	1811

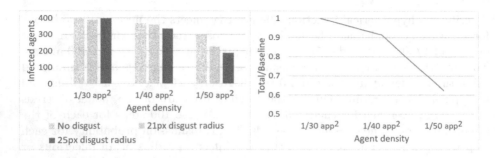

Fig. 2. Total number of infections at different agent densities, totals (left) and compared to baseline (right). The effect decreases as agent density increases.

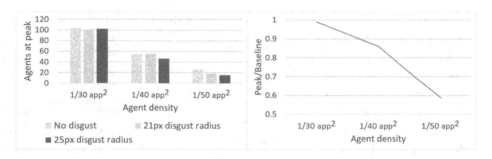

Fig. 3. Peak number of infected agents at different agent densities, totals (left) and compared to baseline (right). The effect decreases as agent density increases.

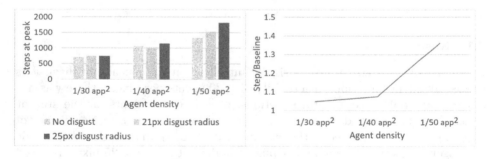

Fig. 4. Simulation step at peak infection at different agent densities, totals (left) and compared to baseline (right). The effect decreases as agent density increases.

infected agents at the peak of infection was reduced by 41.23 % with a disgust radius of 25 pixels, and the peak infection time was slowed by 35.96 % with a disgust radius of 25 pixels.

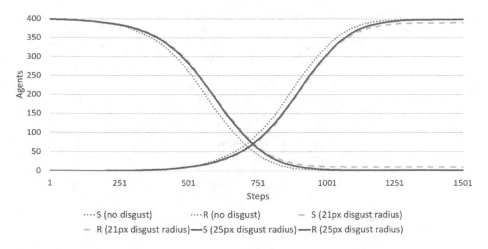

Fig. 5. S and R compartments from epidemic simluations at agent density 1/30 agents per pixels^2

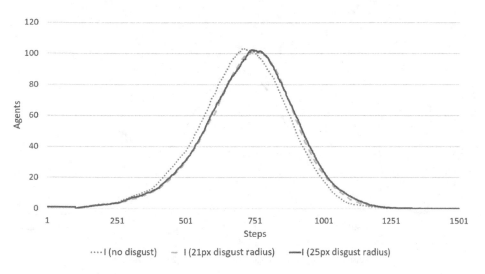

Fig. 6. I compartments from epidemic simulations at agent density 1/30 agents per pixels^2

The protective effect can be seen compared to the baseline in the S and R compartments (Figs. 7 and 9), and also in the I compartments (Figs. 8 and 10). The full results can be seen in Table 2.

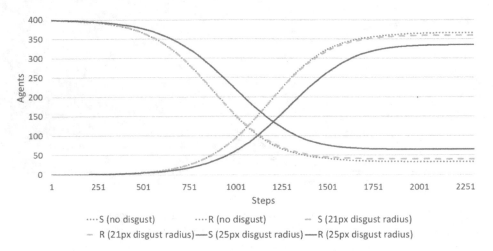

Fig. 7. S and R compartments from epidemic simulations at agent density 1/40 agents per pixels^2

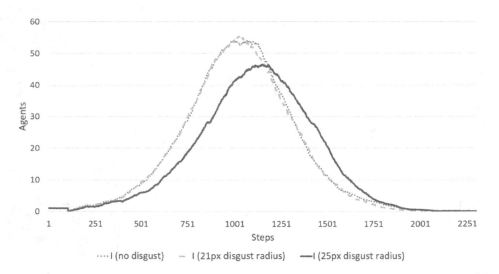

Fig. 8. I compartments from epidemic simulations at agent density 1/40 agents per pixels^2

5   Final Remarks

Our preliminary results using the SLIPR model suggest that disgust can significantly slow and reduce the spread of disease, which is consistent with related work done by Poletti et al. [6]. However, SLIPR further predicts that the effect of disgust is limited in dense populations.

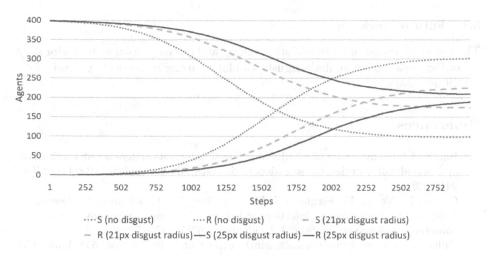

Fig. 9. S and R compartments from epidemic simulations at agent density 1/50 agents per pixels^2

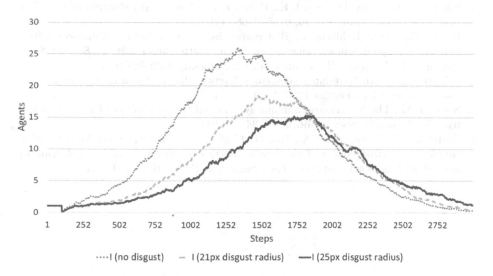

Fig. 10. I compartments from epidemic simulations at agent density 1/50 agents per pixels^2

The results suggest that the reproductive number of an infectious disease, R_0, may vary predictably given the density of a particular population and its known behaviors. This could have implications for forecasting disease risk and preparedness.

5.1 Future Work

The implementation of more realistic walking and disease-avoidant behavior may affect the impact of simulated disgust. Validation using real-world epidemic data is still necessary.

References

1. Bai, Z.: Global dynamics of a SEIR model with information dependent vaccination and periodically varying transmission rate. Math. Meth. Appl. Sci. **38**(11), 2403–2410 (2015)
2. Carrat, F., Vergu, E., Ferguson, N.M., Lemaitre, M., Cauchemez, S., Leach, S., Valleron, A.: Time lines of infection and disease in human influenza: a review of volunteer challenge studies am. J. Epidemiol. **167**(7), 775–785 (2008)
3. Gillis, J.: Correlated random walk. Math. Proc. Camb. Philos. Soc. **51**(4), 639–651 (1955)
4. Kermack, W.O., McKendrick, A.G.: A contribution to the mathematical theory of epidemics. Proc. Roy. Soc. Lond. A **115**, 700–721 (1927)
5. Kiesecker, J., Skelly, D., Beard, K., Preisser, E.: Behavioral reduction of infection risk. Proc. Natl. Acad. Sci. **96**, 9165–9168 (1999)
6. Poletti, P., Ajelli, M., Merler, S.: Risk perception and effectiveness of uncoordinated behavioral responses in an emerging epidemic. Math. Biosci. **238**(2), 80–89 (2012)
7. Popinga, A., Vaughan, T., Stadler, T., Drummond, A.J.: Inferring epidemiological dynamics with bayesian coalescent inference: the merits of deterministic and stochastic models. Genetics **199**(2), 595–607 (2015)
8. Schaller, M.: The behavioural immune system and the psychology of human sociality. Phil. Trans. R. Soc. B **366**, 3418–3426 (2011)
9. Torrens, P.M., Nara, A., Li, X., Zhu, H., Griffin, W.A., Brown, S.B.: An extensible simulation environment and movement metrics for testing walking behavior in agent-based models Computers. Environ. Urban Syst. **36**(1), 1–17 (2012)

Modeling Social Capital as Dynamic Networks to Promote Access to Oral Healthcare

Hua Wang[1](✉), Mary E. Northridge[2], Carol Kunzel[3], Qiuyi Zhang[4],
Susan S. Kum[4], Jessica L. Gilbert[4], Zhu Jin[4], and Sara S. Metcalf[4]

[1] Department of Communication, University at Buffalo,
The State University of New York, 359 Baldy Halll, Buffalo, NY 14260, USA
hwang23@buffalo.edu

[2] Department of Epidemiology and Health Promotion, New York University
College of Dentistry, 433 First Avenue, Room 726, New York, NY 10010, USA
men6@nyu.edu

[3] Section of Population Oral Health, Columbia University College of Dental
Medicine, 630 W. 168th St., New York, NY 10032, USA
ck60@columbia.edu

[4] Department of Geography, University at Buffalo, The State University
of New York, 105 Wilkeson Quad, Buffalo, NY 14261, USA
{qiuyizha, susankum, jlgilber, zhujin,
smetcalf}@buffalo.edu

Abstract. Social capital, as comprised of human connections in social networks and their associated benefits, is closely related to the health of individuals, communities, and societies at large. For disadvantaged population groups such as older adults and racial/ethnic minorities, social capital may play a particularly critical role in mitigating the negative effects and reinforcing the positive effects on health. In this project, we model social capital as both cause and effect by simulating dynamic networks. Informed in part by a community-based health promotion program, an agent-based model is contextualized in a GIS environment to explore the complexity of social disparities in oral and general health as experienced at the individual, interpersonal, and community scales. This study provides the foundation for future work investigating how health and healthcare accessibility may be influenced by social networks.

Keywords: Social capital · Dynamic social network · Agent-based modeling · Oral healthcare accessibility

1 Introduction

Gaps in access to healthcare services, experiences, and outcomes persist in the United States [30]. Health disparities result from a range of economic, social, cultural, and behavioral factors at multiple levels, resulting in persistent and egregious effects on individual and population health. For example, people with fewer resources may suffer from chronic diseases and systemic inflammation, and lose their teeth as they age, compromising their nutrition and social interactions. Racial and ethnic minorities and people living in poor neighborhoods may lack access to healthy food, community-based

© Springer International Publishing Switzerland 2016
K.S. Xu et al. (Eds.): SBP-BRiMS 2016, LNCS 9708, pp. 117–130, 2016.
DOI: 10.1007/978-3-319-39931-7_12

amenities, and quality oral and general healthcare. The result is often delayed treatment until oral pain becomes untenable or complications arise. Diabetes and hypertension are more prevalent in disadvantaged and older populations than in wealthier and younger populations, which can negatively affect both oral health and dental treatment. What is detrimental to the oral and general health of older adults is also financially burdensome to the entire healthcare system [20].

The complexity of such phenomena and the importance of studying the healthcare system as a whole involve breaking through the traditional disciplinary boundaries of medicine, dentistry, and public health [20], and requires more than reductionist scientific approaches such as randomized controlled trials [7]. Systems science is especially useful in understanding complex public health issues [13], studying the dynamic interplays underlying health disparities [9], and informing health policy research [14]. In this collaborative project, we apply social network, system dynamics, and agent-based modeling from the systems science tradition, to understand the mechanisms through which social capital may promote access to oral healthcare and ultimately health equity.

The study context is the *ElderSmile* community outreach program of the Columbia University College of Dental Medicine. Since 2006, *ElderSmile* has conducted health events at senior centers that serve predominantly racial and ethnic minority older adults who live in northern Manhattan or nearby communities. *ElderSmile* offers educational workshops, preventive oral and general health screenings, and referrals for oral healthcare treatment to older adults who frequent the centers. The centers themselves function as "third places" for community gathering and socialization outside the confines of home ("first places") and work ("second places") [19, 21]. This community-based program thereby affords oral healthcare providers with a means of serving older adults' complex social and health needs.

Based on the insights from *ElderSmile* staff and participants, a system dynamics approach is used to articulate our conceptual framework and also to guide group model-building activities. Social capital is examined as dynamic social networks and simulated using an agent-based model contextualized in a GIS environment to explore the complexity of social disparities in oral and general health as experienced at the individual, interpersonal, and community scales. The model is used to explore the role of community-based health promotion that leverages "third places" in stimulating social capital formation, particularly in providing access to referral networks for dental visits to induce positive impacts on health and well-being.

2   Social Capital

The concept of social capital has been studied for decades. Apart from tools and machines in physical capital, skills and talents in human capital, or cash and stocks in financial capital, what people generate, consume, exchange, and accumulate in social capital are human connections and their associated benefits [3]. Social capital is known to be closely related to the health of individuals, communities, and societies at large [12, 24, 28, 31]. It may exert positive impacts on physical and mental health through multiple mechanisms, including by helping people to navigate healthcare systems, providing buffers against stressors, and offering instrumental and emotional support. But conversely, social capital may also exert excessive demands, expectations of conformity, and social exclusion on individuals and groups that present debilitating barriers to health-seeking behaviors and exacerbate healthcare costs [10, 11, 23]. For disadvantaged population groups such as older adults and racial and ethnic minorities, social capital may play a particularly critical role in mitigating the negative effects and reinforcing the positive effects of factors at multiple levels on individuals' health in general [29]. However, previous research on social capital and oral health has been limited. In this project, we extend research on social capital from its connection to individuals' general health to include oral health, healthcare, and health equity, particularly for disadvantaged population groups such as the *ElderSmile* program participants attending the senior centers in predominantly racial and ethnic minority neighborhoods.

In addition, researchers have different approaches to the definition and measurement of social capital [3–5, 10, 12, 25, 32]. The vast majority of the literature has focused on framing social capital as either a cause or an effect [10]. When treated as a cause, social capital is viewed as a network with embedded social structures and properties that can serve as resources available to the members of the network [4, 5, 12]. As an effect, social capital is viewed as a form of social consequences based on interactions between and within networks [8, 32], with the notable example introduced by Putnam [24] of bonding social capital and bridging social capital. As Parks [22] points out, "[R]elationships live in communication. They are made, unmade, and remade in the communicative practices of their participants" (p. 24). Rather than taking a static view of social capital, it may prove more fruitful to examine social capital as dynamic networks with feedback loops and cyclical processes [17, 18]. Social capital can thus be conceived of as social agents and structural conditions for community building and mobilization [25] and encompass communication networks, cultural norms, and trust in relationships [27]. In this project, we acknowledge the evolutionary and emergent capacity of individual actors as well as their social interactions. By modeling social capital as dynamic networks, we hope to better connect the micro factors at the personal level with the macro factors at the societal level, providing useful insights into often intricate and complex phenomena such as oral healthcare for older adults and racial and ethnic disparities in health.

3 Conceptual Framework

The diagram in Fig. 1 illustrates a dynamic hypothesis of how social factors affect access to oral healthcare and ultimately health equity. The relationships in Fig. 1 are interconnected as a causal map (a map of causes and effects) in the tradition of system dynamics such that solid arrows indicate direct causal relationships (positive polarity) whereas the dotted arrows indicate inverse causal relationships (negative polarity). In the modeling tradition of system dynamics, reinforcing (positive) feedback loops can be traced when an even number (including zero) of inverse relationships are encountered in a complete cycle of cause and effect. Balancing (negative) feedback loops are traced when there are an odd number of inverse relationships. These structural relationships lead to amplification of a change in direction for a reinforcing feedback loop, and mitigation of such a change for a balancing feedback loop [26].

The elements of Fig. 1 that are highlighted in yellow reflect aspects of social capital. These include *social connectedness, social support, social engagement, visits to third places, communication about healthcare experience, recommendation for healthcare provider*, and *trust in healthcare provider*. These factors are interrelated with potential leverage points for health promotion that can steer the system toward greater health equity. Elements in Fig. 1 that appear in **boldface** correspond to elements that are represented in the agent-based model described in the following section. As a navigational tool for modeling work, the causal map in Fig. 1 serves as a conceptual framework to guide the design of agent-based models with which we perform experiments that involve heterogeneous individuals and their cross-scalar interactions with each other and the environment.

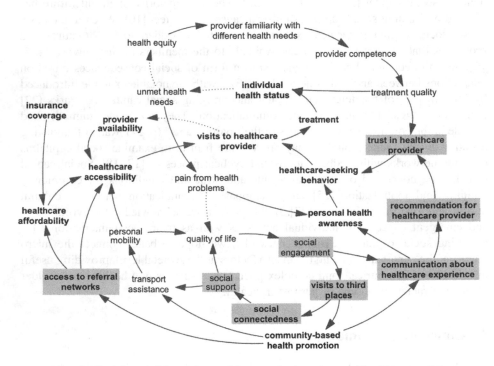

Fig. 1. Social capital formation and impacts in the context of health promotion

As the foundation for pathways toward better access to healthcare, we situate health promotion at the base of Fig. 1. Specifically, *community-based health promotion* activities through programs such as *ElderSmile* provide a novel means of intervention by making healthcare services available at third places such as senior centers. The health education component of community-based health promotion may serve to influence *communication about healthcare experience* and *personal health awareness*. Health promotion may also provide mechanisms for *transport assistance. Access to referral networks* results from *community-based health promotion*. Such access is a critical outcome of older adult participation in the *ElderSmile* preventive screenings: it directly improves both *healthcare affordability* and *healthcare accessibility*. *Healthcare accessibility* and *healthcare-seeking behavior* then combine to influence *visits to provider* for *treatment*.

Insurance coverage is an important leverage point for health promotion at the societal scale, as evidenced by Medicaid expansion under the Affordable Care Act [1]. *Insurance coverage* is a broad term that belies the array of benefits and limitations that arise under different forms of coverage. Although it is shown as an exogenous policy lever in Fig. 1, *insurance coverage* is also influenced by awareness of what services are covered. Limited compensation from Medicaid to healthcare providers for services adversely impacts the effective provider availability and potentially the range of treatment options that are discussed with the patient. Complications of reimbursement processes for Medicaid patients deter utilization of needed healthcare. Metcalf and colleagues [15] discuss Medicaid as it relates to opportunities to promote oral health equity. *Community-based health promotion* programs such as *ElderSmile*, through their outreach and health education arms, help to clarify which providers are available who would accept Medicaid insurance. The relationships drawn in Fig. 1 indicate that the expansion of *insurance coverage* increases *healthcare affordability* by lowering out-of-pocket costs.

Contingent upon sufficient *treatment quality*, *visits to healthcare provider* are presumed to improve a person's *individual health status*, reducing *unmet health needs*. We then move conceptually from the scale of the individual to the scale of society at large, so that *unmet health needs* at the individual scale aggregate to reveal disparities that obstruct *health equity*. Here we note that the translation of *unmet health needs* to *health equity* warrants consideration of the distribution of individual health outcomes across a population, since the reduction of unmet health needs equally across the population would not alone be sufficient to achieve *health equity*.

A path from *health equity* to *provider competence* is hypothesized via *provider familiarity with different health needs*. The logic here is that competence builds with familiarity through exposure to a wider variety of complex patient health needs. Since each provider serves only a subset of the population, the effect of this relationship would be limited by the specific sets of patients that a given provider serves. *Provider competence* then improves the *treatment quality* experienced by the patient. As the *treatment quality* delivered improves, *individual health* status also improves. *Treatment quality* also produces *trust in healthcare provider*, which then leads to *healthcare-seeking behavior*, influencing *visits to provider* for treatment.

At the individual scale, *pain from health problems*, such as dental caries and other oral infections, impinges upon one's *quality of life* and may thereby deter the *social engagement* needed to foster social capital formation. On the other hand, symptomatic pain may also have the beneficial effect of increasing the salience of health issues to promote one's *personal health awareness*, a mechanism for stimulating *health-seeking behavior*. Other symptoms of *unmet health needs*, such as loose teeth and difficulty chewing, may also interfere with *quality of life*.

Older adults' attendance at senior centers, characterized as *visits to third places*, is significant not only because it offers an opportunity to be exposed to *community-based health promotion* through preventive screenings, but also because it offers an opportunity for social interaction and a means of social network formation, thereby enhancing *social connectedness*. *Social engagement* provides opportunities for conversations about relevant matters that may include *communication about healthcare experience*. We consider *communication about healthcare experience* to be an aspect of social capital, as well as a particular consequence of such communication, which is a *recommendation for healthcare provider*. Recommendations serve to instill a sense of *trust in healthcare provider*.

Social connectedness is an outcome of *visits to third places* that is explored further in the model described below. This connectedness, in terms of the network structure and the density of social ties, improves the capacity for *social support* that arises from the network. This support may confer practical advantages such as *transport assistance*. Importantly, *social support* enhances *quality of life* for older adults, making daily challenges easier to manage and reducing social isolation. This effect completes a reinforcing feedback mechanism for maintaining *social engagement*.

We constructed the causal hypothesis in Fig. 1 to explore the complex feedback mechanisms involved in the way that social capital influences health equity through healthcare-seeking behavior at the individual scale, and health promotion at the community scale. These hypothesized relationships have emerged iteratively as a product of dialogues and group model-building activities with the research team. In this way, the conceptual and computational models in our portfolio have emerged from the experiences of the research team in group model-building activities, the *ElderSmile* program, and focused group interviews of ethnic minority seniors living in northern Manhattan. Our modeling approach is broadly in line with the tradition of system dynamics, and we have previously found system dynamics modeling quite useful for examining dynamics of diffusion of healthcare-seeking behavior [16]. However, instead of implementing our model using the stocks (integrals) and flows (rates of change) associated with that tradition, for this study we shift our structural orientation toward the behavior of individual agents. An agent-based modeling platform was chosen for this study because of its capacity to explicitly represent heterogeneous individuals, simulate social network dynamics, and be integrated with a GIS environment. Importantly, this platform provides a means of tracing individual health trajectories and also considering the distribution of oral health needs in the population.

4   Agent-Based Model

The model developed in this study is designed to simulate dynamic social networks so as to operationalize social capital formation and subsequent access to referral networks for dental visits. Dental visits provide an indication of access to oral healthcare and a mechanism for reducing disparities in working toward health equity. This model consists of two agent classes that represent people (i.e., older adult senior center attendees and oral healthcare providers) and two agent classes that represent facilities (i.e., senior centers and dental clinics). In addition to these agents there is a Main class that incorporates all of the other classes in the model, as required in the Java-based AnyLogic software platform used for the model [2]. The Main class contains a GIS environment as the landscape in which people agents can move around and interact with each other. This interactive model has a user interface for customizing parameter settings for population size and the degree threshold to identify hubs that establish the conditions for each simulation run.

An earlier version of this agent-based model was developed without a GIS environment to facilitate discussion and exploration of agents in a group model-building exercise conducted with the research team in May 2015 as part of a workshop on model structure. The model was used to demonstrate agent dynamics that were analogous to those of individuals participating in the *ElderSmile* program. Because the model was also used to set the stage for a cooperative game played by members of the research team in small breakout groups, parameters of the demonstration model were set so that referral delay times were reduced and screening events were held more frequently than in the actual operation of the *ElderSmile* program. During the game, members of the research team acted out agent behaviors using pegs to indicate sick and healthy people attending senior centers and receiving referrals to treatment by oral healthcare providers. For the purpose of engaging collaborators, the model was designed to simulate small numbers of older adult and provider agents in a dental landscape of limited size. Therefore, while the model logic was informed by the operation of the *ElderSmile* program, the model parameterization was simplified for educational and learning purposes. The *ElderSmile* screening program events simulated in the model capture the mechanism of community-based screenings in third places such as senior centers where social networks form and individuals communicate about their experiences with oral healthcare providers. The contribution of the model lies in its possibilities for simulating dynamic social networks that affect access to oral healthcare.

4.1 Model Operation

Upon startup of the simulation, a customizable population of older adult agents is created. Each of these agents has a home location assigned randomly (i.e., with a uniform probability distribution) within the GIS landscape of the northern Manhattan study area. Senior centers are located according to the addresses of centers that are affiliated with the *ElderSmile* program. Dental clinics may be located using information about actual clinic locations [6] or assigned a synthetic location based upon user-specified coordinates to ensure that clinics are in physical proximity to the simulated agents. For the model described here, both empirical and synthetic locations are assigned to simulated clinics.

Each oral healthcare provider is assigned to a dental clinic and also given a random home location in the study area. One of these providers is affiliated with a health promotion program and performs preventive health screenings when not working at the clinic. Screening events occur at one of the senior centers every three days. When attending a screening event, the provider's corresponding dental clinic will be closed for the day. During the screening event, if their actual oral health status is below a diagnostic threshold, participating older adults will receive a referral for treatment at the affiliated provider's dental clinic. The referral is scheduled for the following day.

In the model, we differentiate between perceived and actual oral health status. These are both operationalized as continuous values that can vary between 0 and 1, where 1 implies healthy and 0 implies unhealthy, and are randomly assigned initial values between these bounds. Akin to the notion of personal health awareness in our causal map (see Fig. 1 above), perceived oral health status is contingent upon the experience of symptoms, so that poor perceived oral health induces healthcare-seeking behavior to treat health problems. Actual oral health status is recognized by the provider at preventive screenings and used as a criterion for scheduling follow-up referrals. Although they are initialized differently, both actual and perceived oral health status decline without treatment at the same rate (reduced by 0.1 per 10 days) and are assumed to be exacerbated by the presence of another chronic illness (reduced by 0.15 per 10 days).

Boolean parameters are assigned at random upon initialization to establish whether agents trust oral healthcare providers (i.e., are more disposed to go to oral healthcare providers on their own), whether they can afford treatment, or whether they live with a related chronic illness.

As part of their simulated daily routine, older adult agents either attend their nearest senior center or visit one of the dental clinics. The latter case, in which older adult agents proceed directly to a dental clinic, occurs if they have a need for urgent care (i.e., their perceived oral health status declines below a symptomatic threshold, set to a default value of 0.4), or if they have scheduled a referral appointment with an oral healthcare provider. Referral appointments are scheduled with a time lag of one day, and are held with the same provider that the agent encountered at the preventive screening.

Agents who can afford care will search for dental clinics within a user-specified distance (default value of 1 km) from the home location to identify an affordable dental clinic, e.g. accepts Medicaid insurance or offer sliding scale fees. If an affordable dental clinic is available, the agent will go to that clinic. Otherwise, they will go to the nearest clinic. Agents who cannot afford care will not visit a dental clinic unless they have a need for urgent care. The outcome of treatment is an improvement in oral health status, encoded in a return to the state of healthy teeth. This simplified assumption of a full restoration of health status will be relaxed in future work to account for variation in treatment quality and to render the dynamics of dental health decline less reversible.

If a senior center hosts a preventive health screening event, each older adult at the center is as likely to participate as not (set by a 50 % probability). If a participating older adult agent's actual oral health value is lower than a diagnostic threshold, s/he will be referred to the provider's dental clinic. If the person agent can afford to pay for treatment and also trusts an oral healthcare provider, s/he will bypass the referral process and seek treatment directly, undertaking the same decision process as agents who have an urgent need for care in seeking proximate dental clinics. After each potential activity away from home, older adults then return to their home location.

4.2 Social Networks

Two types of social networks are simulated for agent interactions: (1) the peer social network among older adult senior center attendees; and (2) the patient-provider network between older adults and their oral healthcare providers. The peer social network forms from encounters made at the senior center (under the default assumption, one encounter is sufficient to form a tie) and is attenuated if an older adult no longer frequents the center on a daily basis (ties fade after 3 days). The patient-provider network forms from dental visits.

The visualization of these dynamic networks is facilitated by a user interface enabling selection of either network type for display during simulation. Figure 2 illustrates how the peer network is visualized (at left, with black lines connecting older adults) in contrast to the patient-provider network (at right, with red lines connecting older adults with oral healthcare providers).

The snapshots of the peer and patient-provider social networks shown in Fig. 2 were taken at the same point in the simulation run. All embedded agent classes are visible in Fig. 2: older adults are shown in black, providers are green, the green buildings represent dental clinics, and the blue buildings represent senior centers where screenings may be held. In the panel at right in Fig. 2, the providers are shown working in their clinics and connected through social ties to older adults who are their patients in the study area.

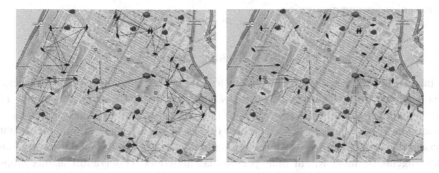

Fig. 2. Simulated peer social network (left) and patient-provider network (right)

Social connectedness in the agent-based model is represented by the degree or number of social ties that each person maintains. In aggregating from the individual level to the community of older adults simulated, we can examine the degree distribution of these connections. Figure 3 depicts the resulting social network degree distribution with a population size of 40 (at top) and 80 (at bottom) older adults at the end of the simulation run (on the 100th day). For each plot, the horizontal (x) axis delineates the degree or number of connections per person, and the vertical (y) axis depicts the percentage of the population with the given degree as the frequency of the degree in the population.

The degree distributions of both experiments stabilized around the 40th day of simulation as social clusters form from activity at the senior centers. The mode is 4 connections for a population of 40 versus 10 connections for a population of 80. This comparison of social network structure for these different population sizes indicates that a larger population in the same geographic area leads to a greater density of older adults who therefore have more chances to encounter each other at third places such as

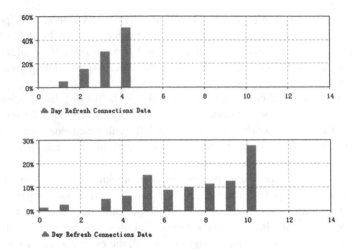

Fig. 3. Emergent degree distributions for 40 people (top) and 80 people (bottom)

senior centers. These encounters lead to the formation of social ties that provide opportunities for communication.

4.3   Social Influence on Healthcare-Seeking Behavior

For this study we created a set of scenarios to contrast the simulation results for dental visits achieved through mechanisms of *seeking care through a referral* with *seeking care through a trusted health provider*. An older adult agent's trust in oral healthcare providers is spread through the social network as opinion leaders communicate with others in their social network. A threshold parameter is used in the model to designate certain agents as hubs within their peer social networks. Older adult agents whose degree is greater than the threshold will act as opinion leaders and send a "trust" message to connected agents through the peer social network, exerting a social influence toward trust of oral healthcare providers. This mechanism functions much as a recommendation for an oral healthcare provider as expressed in the causal map of Fig. 1. However, here trust extends beyond a particular provider. After receiving a trust message, the connected older adults are considered to have trust in oral healthcare providers, indicated by a change in their status from "no trust" to "trust." A forgetting time is applied as a balancing mechanism, with a default value of 30 days, after which the status returns to "no trust." Because it can be lost through this "forgetting," trust functions much like an awareness of or inclination toward care, and can similarly be socially influenced. Trust thereby emerges from the peer network to influence the construction of the provider-patient network. If an agent trusts oral healthcare providers, they may seek treatment on their own, without waiting for an appointment by referral.

The dynamics of dental visits are illustrated in Fig. 4 for the clinic that accommodates referrals made during the preventive screenings offered by the oral healthcare provider who works there. The left-hand side of Fig. 4 indicates the number of visits under the scenario of no social influence. In this scenario, no agents can function as

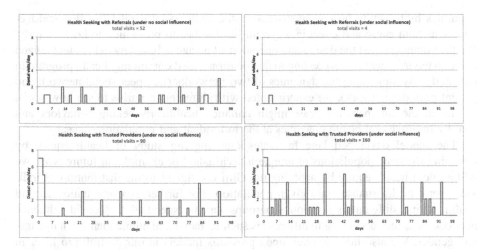

Fig. 4. Effect of social influence (right) on dental visits with referrals (top) and trusted providers (bottom)

opinion leaders, which is achieved by setting the parameter for hub degree threshold to 100 % of the population size. In contrast, the right-hand side of Fig. 4 indicates the number of dental visits attained in a scenario of social influence, in which the hub degree threshold is set to 10 % of the population size. For the default population of 40 older adults, opinion leaders are those agents who have more than 4 peer connections.

The dynamics of dental visits for the experiments charted in Fig. 4 are arrayed so that the dental visits resulting from referrals established during a preventive screening event appear in the top row, while visits resulting from an older adult agent's own initiative to seek healthcare with trusted providers appear in the bottom row. The upper right scenario highlights that as social influence takes effect through opinion leader dynamics in the social network, all agents elect to go directly to a trusted oral healthcare provider (shown at bottom right) instead of waiting for a referral appointment to be arranged.

Under the social influence scenario, the simulated results indicate an increase in overall dental visits to the referral clinic as well as the other clinics, demonstrating the potential influence of social networks on healthcare behaviors. Because treatment improves oral health, these visits translate to improved oral health outcomes. In the absence of social influence, cumulative dental visits to the referral clinic are less than visits to other clinics. This gap has been explored in the social influence simulation where hubs have more social influence than others via word of mouth communication.

5   Discussion

This modeling effort was designed to explore how social capital might help mobilize older adults to utilize community-based health promotion and healthcare services. A conceptual framework was first developed as a causal map in the tradition of system dynamics highlighting how elements associated with social capital function endogenously as both cause and effect in a system linking health promotion to health equity. We then demonstrated the design and operation of an agent-based model contextualized in a GIS environment corresponding to the location-based community health program in this study. By simulating the dynamics of two networks (i.e., peer social network and patient-provider network), we experimented with different scenarios to see how social factors at the interpersonal scale might enhance healthcare-seeking behaviors and thereby improve oral health outcomes at the individual and population scales.

The model developed here facilitates our ongoing study of health equity by reflecting dynamic populations of unique individuals. For example, in future work we will explore health equity by applying the Gini coefficient to the distribution of individual oral health statuses in a given population. For further experimentation around health equity, different population subgroups can be specified that could influence network dynamics according to differences such as racial and ethnic identity, language, and gender. We also hope to draw upon previous models in our portfolio to layer in other elements of network dynamics such as communication about health promotion, healthcare affordability, and transport accessibility as a mediator of healthcare accessibility [16]. Alternative mechanisms for the formation of trust can be designed to consider factors such as treatment quality as well as the recommendation for a particular healthcare provider. This study expanded our portfolio and can help inform future modeling efforts as well as policy research on community-based oral public health to promote health equity.

Acknowledgements.   Work on this study was supported by NIH (NIDCR/OBSSR) R01 award DE023072, "Integrating Social and Systems Science Approaches to Promote Oral Health Equity."

References

1. Affordable Care Act: The Patient Protection and Affordable Care Act (PPACA), Pub. L. No. 111–148, 124 Stat. 119 (2010)
2. AnyLogic: Multimethod Simulation Software. Version 7.2.0. XJ Technologies, St. Petersburg (2016)
3. Bourdieu, P.: The Forms of Social Capital. Handbook of Theory and Research for the Sociology of Education, Green- wood, New York (1986)
4. Burt, R.S.: Brokerage and closure: An introduction to social capital. Oxford University Press, Oxford, New York (2005)
5. Coleman, J.S.: Social capital in the creation of human capital. Am. J. Sociol. **94**, S95–S121 (1988)
6. DFTA: Oral care provider directory: A resource directory for use by New York City Department for the Aging partner agencies to help older adults access dental care providers. NYC Department for the Aging (DFTA) and Columbia University College of Dental Medicine, New York (2014)
7. Diez Roux, A.V.: Complex systems thinking and current impasses in health disparities research. Am. J. Public Health **101**(9), 1627–1634 (2011)

8. Ellison, N., Steinfield, C., Lampe, C.: The benefits of Facebook "friends:" Social capital and college students' use of online social network sites. J. Comput. Mediated Commun. **12**(3), 434–445 (2007). Article 1
9. Frerichs, L., Lich, K.H., Dave, G., Corbie-Smith, G.: Integrating systems science and community-based participatory research to achieve health equity. Am. J. Public Health **106**, 215–222 (2016)
10. Kawachi, I.: Social capital and health - Making the connections one step at a time. Int. J. Epidemiol. **35**(4), 989–993 (2006)
11. Kawachi, I., Subramanian, S.V., Kim, D.: Social capital and health: A decade of progress and beyond. In: Kawachi, I., Subramanian, S.V., Kim, D. (eds.) Social capital and health. Springer, New York (2008)
12. Lin, N.: Social capital: A theory of social structure and action. Cambridge University Press, New York (2001)
13. Luke, D.A., Stamatakis, K.A.: Systems science methods in public health: Dynamics, networks, and agents. Ann. Rev. Public Health **33**, 357–376 (2012)
14. Mabry, P.L., Marcus, S.E., Clark, P.I., Leischow, S.J., Mendez, D.: Systems science: A revolution in public health policy research. Am. J. Public Health **100**(7), 1161–1163 (2010)
15. Metcalf, S.S., Birenz, S.S., Kunzel, C., Wang, H., Schrimshaw, E.W., Marshall, S.E., Northridge, M.E.: The impact of Medicaid expansion on oral health equity for older adults: A systems perspective. J. Calif. Dental Assoc. **43**, 369–377 (2015)
16. Metcalf, S., Northridge, M., Widener, M., Chakraborty, B., Marshall, S., Lamster, I.: Modeling social dimensions of oral health among older adults in urban environments. Health Educ. Behav. **40**(1S), 63S–73S (2013)
17. Monge, P.S., Contractor, N.S.: Theories of communication networks. Oxford University Press, Oxford (2003)
18. Newton, K.: Social capital and democracy. Am. Behav. Sci. **40**(5), 575–586 (1997)
19. Northridge, M.E., Nye, A., Zhang, Y., Jack, G., Cohall, A.T.: "Third places" for healthy aging: Online opportunities for health promotion and disease management in adults in Harlem. J. Am. Geriatr. Soc. **59**(1), 175–176 (2011)
20. Northridge, M.E., Yu, C.C., Chakraborty, B., GreenBlatt, A.P., Mark, J., Golembeski, C., Cheng, B., Kunzel, C., Metcalf, S.S., Marshall, S.E., Lamster, I.B.: A community-based oral public health approach to promote health equity. Am. J. Public Health **105**(S3), S459–S465 (2015)
21. Oldenburg, R.: Celebrating the third place: Inspiring stories about the "great good places" at the heart of our communities. Da Capo Press, Boston (2000)
22. Parks, M.R.: Personal relationships and personal networks. Routledge, New York (2006)
23. Portes, A.: Social capital: Its origins and application in modern sociology. Ann. Rev. Sociol. **24**, 1–24 (1998)
24. Putnam, R.: Bowling alone: The collapse and revival and American community. Simon & Schuster, New York (2000)
25. Putnam, R., Feldstein, L.M., Cohen, D.: Better together: Restoring the American community. Simon & Schuster, New York (2003)
26. Richardson, G.: Feedback thought in social science and systems theory. Pegasus, Waltham (1991)
27. Singhal, A., Papa, M.J., Sharma, D., Pant, S., Worrell, T., Muthuswamy, N., et al.: Entertainment-education and social change: The communication dynamics of social capital. J. Creative Commun. **1**(1), 1–18 (2006)
28. Song, L., Son, J., Lin, N.: Social capital and health. In: Cockerham, W. (ed.) The new companion to medical sociology. Wiley-Blackwell, Oxford (2010)

29. Uphoff, E.P., Pickett, K.E., Cabieses, B., Small, N., Wright, J.: A systematic review of the relationships between social capital and socioeconomic inequalities in health: A contribution to understanding the psychosocial pathway of health inequalities. Int. J. Equity Health **12**, 54 (2013)

30. US Institute of Medicine: How far have we come in reducing health disparities? Progress since 2000: Workshop summary. National Academies Press, Washington, DC (2012)

31. Valente, T.W.: Social networks and health: Models, methods, and applications. Oxford University Press, New York (2010)

32. Williams, D.: On and off the net: Scales for social capital in an online era. J. Comput. Mediated Commun. **11**(2), 593–628 (2006)

Information, Systems, and Network Sciences

The Implications for Network Structure of Dynamic Feedback Between Influence and Selection

Ran Xu[✉] and Kenneth A. Frank

Michigan State University, East Lansing, USA
{ranxu, kenfrank}@msu.edu

Abstract. Although network structures shape diffusion and ultimately systemic performance, the underlying dynamics generating different network structures during diffusion are not well understood. To explore these dynamics we present a set of models in terms of a common drive for relational balance – the inclination to align the attributes of one's network members with one's own attributes either by adopting the attributes of network members as in the diffusion process or by selecting to interact with similar others. Agent based models show that the models generate modular (clustering) structure for high levels of relational balance. Moreover, when network members have strong influence over one another the transition to modularity is delayed but then drastic, in the extreme creating a phase transition. Thus, the rate of influence amplifies the attractiveness of a particular modularity state of the system.

Keywords: Modular networks · Selection · Diffusion · Amplification

1 Introduction

It is known that network structures have important implications for diffusion as well as the performance of the system [1–5]. While how these network structures are generated through different selection processes have been extensively studied [6, 7], there has been relatively little study of how network structures are indirectly altered as actors influence one another during the diffusion process. A small number of empirical studies have found greater support for the rate of selection relative to the rate at which actors influence one another during diffusion [8–10]. But while these empirical studies have used agent based models to simultaneously estimate the effects of influence and selection, agent based models have not been as extensively used to explore the long term systemic implications of simultaneous influence and selection.

In this study we use agent based models to explore how the process of influence dynamically complements the process of selection during the diffusion process. We first specify models of network selection that can generate different social structures depending on a single parameter governing actors' relative preferences for homophily in selecting with whom to interact. To investigate how networks are shaped by the diffusion process we then use the same parameter to characterize how the actors are influenced as they change their beliefs/behaviors in response to the mean belief/behavior

© Springer International Publishing Switzerland 2016
K.S. Xu et al. (Eds.): SBP-BRiMS 2016, LNCS 9708, pp. 133–141, 2016.
DOI: 10.1007/978-3-319-39931-7_13

of others in their specific networks. This influence process is the engine of diffusion; behaviors diffuse through a system as actors adopt the behaviors of those in their network. In turn, the changes in behaviors then affect the selection process. Thus our study synthesizes the diffusion of innovation and network evolution literatures as we examine how the factors that affect the formation of network structures are altered during diffusion.

1.1   The Micro-Dynamic Drivers of Influence and Selection

The micro-dynamics of our system are driven by actors' choices of network members (selection) and how to respond to norms (average behavior or belief) among network members (influence). In our framework the choices are driven by two utilities, each of which is a function of relational balance – the inclination to align one's own attributes with those of one's network members. Relational balance defined here is grounded in Heider's balance theory [11], which states that people are motivated to achieve cognitive consistency by balancing their beliefs or behaviors with those with whom they interact. Furthermore, people can achieve relational balance either by choosing to interact with similar others or adopting the beliefs or behaviors of those with whom they interact.

Selection Process. In the selection model actors pursue relational balance by seeking ties with others with similar attributes, known as homophily – birds of a feather flocking together [12, 13]. Formally, we parameterize the pursuit of relational balance through homophily as α, where homophily is defined in terms of the absolute value of the difference in an attribute y between actor i and j at time t: $|y_{it} - y_{jt}|$. The pursuit of homophily is offset against the pursuit of others based on other characteristics, g (e.g., the control of resources, popularity, control of information) by 1-α.[1] Thus the selection utility, U, of actor i choosing to interact with actor j at time t:

$$U_{ijt} = (1-\alpha)g_{jt-1} - \alpha|y_{it-1} - y_{jt-1}| \tag{1}$$

where y_{it-1} represents an attribute possessed by actor i at time t-1.[2] α ranges from 0 to 1, the larger the value of α the more the agents pursue relational balance through homophily based on the similarity of attributes $|y_{it-1} - y_{jt-1}|$. The smaller the value of α the more agents seek others based on criteria other than homphily. Note that in the following experiments we will choose a particular form of g based on the pursuit of popular others (in future work we will consider of alternatives to homophily, such as transitivity [13]). Furthermore, note that here ties are directional so that in general $U_{ij} \neq U_{ji}$.

---

[1] Though it would be more realistic to represent preference of other characteristics g with another parameter, for parsimony we parameterize it as 1-α to offset homophily.

[2] For simplicity we only consider one attribute in the model. However this attribute can also be represented as aggregate functions of multiple attributes.

Influence Process. In the influence process actors may seek relational balance by conforming to the behaviors/beliefs of those with whom they interact. Here we parameterize relational balance as γ, offsetting it against the retention of one's own beliefs/behaviors associated with 1- γ. In this sense agents tolerate relational imbalance, which we parameterize by 1- γ.

Starting with a utility function [14], one can show that to maximize utility agent i should choose behavior y at time t according to:

$$y_{it} = (1 - \gamma)y_{it-1} + \gamma \frac{\sum w_{ijt-1}y_{jt-1}}{\sum w_{ijt-1}} \tag{2}$$

where w_{ijt-1} is 1 if actor i interacts with actor j at time t-1, 0 otherwise. Therefore $\frac{\sum w_{ijt-1}y_{jt-1}}{\sum w_{ijt-1}}$ represents the mean behavior of i's network members at time t-1,[3] In Eq. (2), the larger the value of γ, the more the actor pursues relational balance by conforming to network members. Thus the larger the value of γ the more readily can behavior or belief diffuse through networks within the system.[4]

Equations (1) and (2) can be re-expressed to represent the relative rates of the pursuit of relational balance through influence and selection. In particular, define k as the rate of the pursuit of relational balance in the influence process relative to the selection process: $k = \gamma/\alpha$. Substituting $k\alpha$ for γ in Eq. (2) yields:[5]

$$y_{it} = (1 - k\alpha)y_{it-1} + k\alpha \frac{\sum w_{ijt-1}y_{jt-1}}{\sum w_{ijt-1}} \tag{3}$$

Generally, when $k \to 0$ actors retain only their previous beliefs or behaviors – influence occurs slowly relative to selection; as k increases the process of influence occurs faster relative to selection and when $k \to 1$ influence occurs at the same rate as selection.[6] For example, α might represent the general tendency for an adolescent to have friends who engage in academics as much as she does, while k would represent the relative rates at which the adolescent pursues that balance by conforming to her friends' behaviors versus selecting friends who engage in academic behaviors similar to her own. If k is small, then an actor will pursue relational balance more through selecting members who engage in similar behaviors than by conforming in behavior to those of network members. As k increases the actor pursues relational balance more by conforming to the norms of network members and less by selecting network members

---

[3] We use mean behavior of network members to represent norm-based influence. Other functional forms can also be used to represent different forms of influence.

[4] Different from Centola [15] we assume a fairly simple and direct influence process in which actors respond to the attributes of network members without requiring repeated exposure.

[5] Alternatively we could substitute: γ/k for α in Eq. (1) and express relational balance generally in terms of γ.

[6] If selection does not occur and therefore the network does not change, the system converges to the same end point for $0 < k < 1$ [16].

who engage in similar behaviors. In this sense our models allow us to express the system in terms of relational balance (α) and the rate of influence relative to the rate of selection (k).

We evaluate our models using simulations in which actors can select with whom they interact and can change their behaviors/beliefs based on with whom they interact. Specifically we perform two sets of simulations, one in which the baseline networks are determined at random, and the other in which scale-free networks are established at baseline, allowing us to evaluate robustness of our results to different initial conditions. For each set of simulations we vary α and k to evaluate the relative effects of relational balance and the rate of influence on the occurrence of modular social structures, and how the two factors interact with one another. Here we express our results in terms of changes in behavior which drive the diffusion process, but our models pertain to changes in any attribute, such as taste or belief.

2   Results

2.1   Initiated with Random Networks

In Fig. 1 each data point represents the results of 200 replications for agent based models based on Eqs. (1) and (3) for different values of relational balance (α) and rate of influence relative to selection (k). The solid lines represent conditions in which actors are not influenced by their network members (k = 0). For $\alpha < .5$ the system exhibits near zero modularity. The system exhibits moderate modularity for $.5 < \alpha < .8$ as actors seek others with similar behaviors as much or more as they seek already popular others. In the extreme, highly modular structures emerge for $\alpha > .8$. For these high values of relational balance, homophily generates groups with distinct behaviors. Consistent with the formation of groups represented by the measures of modularity, in Fig. 1B the relative proportion of variation in behavior within groups decreases as α exceeds .5.

Examples of equilibria exhibiting the network structures as well as the behaviors of the actors for $\alpha = .2$ and $\alpha = .7$ are shown in Fig. 2A and B. One can observe modularity for high values of α (relational balance) in 2B relative to 2A. Correspondingly, diffusion would be more rapid in 2A than in 2B as the core members in 2A can directly diffuse behaviors or information to others.

Critically, Fig. 1 shows that modularity depends on the relative rate of influence. For a modest level of influence (dashed line with triangle dots, k = .5), modularity emerges at higher levels of relational balance ($\alpha = .75$) than when there is no influence ($\alpha = .5$ when k = 0). Our explanation is that with moderate influence the overall variance in behaviors is reduced as actors' behaviors become more similar to one another. As a result, the pursuit of homophilous others becomes relatively less important in the selection model.

Figure 1 also shows that the transition to modularity is more rapid when influence is present than when influence is absent (k = 0) as indicated by the steeper slopes for the dashed and dotted lines than for the solid line in Fig. 1A. Because relational balance operates through influence and selection, behaviors within nascent groups become more similar through influence as actors conform in their behaviors to their network members and ties become denser. The dual forces through influence and selection accelerate the formation of groups once they begin to emerge.

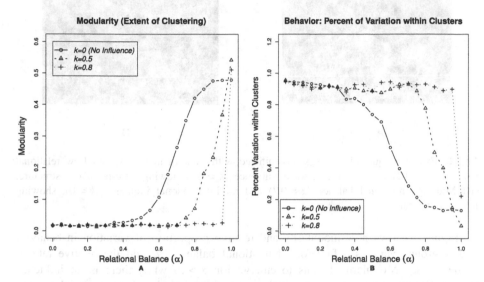

Fig. 1. Simulations initiated with random networks for different rates of interpersonal influence. Modularity increases with relational balancing, amplified with influence. Results show more dramatic changes at higher levels of interpersonal influence. (A) Mean modularity increases with relational balance (inclination to align individual behaviors with network members); (B) Average proportion of total behavior variation within clusters decreases with relational balance; Each data point represents 200 simulations.

The trends in Figs. 1 and 2 indicate that influence amplifies the effects of selection, decreasing the attraction of modularity for low values of α and then increasing the attraction of modularity for high values of α. The amplification effects increase with k, to the point that there is a near phase transition as indicated by the steep slope for the red lines for k = .8. Thus when relational balance is high and the rate of influence is high, the system can transition rapidly from one able to diffuse information to one not able to diffuse information, challenging a static characterization of the system as either supporting diffusion or not.

2.2 Initiated with Scale-Free Networks

The previous agent based models characterized the effects of relational balance (α) and relative rate of influence (k) on the emergence of modularity from random networks. Of course, different initial conditions may also contribute to modularity in networks, such as different initial network structures. Therefore we explore the effects of relational balance and relative rate of influence on networks with initial structures in which few nodes are highly connected to others while most nodes only possess few links [17].

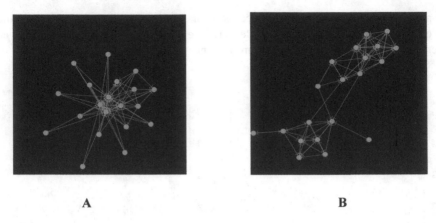

A   B

Fig. 2. Example equilibria for systems initiated with random networks: (A) Low relational balance ($\alpha = 0.2$) and no interpersonal influence ($k = 0$), showing core-periphery structure; (B) Mid-high relational balance ($\alpha = 0.7$) and no interpersonal influence ($k = 0$), showing modular structure.

Figure 3 shows how modularity and relative proportion of variation in behavior within groups vary as a function of relational balance (α) and the relative rate of influence (k). Modularity begins to emerge for $\alpha > .6$ when there is no influence ($k = 0$). But when influence is present the emergence of modularity is more delayed but then drastic. Furthermore, modularity emerges at higher levels of relational balance than when we start from random networks. Our explanation is that for modularity to

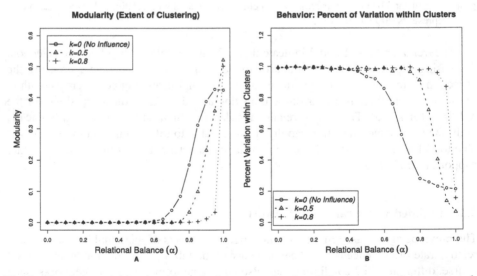

Fig. 3. Simulations initiated with scale-free networks for different rates of interpersonal influence. (A) Mean modularity increases with relational balance (inclination to align individual behaviors with network members); (B) Average proportion of total behavior variation within clusters decreases with relational balance; Each data point represents 200 simulations.

emerge in a scale-free network, actors need to have higher relational balance to offset the utility gained by connecting to the high-degree nodes. Furthermore, when actors influence one another central members are able to influence others to adopt behaviors similar to their own. In turn, the peripheral members continue to engage in connections with the central members because of the similarity in behaviors even for relatively high levels of relational balance.

3 Discussion

We have expressed selection and influence as a function of a common parameter representing the relational balance of a social system. Thus, we were able to express the emergence of modular structures relative to the intensity of relational balance. Generally, we show that for very high levels of relational balance, clusters emerge as actors prefer to interact with similar others (homophily).

The unexpected finding is that the rate of transition to modularity is affected by the rate at which actors influence one another. The greater the rate of influence of network partners relative to selection of network partners, the more delayed, but then ultimately dramatic, the emergence of modularity. In general, the rate of influence amplifies the attractiveness of a particular state of the system. When influence is strong, the system is attracted either to a low modularity structure or extreme modularity, with few other states or options in between.

Correspondingly, change agents must be aware that the interplay between influence and selection changes the dynamics of diffusion. Ignoring this interplay by characterizing diffusion as though the network is static can lead to markedly misleading predictions about the diffusion process. Attention to this interplay can lead to more nuanced approaches to leveraging networks as well as an understanding of how social networks evolve during the diffusion process.

4 Materials and Methods

4.1 Simulation Process

After defining our models we perform agent based simulations in Netlogo 5.2.0 [18]. Specifically, in each round: 1. Each actor evaluates all other actors in the system and calculates the utility of establishing a tie with each other actor based on the selection equation in (1). In particular, we chose in-degree of the alter as g in Eq. (1), the offset to homophily, representing actors' tendencies to seek interactions with popular others who might confer status or information [7, 17]; 2. Each actor then establishes a tie with other actors with highest utilities, holding the out-degree (number of others identified as network ties) for each actor constant. For example if an actor starts with 3 out-going ties, then each round it will terminate all previous out-going ties and select the 3 actors who yield the highest utilities; 3. As actors select with whom they interact they are influenced by their network neighbors and adjust their behaviors based on the influence Eq. (3). For each experiment described below we varied relational balance (α) from

0 to 1 by intervals of 0.05 and chose the relative rate of influence (k) to be 0, 0.5 or 0.8. In each configuration we simulated 200 times, with a total of 200*21*3 = 12600 simulations.

First Experiment. We initialized a random network as follows: 1. 20 actors with normally distributed behavior state with mean 20 and standard deviation 4. We chose a behavior state from a normal distribution with mean of 20 so that behavior states remained positive throughout the simulation. We used a standard deviation of 4 to scale the behavior state relative to in-degree. If the standard deviation were smaller α would have to be larger to generate the effects; modularity or polarization would occur for larger values of α; 2. Network density of 0.2, which generated 76 ties on average.

Second Experiment. We initialized a scale-free network as follows: 1. 20 actors with normally distributed behavior state with mean 20 and standard deviation 4; 2. We assign a scale-free in-degree distribution to the 20 actors, while keeping total number of ties and average out-degrees similar to that of the first experiment.

4.2 Key Outcome Measures

In both experiments we stopped each simulation after 50 rounds. To measure the modularity of the network we first used Kliquefinder [19, 20] to identify cohesive groups and calculated the modularity of ties within versus between groups. We chose KliqueFinder because it maximizes the odds of a tie occurring within groups relative to between groups as expressed in Exponential Random Graph Models [19]. This odds ratio is very similar to the modularity index. Moreover, KliqueFinder identifies the number of groups on the fly without a priori or posteriori input from the user, allowing the process to be fully automated. To measure the percent of variation in behavior within clusters we calculate the ratio between sum of squares for actors' behaviors within each cluster (as identified by Kliquefinder) and the total sum of squares for actors' behaviors.

References

1. Kozma, B., Hastings, M.B., Korniss, G.: Diffusion processes on power-law small-world networks. Phys. Rev. Lett. **95**(1), 18701 (2005)
2. Watts, D.J.: A simple model of global cascades on random networks. Proc. Natl. Acad. Sci. U.S.A. **99**(9), 5766–5771 (2002)
3. Weng, L., Ratkiewicz, J., Perra, N., Gonçalves, B., Castillo, C., Bonchi, F., Flammini, A.: The role of information diffusion in the evolution of social networks. In: Proceedings of the 19th ACM SIGKDD International Conference on Knowledge Discovery and Data Mining, pp. 356–364. ACM (2013)
4. Newman, M.E.: The structure and function of complex networks. SIAM Rev. **45**(2), 167–256 (2003)
5. Krause, A., Frank, K.A., Mason, D.M., Ulanowicz, R.E., Taylor, W.M.: Compartments exposed in food-web structure. Nature **426**, 282–285 (2003)

6. Macy, M.W., Kitts, J.A., Flache, A., Benard, S.: Polarization in dynamic networks: A Hopfield model of emergent structure. Dyn. Soc. Netw. Model. Anal. **24**, 162–173 (2003)
7. Barabási, A.L., Albert, R.: Emergence of scaling in random networks. Science **286**(5439), 509–512 (1999)
8. Müller, M., Buchmann, T., Kudic, M.: Micro strategies and macro patterns in the evolution of innovation networks: an agent-based simulation approach. In: Gilbert, N., Ahrweiler, P., Pyka, A. (eds.) Simulating Knowledge Dynamics in Innovation Networks, pp. 73–95. Springer, Heidelberg (2014)
9. Lewis, K., Gonzalez, M., Kaufman, J.: Social selection and peer influence in an online social network. Proc. Natl. Acad. Sci. U.S.A. **109**(1), 68–72 (2012)
10. Powell, W.W., White, D.R., Koput, K.W., Owen-Smith, J.: Network dynamics and field evolution: The growth of interorganizational collaboration in the life sciences. Am. J. Sociol. **110**(4), 1132–1205 (2005)
11. Heider, F.: The psychology of interpersonal relations. Psychology Press, New York (2013)
12. McPherson, M., Smith-Lovin, L., Cook, J.M.: Birds of a feather: Homophily in social networks. Annu. Rev. Sociol. **27**, 415–444 (2001)
13. Rivera, M.T., Soderstrom, S.B., Uzzi, B.: Dynamics of dyads in social networks: Assortative, relational, and proximity mechanisms. Annu. Rev. Sociol. **36**, 91–115 (2010)
14. Frank, K.A., Kim, C., Belman, D.: Utility theory, social networks, and teacher decision making. Soc. Netw. Theor. Educ. Change **49**, 223–242 (2010)
15. Centola, D.: The spread of behavior in an online social network experiment. Science **329** (5996), 1194–1197 (2010)
16. Frank, K.A., Fahrbach, K.: Organization culture as a complex system: balance and information in models of influence and selection. Organ. Sci. **10**(3), 253–277 (1999)
17. Newman, M.E.: Power laws, Pareto distributions and Zipf's law. Contemp. Phys. **46**(5), 323–351 (2005)
18. Wilensky, U.: NetLogo (1999)
19. Frank, K.A.: Identifying cohesive subgroups. Soc. Netw. **17**(1), 27–56 (1995)
20. Frank, K.A.: Mapping interactions within and between cohesive subgroups. Soc. Netw. **18** (2), 93–119 (1996)

Leveraging Network Dynamics for Improved Link Prediction

Alireza Hajibagheri[1], Gita Sukthankar[1(✉)], and Kiran Lakkaraju[2]

[1] University of Central Florida, Orlando, FL, USA
{alireza,gitars}@eecs.ucf.edu
[2] Sandia National Laboratories, Albuquerque, NM, USA
klakkara@sandia.gov

Abstract. The aim of link prediction is to forecast connections that are most likely to occur in the future, based on examples of previously observed links. A key insight is that it is useful to explicitly model *network dynamics*, how frequently links are created or destroyed when doing link prediction. In this paper, we introduce a new supervised link prediction framework, RPM (Rate Prediction Model). In addition to network similarity measures, RPM uses the predicted rate of link modifications, modeled using time series data; it is implemented in Spark-ML and trained with the original link distribution, rather than a small balanced subset. We compare the use of this network dynamics model to directly creating time series of network similarity measures. Our experiments show that RPM, which leverages predicted rates, outperforms the use of network similarity measures, either individually or within a time series.

Keywords: Link prediction · Network dynamics · Time series · Supervised classifier

1 Introduction

Many social networks are constantly in flux, with new edges and vertices being added or deleted daily. Fully modeling the dynamics that drive the evolution of a social network is a complex problem, due to the large number of individual and dyadic factors associated with link formation. Here we focus on predicting one crucial variable–the *rate* of network change. Not only do different networks change at different rates, but individuals within a network can have disparate tempos of social interaction. This paper describes how modeling this aspect of network dynamics can ameliorate performance on link prediction tasks.

Link prediction approaches commonly rely on measuring topological similarity between unconnected nodes [1–3]. It is a task well suited for supervised binary classification since it is easy to create a labeled dataset of node pairs; however, the datasets tend to be extremely unbalanced with a preponderance of negative examples where links were not formed. Topological metrics are used

© Springer International Publishing Switzerland 2016
K.S. Xu et al. (Eds.): SBP-BRiMS 2016, LNCS 9708, pp. 142–151, 2016.
DOI: 10.1007/978-3-319-39931-7_14

to score node pairs at time t in order to predict whether a link will occur at a later time $t'(t' > t)$. However, even though these metrics are good indicators of future network connections, they are less accurate at predicting *when* the changes will occur (the exact value of t'). To overcome this limitation, we explicitly learn link formation rates for all nodes in the network; first, a time series is constructed for each node pair from historic data and then a forecasting model is applied to predict future values. The output of the forecasting model is used to augment topological similarity metrics within a supervised link prediction framework. Prior work has demonstrated the general utility of modeling time for link prediction (e.g., [4–6]); our results show that our specific method of rate modeling outperforms the use of other types of time series.

RPM is implemented using Spark MLlib machine learning library. Using Spark, a general purpose cluster computing system, enables us to train our supervised classifiers with the full data distribution, rather than utilizing a small balanced subset, while still scaling to larger datasets. Moreover, we evaluate the classifiers with a full test dataset, so the results are representative of the performance of the method "in the wild". Our experiments were conducted with a variety of datasets, in contrast to prior work on link prediction that has focused on citation or collaboration networks [7]. In addition to a standard co-authorship network (hep-th arXiv [8]), we analyze the dynamics of an email network (Enron [9]) and two player networks from a massively multiplayer online game (Travian [10]). Networks formed from different types of social processes may vary in their dynamics, but our experiments show that RPM outperforms other standard approaches on all types of datasets.

2   Background

Approaches to the link prediction problem are commonly categorized as being *unsupervised* [4,7,11–13] or *supervised* [8,14–16]. In unsupervised approaches, pairs of non connected nodes are initially ranked according to a chosen similarity metric (for instance, the number of common neighbors) [17,18]. The top k ranked pairs are then assigned as the predicted links. The strength of this paradigm is that it is simple and generalizes easily to many types of data, but there are some limitations: for instance, how to *a priori* select the cutoff threshold for link assignment? Implicitly, these approaches assume that the links with the highest scores are most likely to occur and form the earliest; however this is often not the case in many dynamic networks [18]. If the rank correlation between the selected metric and the sequence of formed links is poor, the accuracy of this approach suffers.

Supervised approaches have the advantage of being able to (1) simultaneously leverage multiple structural patterns and (2) accurately fit model parameters using training data. In this case, the link prediction task is treated as a classification problem, in which pairs of nodes that are actually linked are assigned to class 1 (positive class), whereas the non-connected ones are assigned to class 0 (negative class). The standard model assumes that feature vectors encapsulating the current network structure at time t are used to predict links formed at $t+1$; in some sense, this model is "amnesiac", ignoring the past connection history of individual nodes. To address this issue, our proposed method, RPM represents the network with *time series*. A forecasting model is then used to predict the next value of the series; this value is in turn used to augment the input to the supervised learning process.

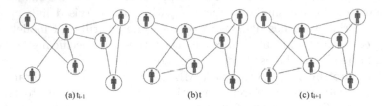

(a) t_{i-1} (b) t (c) t_{i+1}

Fig. 1. Evolution of a network over time. Blue nodes have higher *rates* of link formation. This behavior can only be captured by taking temporal information into account; RPM identifies these nodes through the use of time series. (Color figure online)

2.1   Time Series

To construct the time series, the network G observed at time t must be split into several time-sliced snapshots, that is, states of the network at different times in the past. Afterwards, a window of prediction is defined, representing how further in the future we want to make the prediction. Then, consecutive snapshots are grouped in small sets called frames. Frames contain as many snapshots as the length of the window of prediction. These frames compose what is called Framed Time-Sliced Network Structure (S) [8]. Let G_t be the graph representation of a network at time t. Let $[G_1, G_2, ..., G_T]$ be the frame formed by the union of the graphs from time 1 to T. Let n be the number of periods (frames) in the series. And let w be the window of prediction. Formally, S can be defined as:

$$S = \{[G_1, ..., G_w], [G_{w+1}, ..., G_{2w}], ...[G_{(n-1)w+1}, ..., G_{nw}]\}$$

For instance, suppose that we observed a network from day 1 to day 9, and our aim is to predict links that will appear at day 10. In this example, the forecast horizon (window of prediction) is one day. Our aim here is to model how the networks evolve every day in order to predict what will happen in the forecast horizon. Figure 1 shows an example of the evolution of network over time.

2.2   Network Similarity Metrics

In this paper, we use a standard set of topological metrics to assign scores to potential links:

1. Common Neighbors (CN) [19] is defined as the number of nodes with direct relationships with both members of the node pair: $CN(x,y) = |\Gamma(x) \cap \Gamma(y)|$ where $\Gamma(x)$ is the set of neighbors of node x.

2. Preferential Attachment (PA) [7,20] assumes that the probability that a new link is created is proportional to the node degree $|\Gamma(y)|$. Hence, nodes that currently have a high number of relationships tend to create more links in the future: $PA(x,y) = |\Gamma(x)| \times |\Gamma(y)|$.
3. Jaccard's Coefficient (JC) [21] assumes higher values for pairs of nodes that share a higher proportion of common neighbors relative to total number of neighbors they have: $JC(x,y) = \frac{|\Gamma(x) \cap \Gamma(y)|}{|\Gamma(x) \cup \Gamma(y)|}$.
4. Adamic-Adar (AA) [22], similar to JC, assigns a higher importance to the common neighbors that have fewer total neighbors. Hence, it measures exclusivity between a common neighbor and the evaluated pair of nodes:

$$AA(x,y) = \sum_{z \in |\Gamma(x) \cap \Gamma(y)|} \frac{1}{log(|\Gamma(z)|)}.$$

These metrics serve as (1) unsupervised baseline methods for evaluating the performance of RPM and (2) are also included as features used by the supervised classifiers.

2.3 Datasets

For our analysis, we selected three datasets: player communication and economic networks from the Travian massively multiplayer online game [10], the popular Enron email dataset [9], and the co-authorship network from arXiv hep-th [8]. Table 1 gives the network statistics for each of the datasets:

1. **Enron email dataset** [9]: This email network shows the evolution of the Enron company organizational structure over 24 months (January 2000 to December 2001).
2. **Travian MMOG** [10]: We used the communication and trading networks of users playing the Travian massively multiplayer online game. Travian is a browser-based, real-time strategy game in which the players compete to create the first civilization capable of constructing a Wonder of the World. The experiments in this paper were conducted on a 30 day period in the middle of the Travian game cycle (a three month period). Figure 2 indicates that Travian is a highly dynamic dataset, with over 90 % of the edges changing between snapshots.
3. **co-authorship network hep-th arXiv** [8]: This co-authorship network shows the evolution in co-authorship relationships extracted from the arXiv High Energy Physics (Theory) publication repository between 1991 and 2010.

3 Method

RPM treats the link prediction problem as a supervised classification task, where each data point corresponds to a pair of vertices in the social network graph. This

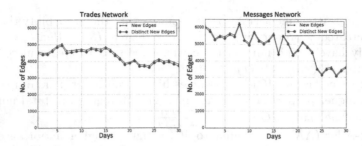

Fig. 2. Dynamics of the Travian network (trades: left and messages: right). The blue line shows the new edges added, and the red line shows edges that did not exist in the previous snapshot. (Color figure online)

Table 1. Dataset Summary

Data	Enron	Travian (Messages)	Travian (Trades)	hep-th
No. of nodes	150	2,809	2,466	17,917
Link (Class 1)	5,015	44,956	87,418	59,013
No Link (Class 0)	17,485	7,845,525	5,993,738	320,959,876
No. of snapshots	24	30	30	20

is a typical binary classification task that could be addressed with a variety of classifiers; we use the Spark support vector machine (SVM) implementation. All experiments were conducted using the default parameters of the Spark MLlib package: the SVM is defined with a polynomial kernel and a cost parameter of 1. Algorithms were implemented in Python and executed on a machine with Intel(R) Core i7 CPU and 24 GB of RAM. We have made our code and some example datasets available at: http://ial.eecs.ucf.edu/travian.php.

In order to produce a labeled dataset for supervised learning, we require timestamps for each node and edge to track the evolution of the social network over time. We then consider the state of the network for two different time periods t and t' (with $t < t'$). The network information from time t is used to predict new links which will be formed at time t'. One of the most important challenges with the supervised link prediction approach is handling extreme class skewness. The number of possible links is quadratic in the number of vertices in a social network, however the number of actual edges is only a tiny fraction of this number, resulting in large class skewness.

The most commonly used technique for coping with this problem is to balance the training dataset by using a small subset of the negative examples. Rather than sampling the network, we both train and test with the original data distribution and reweight the misclassification penalties. Let $G(V, A)$ be the social network of interest. Let $G[t]$ be the subgraph of G containing the nodes and edges recorded at time t. In turn, let $G[t']$ be the subgraph of G observed at time t'. In order to generate training examples, we considered all pairs of nodes in $G[t]$. Even though this training paradigm is more computationally demand-ing it avoids the concern that the choice of sampling strategy is distorting the classifier performance [16].

Selecting the best feature set is often the most critical part of any machine learning implementation. In this paper, we supplement the standard set of fea-tures extracted from the graph topology (described in the previous section), with features predicted by a set of time series. Let $F_t(t = 1,...,T)$ be a time series with T observations with A_t defined as the observation at time t and F_{t+1} the time series forecast at time $t + 1$. First, we analyze the performance of the following time series forecasting models for generating features:

1. **Simple Mean**: The simple mean is the average of all available data:

$$F_{t+1} = \frac{A_t + A_{t-1} + ... + A_{t-T}}{T}$$

2. **Moving Average**: This method makes a prediction by taking the mean of the n most recent observed values. The moving average forecast at time t can be defined as:

$$F_{t+1} = \frac{A_t + A_{t-1} + ... + A_{t-n}}{n}$$

3. **Weighted Moving Average**: This method is similar to moving average but allows one period to be emphasized over others. The sum of weights must add to 100 % or 1.00:

$$F_{t+1} = \sum C_t A_t$$

4. **Exponential Smoothing**: This model is one of the most frequently used time series methods because of its ease of use and minimal data requirements. It only needs three pieces of data to start: last period's forecast (F_t), last period's actual value (A_t) and a value of smoothing coefficient,α, between 0 and 1.0. If no last period forecast is available, we can simply average the last few periods:

$$F_{t+1} = \alpha A_t + (1 - \alpha)F_t$$

We identify which time series prediction model produces the best rate esti-mate, according to the AUROC performance of its RPM variant. Parameters of weighted moving average and exponential smoothing were tuned to maximize performance on the training dataset. Figure 3 shows that the best performing model was Weighted Moving Average with $n = 3$ and parameters C_1, C_2 and C_3 set to 0.2,0.3, and 0.5 respectively.

3.1   Results

Our evaluation measures receiver operating characteristic (ROC) curves for the different approaches. These curves show achievable true positive rates (TP) with respect to all false positive rates (FP) by varying the decision threshold on probability estimations or scores. For all of our experiments, we report area

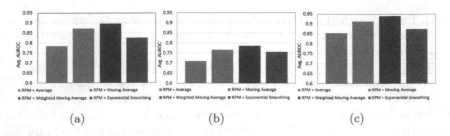

(a)   (b)   (c)

Fig. 3. Performance of RPM using different forecasting models on (a) Travian Messages (b) hep-th (c) Enron. Weighted Moving Average is consistently the best performer across all datasets and is used in RPM.

Table 2. AUROC Performance

Algorithms / Networks	Travian(Messages)	Travian(Trades)	Enron	hep-th
RPM	**0.8970**	**0.7859**	**0.9399**	**0.7834**
Supervised-MA	0.8002	0.6143	0.8920	0.7542
Supervised	0.7568	0.7603	0.8703	0.7051
Common Neighbors	0.4968	0.5002	0.7419	0.5943
Jaccard Coefficient	0.6482	0.4703	0.8369	0.5829
Preferential Attachment	0.5896	0.5441	0.8442	0.5165
Adamic/Adar	0.5233	0.4962	0.7430	0.6696

under the ROC curve (AUROC), the scalar measure of the performance over all thresholds. Since link prediction is highly imbalanced, straightforward accuracy measures are well known to be misleading; for example, in a sparse network, the trivial classifier that labels all samples as missing links can have a 99.99 % accuracy.

In all experiments, the algorithms were evaluated with stratified 10-fold cross-validation. For more reliable results, the cross-validation procedure was executed 10 times for each algorithm and dataset. We benchmark our algorithm against **Supervised-MA** [8]. Supervised-MA is a state of the art link prediction method that is similar to our method, in that it is supervised and uses moving average time series forecasting. In contrast to RPM, Supervised-MA creates time series for the unsupervised metrics rather than the link formation rate itself. **Supervised** is a baseline supervised classifier that uses the same unsupervised metrics as features without the time series prediction model. As a point of reference, we also show the unsupervised performance of the individual topological metrics: (1) **Common Neighbors**, (2) **Preferential Attachment**, (3) **Jaccard Coefficient**, and (4) **Adamic-Adar**. Table 2 presents results for all methods on Travian (communication and trade), Enron, and hep-th networks. Results for our proposed method are shown using bold numbers in the table; in all cases, RPM outperforms the other approaches. Two-tailed, paired t-tests across multiple network snapshots reveal that the RPM is significantly better ($p < 0.01$) on all four datasets when compared to Supervised-MA.

We discover that explicitly including the rate feature (estimated by a time series) is decisively better than the usage of time series to forecast topological metrics. The rate forecast is useful for predicting the source node of future links, hence RPM can focus its search on a smaller set of node pairs. We believe a combination of topological metrics is useful for predicting the destination node, but that relying exclusively on the topological metrics, or their forecasts, is less discriminative.

4   Related Work

The performance of RPM relies on three innovations: (1) explicit modeling of link formation rates at a node level, (2) the usage of multiple time series to leverage information from earlier snapshots, (3) training and testing with the full data distribution courtesy of the Spark fast cluster computing system. Rate is an important concept in many generative network models, but its usage has been largely ignored within discriminative classification frameworks. For instance, the stochastic actor-oriented model of network dynamics contains a network rate component that is governed by both the time period and the actors [23]. RPM does not attempt to create a general model of how the rate is affected by the properties of the actor (node), but instead predicts the link formation rate of each node with a time series.

Time series are useful because they enable us to track the predict future network dynamics, based on the past changes. Soares and Prudêncio [8] investigated the use of time series within both supervised and unsupervised link prediction frameworks. The core concept of their approach is that it is possible to predict the future values of topological metrics with time series; these values can either be used in an unsupervised fashion or combined in a supervised way with a classifier. In this paper, we compare RPM to the best performing version of their methods, Supervised-MA (Supervised learner with Moving Average predictor), that we reimplemented in Spark and evaluated using our full test/train distribution paradigm, rather than their original sampling method. Predicting the rate directly was more discriminative that predicting the topological metrics. We predict the rate of the source node's link formation using a time series, in contrast to Huang et al. [4] who used a univariate time series model to predict link probabilities between node pairs. In our work, we use a supervised model to assign links, rather than relying on the time series alone.

Feature selection is especially critical to the performance of a supervised classifier. For co-authorship networks, Hasan et al. [14] identified three important categories of classification features: (1) proximity (for comparing nodes) (2) aggregated (for summing features across nodes), and (3) topological (network-based). In our work, we only use network-based features, since those are the easiest to generalize across different types of networks; both proximity and aggregated features require more feature engineering to transfer to different datasets. Wang and Sukthankar [11] promoted the importance of social features in both supervised and unsupervised link prediction; social features aim to express the community membership of nodes and can be used to construct alternate distance metrics. However we believe that rate generalizes better across different types of dynamic networks; moreover it can be easily combined with dataset-specific feature sets.

5   Conclusion and Future Work

In this paper, we introduce a new supervised link prediction method, RPM (Rate Prediction Model), that uses time series to predict the rate of link formation. By accurately identifying the most active individuals in the social network, RPM achieves statistically significant improvements over related link prediction methods. Unlike the preferential attachment metric which identifies active individuals based on the degree measure of a single snapshot, RPM measures time-sliced network structure and finds individuals whose influence is rapidly rising. Our experiments were performed on networks created by a variety of social processes, such as communication, collaboration, and trading; they show that the rate of link generation varies with the type of network. In future work, we plan to extend this method to do simultaneously link prediction on different layers of multiplex networks, such as Travian, by modeling the relative rate difference between network layers.

Acknowledgments. Research at University of Central Florida was supported with an internal Reach for the Stars award. Sandia National Laboratories is a multi-program laboratory managed and operated by Sandia Corporation, a wholly owned subsidiary of Lockheed Martin Corporation, for the U.S. Department of Energy's National Nuclear Security Administration under contract DE-AC04-94AL85000.

References

1. Al Hasan, M., Zaki, M.J.: A survey of link prediction in social networks. In: Aggarwal, C.C. (ed.) Social Network Data Analytics, pp. 243–275. Springer, Heidelberg (2011)
2. Getoor, L., Diehl, C.P.: Link mining: a survey. ACM SIGKDD Explor. Newslett. **7**(2), 3–12 (2005)
3. Wang, C., Satuluri, V., Parthasarathy, S.: Local probabilistic models for link prediction. In: IEEE International Conference on Data Mining, pp. 322–331 (2007)
4. Huang, Z., Lin, D.K.: The time-series link prediction problem with applications in communication surveillance. INFORMS J. Comput. **21**(2), 286–303 (2009)
5. Berlingerio, M., Bonchi, F., Bringmann, B., Gionis, A.: Mining graph evolution rules. In: Buntine, W., Grobelnik, M., Mladenić, D., Shawe-Taylor, J. (eds.) ECML PKDD 2009, Part I. LNCS, vol. 5781, pp. 115–130. Springer, Heidelberg (2009)
6. Potgieter, A., April, K.A., Cooke, R.J., Osunmakinde, I.O.: Temporality in link prediction: Understanding social complexity. Emergence Complex. Organ. (E: CO) **11**(1), 69–83 (2009)
7. Liben-Nowell, D., Kleinberg, J.: The link prediction problem for social networks. In: Proceedings of the International Conference on Information and Knowledge Management, pp. 556–559 (2003)

8. Soares, P.R.D.S., Prudêncio, R.B.C.: Time series based link prediction. In: International Joint Conference on Neural Networks, IEEE, pp. 1–7 (2012)
9. Cohen, W.W.: Enron email dataset (2009). http://www.cs.cmu.edu/enron/
10. Hajibagheri, A., Lakkaraju, K., Sukthankar, G., Wigand, R.T., Agarwal, N.: Conflict and communication in massively-multiplayer online games. In: Agarwal, N., Kevin, X., Osgood, N. (eds.) Social Computing, Behavioral-Cultural Modeling, and Prediction. LNCS, pp. 65–74. Springer, Heidelberg (2015)
11. Wang, X., Sukthankar, G.: Link prediction in heterogeneous collaboration networks. In: Missaoui, R., Sarr, I. (eds.) Social Network Analysis: Community Detection and Evolution. Lecture Notes in Social Networks, pp. 165–192. Springer, Heidelberg (2014)
12. Beigi, G., Tang, J., Liu, H.: Signed link analysis in social media networks. In: International AAAI Conference on Web and Social Media (ICWSM) (2016)
13. Davoudi, A., Chatterjee, M.: Modeling trust for rating prediction in recommender systems. In: SIAM Workshop on Machine Learning Methods for Recommender Systems, SIAM, pp. 1–8 (2016)
14. Hasan, M.A., Chaoji, V., Salem, S., Zaki, M.: Link prediction using supervised learning. In: Proceedings of the SDM Workshop on Link Analysis, Counterterrorism and Security (2006)
15. Wang, X., Sukthankar, G.: Link prediction in multi-relational collaboration networks. In: Proceedings of the IEEE/ACM International Conference on Advances in Social Networks Analysis and Mining, Niagara Falls, Canada, pp. 1445–1447, August 2013
16. Lichtenwalter, R.N., Lussier, J.T., Chawla, N.V.: New perspectives and methods in link prediction. In: Proceedings of the 16th ACM SIGKDD International Conference on Knowledge Discovery and Data Mining, pp. 243–252. ACM (2010)
17. Lü, L., Zhou, T.: Role of weak ties in link prediction of complex networks. In: Proceedings of the ACM International Workshop on Complex networks Meet Information & Knowledge Management, pp. 55–58. ACM (2009)
18. Murata, T., Moriyasu, S.: Link prediction based on structural properties of online social networks. New Gener. Comput. 26(3), 245–257 (2008)
19. Newman, M.E.J.: Clustering and preferential attachment in growing networks. Phys. Rev. E 64, 025102 (2001)
20. Barabási, A.L., et al.: Scale-free networks: a decade and beyond. Science 325(5939), 412 (2009)
21. Tan, P.N., Steinbach, M., Kumar, V.: Introduction to Data Mining, 1st edn. Addison-Wesley Longman Publishing Co. Inc., Boston (2005)
22. Adamic, L.A., Adar, E.: Friends and neighbors on the web. Soc. Netw. 25(3), 211–230 (2003)
23. Snijders, T., van de Bunt, G., Steglich, C.E.G.: Introduction to actor-based models for network dynamics. Soc. Netw. 32, 44–60 (2010)

Dynamic Directed Influence Networks:
A Study of Campaigns on Twitter

Brandon Oselio$^{(\boxtimes)}$ and Alfred Hero

Department of Electrical Engineering and Computer Science, University of Michigan,
Ann Arbor, MI, USA
{boselio,hero}@umich.edu

Abstract. Studying the flow of influence in social media can allow insight into the nature of the agents involved and the corresponding actions that they take. In this paper, we study the influence of content among social media users with a concept called directed information (DI). Originally found in information theory, DI measures the amount of causal influence that an agent's actions have on others. By estimating these quantities, influence networks are built that show the leaders and followers of a social circle. In order to demonstrate this technique, we extract tweets from the US presidential candidates and build an influence network. The time-varying influence is extracted using an extension of DI called adaptive directed influence (ADI), which is able to identify changes in influence over different timescales. Using the example of presidential candidates, we are able to show the power of building an influence network using DI and ADI and we compare and contrast with other relevant metrics.

1 Introduction

Often times, we are interested in characterizing the interactions between a set of agents. We are specifically interested in the case where we do not assume any time invariant structure - we assume that these agents change their behaviors, actions, and influences over time. Thus, a metric is needed that incorporates this inherent time sensitivity into its calculations. In this paper, directed information (DI), a metric that finds its origins in information theory, is used to measure the influence of one agent on another, and vice versa. Using these pairwise calculations, one is then able to create an influence network; this allows for the end user to quickly and efficiently understand the influence topology between all of the agents in a social circle.

Directed information, however, is not enough to efficiently understand the time-varying structure of influence. Thus, adaptive directed information (ADI) is introduced to fill this gap. ADI can be thought of as a windowed version of DI, so that the analyst has control over the timescale that is of interest to them. Using ADI, it is possible to see slow trends or changes in influence in the data, as well as event-based changes in influence that are on a shorter timescale.

This work was partially supported by ARO under grant #W911NF-12-1-0443.

© Springer International Publishing Switzerland 2016
K.S. Xu et al. (Eds.): SBP-BRiMS 2016, LNCS 9708, pp. 152–161, 2016.
DOI: 10.1007/978-3-319-39931-7_15

In order to illustrate the effectiveness of DI and ADI, they are applied to a Twitter dataset of the current US presidential candidates, whose names and abbreviations are shown in Table 1. Using this data, relevant information about the influence among candidates can be extracted using a combination of DI, ADI, and a multinomial model that encapsulates the interaction between the content of the tweets.

The paper is structured as follows: Sect. 2 describes the concept and calculation of DI and ADI. It also describes the assumptions we make about the data to perform tractable estimation. Section 3 gives an overview of the Twitter dataset extracted from the REST API. Section 4 shows the application of DI and ADI to the datasets, and we discuss some interesting results that appear. Section 5 describes some related work on DI, including references that describe the underlying theory, as well as applications that have been seen in the past. Finally, Sect. 6 concludes with a summary of the paper and future directions for this work.

2 Methods

2.1 Directed Information

We are interested in extracting influences in content among users. In order to do this, we introduce a notion called directed information (DI). DI was originally used in an information theoretic context [5], where the goal was to measure causality in a channel with feedback. Before mathematically defining what DI is, an example is needed to give the reader a qualitative idea of what DI is measuring.

Assume that there are two agents, V_x and V_y (see Fig. 1). For each time step $m = 1, \ldots, M$, we are able to observe two feature vectors, X_m, Y_m, corresponding to users V_x, V_y. In the case of Fig. 1, the content of speech can be encapsulated

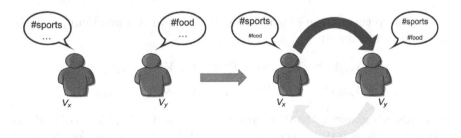

Fig. 1. Influence over time of two agents. In this diagram, we see initially that agent V_x is talking about sports, while agent V_y is talking about food. As they continue talking in a shared space, they influence each other. V_x influences V_y to a greater degree, so by the next time step V_y is mostly talking about sports. However, we see that V_x is talking a little about food due to the small influence that V_y has had on them.

into a discrete feature vector. Using these feature vectors, and knowledge of their underlying distributions, the objective is to measure the influence the two users have over one another. DI attempts to capture this influence.

The DI between two users V_x, V_y is a measure of the causal information flow from V_x to V_y. DI can also be thought of as a generalization of mutual information, which is a commonly used metric in information theory and other fields. Let us now mathematically define DI:

$$I(V_x \rightarrow V_y) = \sum_{m=1}^{M} I(X^{(m)}; Y_m | Y^{(m-1)}), \qquad (1)$$

where $I(A; B|C)$ is conditional mutual information. It is easy to show that this measure is asymmetric. With basic principles, it is possible to write DI as a sum of conditional Shannon entropies:

$$I(V_x \rightarrow V_y) = \sum_{m=1}^{M} \left[H(Y_m | Y^{(m-1)}) - H(Y_m | Y^{(m-1)}, X^{(m)}) \right]. \qquad (2)$$

Here, the notation $X^{(j)} = [X_1 \ldots X_j]$ is used to represent the random features from V_x up to and including timestep j. Note that in order to calculate the DI between users V_x and V_y (or vice-versa), some prior knowledge of the joint distributions of $[X^{(m)}, Y^{(m)}]$ is necessary, or at least the ability to estimate such quantities. In this paper, we assume that each feature vector is distributed as a discrete multinomial vector. Estimation of these quantities can be a challenging problem, especially when dealing with discrete variables - the number of parameters to estimate can explode quickly with the time parameter M and feature length k. In the next sections, we introduce some assumptions that help alleviate this issue.

2.2 Markovian Assumptions

In order to keep the number of parameters to estimate at a minimum, a Markov assumption is introduced:

$$P(X_m | Y^{(m)}, X^{(m-1)}) = P(X_m | Y_m, Y_{m-1}, X_{m-1}), \qquad (3)$$

$$P(Y_m | X^{(m)}, Y^{(m-1)}) = P(Y_m | X_m, X_{m-1}, Y_{m-1}). \qquad (4)$$

In words, it's assumed that the current feature vector from V_x is only dependent on its previous feature vector and the current and previous feature vector from V_y. This allows us to model the joint distribution of the feature vectors simply, while still capturing some dependence structure. In particular, the calculation of DI is now possible with the following formula:

$$I(V_x \rightarrow V_y) = H(X_{m-1}, Y_m, Y_{m-1}) - H(X_m, X_{m-1}, Y_m, Y_{m-1}) \qquad (5)$$
$$+ H(Y_m, Y_{m-1}) - H(Y_{m-1}).$$

2.3 Modeling Tweets as a Concatenated Multinomial Model

With Twitter, incoming tweets are modeled as multinomial feature vectors over a set vocabulary of words. In order to calculate the DI, there must be some assumption about the joint interaction of the random multinomial vectors, whether it be an assumption on their conditional or joint distributions. This is necessary in order to calculate the directed information. In this case, the random vectors are modeled using a concatenated multinomial model. With the Markov assumption, it is only necessary to estimate the joint distribution of four multinomial vectors. We model the joint distribution of vectors W_1, W_2, W_3, W_4 as:

$$P(W_1 = w_1, W_2 = w_2, W_3 = w_3, W_4 = w_4) = \frac{n!}{\prod_{i=1}^{4} \prod_{j=1}^{k} w_{ij}!} \prod_{i=1}^{4} \prod_{j=1}^{k} \theta_{ij}^{w_{ij}}, \quad (6)$$

where $n = n_1 + n_2 + n_3 + n_4$, and n_i is the number of trials for each separate multinomial vector. From this assumed model, it is possible to produce the appropriate condensed joint and marginal distributions necessary to calculate DI. It should be noted that this introduces an implicit dependence among the multinomials, which decays with $1/k$, the length of the feature vectors.

Given this model, the task is now to estimate the parameters. Specifically, it is necessary to estimate $\theta_{ij}, i = 1, \ldots, 4, j = 1, \ldots, k$. A simple MLE estimator is used for each parameter:

$$\hat{\theta}_{ij} = \frac{w_{ij}}{n}. \quad (7)$$

This is then used to obtain an estimate of the joint entropy:

$$\hat{H}(W_1, W_2, W_3, W_4) = -n \left(\sum_{i=1}^{4} \sum_{j=1}^{k} \hat{\theta}_{ij} \log_2 \hat{\theta}_{ij} \right). \quad (8)$$

2.4 Adaptive Directed Information

In order to observe the time-varying aspect of the data, the concept of adaptive directed information (ADI) is introduced. This takes the previously calculated DI and computes a moving average, or convolution, with a specified window:

$$I_m(V_x \to V_y) = \sum_{i=1}^{m} g(m, i) I(X^{(i)}; Y_i | Y^{(i-1)}). \quad (9)$$

Two examples of a sliding window include the uniform window $g(t, i) = 1/T, |t - i| \leq T$, and the exponential window $g(t, i) = e^{(t-1)\lambda} c_t, i \leq t, \lambda > 0$ and $c_t = (1 - e^{-\lambda})/(1 - e^{-(t+1)\lambda})$.

3   Presidential Candidate Twitter Dataset

In order to test this model on real data, tweets were collected using the REST API system. In particular, we study the influences of the current US presidential candidates. From October 1st to January 13th, all the tweets from each presidential candidate's Twitter account were collected, totaling 12132 tweets over the three and a half month period. Figure 2 shows the cumulative number of tweets for each candidate, with one standard deviation around the mean shaded. Only two candidates fall outside this area. Senator Rick Santorum tweets less than other candidates, and is generally less than one standard deviation below the mean. The large outlier in terms of volume, however, is Mr. Donald Trump, who has over 700 tweets more than the next highest candidate, and almost triple the mean. One aspect that the influence graphs will allow us to explore is the inverse correlation between the volume of tweets and the amount of influence on other candidates.

Before performing the directed information analysis, some preprocessing was performed. The tweets were aggregated into 12 h windows. This allows the candidates with a low volume of tweets to still have data at most timesteps. After binning the data, the tweets were parsed, while eliminating confounding entities such as URLs and mentions. Further, the remaining data was stemmed using the Porter stemmer, and words that had a document frequency less than 10 were discarded. Similarly, words that had appeared in over 50 % of the binned documents were thrown out. This left 1554 words to be used as features.

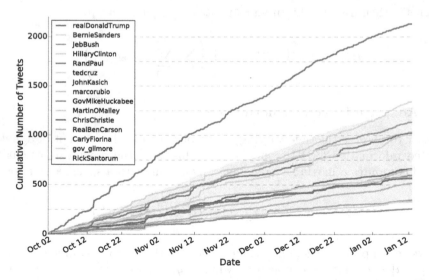

Fig. 2. Volume plot of Twitter dataset. The plot shows Mr. Donald Trump as the major outlier in terms of volume of tweets. There is also a bimodal distribution between candidates who heavily employ social media, such as Sen. Bernie Sanders, Sec. Hillary Clinton, and Gov. Jeb Bush, and those who do not, such as Gov. John Gilmore, Mrs. Carly Fiorina, and Sen. Rick Santorum.

4 Results

Figure 3 shows the relative DI networks over the entire time period. The relative DI of two users V_x and V_y is the difference between their DI w.r.t. one another. The direction of the arrow shows the direction of influence as measured by relative DI. The network on the left shows all the connections that were above the mean relative DI, while the network on the right shows those edges whose z-scores were above 1.645, which corresponds to a p-value of 0.05.

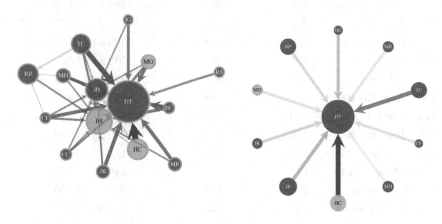

Fig. 3. DI graphs. The graph on the left shows all connections that are above the mean in value for relative DI. The graph on the right shows the connections whose z-scores were above 1.645, corresponding to a p-value of 0.05. In both graphs, the node size is relative to the total volume of tweets, and the shade and width of the edges correspond to the magnitude of relative DI.

From the networks and Table 1, we see that volume of tweets is inversely correlated with the influence over other candidates. Indeed, the two candidates with the largest volume of tweets (Mr. Trump and Sen. Sanders) are the most influenced. We see, however, that this relationship is more subtle. For instance, Sec. Hillary Clinton is an influencer in the network on the left of Fig. 3, despite her having a relatively large tweet count. Although her total DI is negative, her most significant connections are her influencing others.

We also see from the network on the left that the most significant edges are those connected to Mr. Trump, as he is being heavily influenced overall. One possible explanation is that those who tweet at a high frequency are both repetitive as well as responding to content that other candidates are creating. For comparison, we provide another network based on hashtag co-usage among candidates. This network was created by counting the number of times that two candidates used the same hashtag (Fig. 4).

In the hashtag networks, especially in the case of the thresholded graph with a p-value of 0.05, there is community structure along party lines. This suggests that there are some hashtags that are used within the parties and primaries more

Table 1. Table of candidates, their names, total DI, and total tweet volume.

Candidate	Abbreviation	Total Tweets	Total DI	DI per Tweet
Mr. Donald Trump	DT	2131	−34046.9	−15.98
Sen. Bernie Sanders	BS	1337	−19486.1	−14.57
Gov. Jeb Bush	JB	1131	−12185.2	−10.77
Sec. Hillary Clinton	HC	1033	−4959.0	−4.80
Sen. Rand Paul	RP	1022	−2598.8	−2.54
Sen. Ted Cruz	TC	1003	3225.5	3.21
Gov. John Kasich	JK	658	3840.3	5.84
Sen. Marco Rubio	MR	651	5976.7	9.18
Gov. Mike Huckabee	MH	594	7090.3	11.94
Gov. Martin O'Malley	MO	574	6111.8	10.65
Gov. Chris Christie	CC	564	7345.2	13.02
Dr. Ben Carson	BC	512	9005.4	17.59
Mrs. Carly Fiorina	CF	342	11147.0	32.59
Gov. John Gilmore	JG	326	9899.6	30.36
Sen. Rick Santorum	RS	254	9634.1	37.93

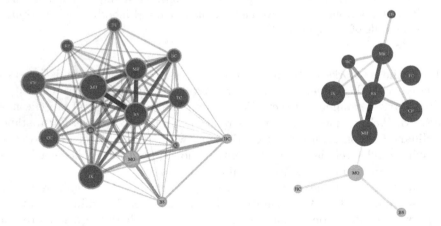

Fig. 4. Hashtag networks. These networks were built using co-usage of hashtags. The network on the left shows the connections whose weights were above the mean, while the network on the right shows the connections whose z-scores were above 1.645, corresponding to a p-value of 0.05. In both networks, size of the node is associated with the amount of hashtag usage in general, while the shade and width of the edge is associated with the number of co-usages of hashtags.

frequently than they are used elsewhere. There are also a few strong connections among candidates that are not particularly active Twitter users, such as Mike Huckabee and Rand Paul.

While the advantage of the hashtag network is that it allows us some insight into community structure by leveraging the usage of commonly used hashtags, its disadvantage is that it does not take into account time. If one candidate repeats everything that another candidate said a day later, then that candidate has been heavily influenced. The naïvely constructed hashtag network would not see this one-sided flow of information.

We are also interested in time varying phenomena. Figure 5 uses adaptive directed information (ADI) and the ego network of Sec. Clinton to show how a candidate's influence varies over time, from an influencer to being heavily influenced and vice versa. The calculations for ADI were made with an exponential window with $\lambda = 0.07$, corresponding to an approximately 20 day period. The edge weights correspond to the magnitude of the relative ADI, while the direction corresponds to the sign.

Near the beginning of the dataset, we see that Sec. Clinton is heavily influenced by nodes in the network. More recently, however, we see that she is an influencer on others, and in fact she heavily influences the largest nodes in the network, Sen. Sanders and Mr. Trump.

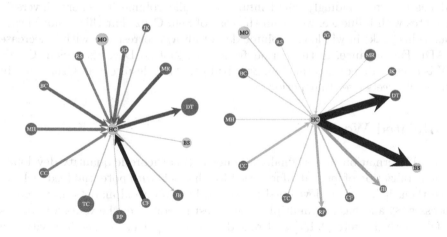

Fig. 5. Ego networks of adaptive directed information (ADI) for two time periods. These two networks represent the influence of Sec. Clinton. The network on the left was taken from October 24th, while the second is from December 19th. The ADI was calculated with an exponential window with $\lambda = 0.07$, approximately corresponding to a 20 day period. We notice that at the beginning of October, Sec. Clinton was heavily influenced by the rest of the network. More recently, however, she is more of an influencer than she is influenced by others.

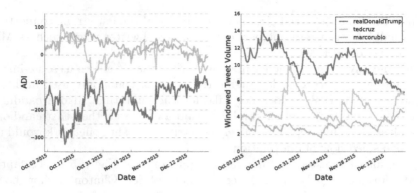

Fig. 6. A plot of ADI and windowed tweet volume for three Republican candidates. We see that while there is an inverse correlation between volume of tweets and ADI, this correlation is not always perfect. For instance, in some timespans, while Sen. Cruz's volume rises, his total ADI stays approximately equivalent to that of Sen. Rubio's.

Figure 6 shows a plot of total ADI and corresponding windowed tweet volume over time for three Republican candidates: Sen. Rubio, Mr. Trump, and Sen. Cruz. Total ADI is calculated by summing all ADI of a candidate, both outgoing (positive) and incoming (negative) influence. The windowed tweet volume is calculated by a convolution with the same window used to calculate ADI. We see that Mr. Trump tweets a significant amount more than the others over time, and has a correspondingly small influence. While volume in general inversely correlates with influence, we see in the case of Sen. Cruz that this is not always true, as his peaks in windowed volume do not always correspond with a decrease in ADI. For instance, in the period from Nov. 21st to Dec. 12th, Sen. Cruz's ADI is comparable or greater than Sen. Rubio's ADI, despite Sen. Cruz's rise in volume of tweets over that period.

5   Related Work

Directed information was originally an information theoretic quantity developed as a generalization of mutual information for channels with potential feedback [5]. Since then, it has been developed for countably infinite alphabets and ergodic processes [8], and has had multiple types of estimators over the years for various models, both discrete [2,4,7] and mixed regimes [3]. DI's connection with the concept of Granger causality has also been explored [1]. To the best of our knowledge, adaptive directed information is a new concept.

The applications of DI in its various forms have included video indexing [2], EEG analysis [3], portfolio theory [6] and social network analysis [7]. Our analysis greatly differs from [7], as they use a keyword based time-arrival model; we are less interested in arrival times and more interested in the content of messages.

6 Conclusion and Future Work

This work examined the influence of people over time on Twitter. Specifically, tweets were extracted from the US presidential candidates, and the influence they have on each other was measured. Using the concept of directed information, time-varying influence and dependence was visualized in a network structure. Using this technique, nodes with heavy outdegree were identified, which correspond to heavy influencers, as well as nodes with heavy indegree, which are candidates who follow the discussion, but don't lead it. Using ego networks on a particular candidate and the novel concept of adaptive directed information, we were also able to see that a node's role may change over time. Finally, total ADI was used to explore the change in influence over time of certain candidates.

Moving forward with this work, we would like to understand the theoretical aspects of DI and ADI better, and be able to understand some finite sample error bounds that might be applied to this model. In addition, regularization might be useful to lower the MSE of the DI estimator. Finally, early tests have shown this method to be useful on much larger datasets - this would increase the applicability of this method greatly.

References

1. Amblard, P.O., Michel, O.J.J.: The relation between granger causality and directed information theory: A review. Entropy **15**, 113–143 (2013)
2. Chen, X., Hero, A.O., Savarese, S.: Multimodal video indexing and retrieval using directed information. IEEE Trans. Multimedia **14**(1), 3–16 (2012)
3. Chen, X., Syed, Z., Hero, A.: EEG spatial decoding with shrinkage optimized directed information assessment. In: ICASSP 2012 Proceedings, pp. 577–580 (2012)
4. Jiao, J., Permuter, H.H., Zhao, L., Kim, Y.H., Weissman, T.: Universal estimation of directed information. IEEE Trans. Inf. Theor. **59**(10), 6220–6242 (2013)
5. Massey, J.: Causality, feedback and directed information. In: Proceedings of 1990 International Symposium on Information Theory and its Applications, pp. 303–305 (1990)
6. Permuter, H.H., Kim, Y.H., Weissman, T.: Interpretations of directed information in portfolio theory, data compression, and hypothesis testing. IEEE Trans. Inf. Theor **57**(6), 3248–3259 (2011)
7. Quinn, C., Kiyavash, N., Coleman, T.P.: Directed information graphs. IEEE Trans. Inf. Theor. **61**(12), 6887–6909 (2015). coleman.ucsd.edu
8. Weissman, T., Kim, Y.H., Permuter, H.H.: Directed information, causal estimation, and communication in continuous time. IEEE Trans. Inf. Theor. **59**(3), 1271–1287 (2012)

Link Prediction via Multi-hashing Framework

Mengdi Wang[1]([✉]) and Yu-Ru Lin[2]

[1] Intelligent System Program, University of Pittsburgh, Pittsburgh 15260, USA
mew133@pitt.edu
[2] School of Information Sciences, University of Pittsburgh, Pittsburgh 15260, USA

Abstract. Link prediction is crucial in various real world applications such as social network analysis and recommendation systems. For example, in social networks, where social actors and their ties (friendship or collaboration) are represented as nodes and links, link prediction can help anticipate future social tie formation. This problem has generally been tackled through computing a "similarity" – measured through graph topological structure or various node attributes and relationships among them (e.g. researcher's affiliation or research interest). However, when considering multiple relationships, existing link prediction methods often ignored that similarities across different relationships may be "non-transitive", i.e., they are not necessarily consistent with each other. Here, we develop a semi-supervised link prediction method via a *Multi-Component Hashing* framework. We derive multiple hashing tables for nodes in a network with each hash table corresponding to a particular type of non-transitive similarity aspect such as prior collaboration experience or topical interest. New links are predicted based on whether nodes are closer in the hashing tables. Results on three co-authorship networks show that our approach outperforms the state-of-the-art unsupervised and supervised methods. The results also show the superiority of our method in *cold-start* link prediction setting, where no or little knowledge about the nodes' network positions is given in the training phase.

1 Introduction

Network has become an increasingly popular way to model many phenomena in the world, which represents entities as nodes and relationships as links. The relationships can be friendships among people, collaborations among researchers, and interactions between proteins, etc. However, networks are highly dynamic, since they grow and evolve over time with the addition of new edges signifying the formation of new interactions between nodes [9]. Therefore, predicting possible links in a network is an interesting but challenge issue and has been attracted more attention recently.

In the *link prediction* problem, we are given a snapshot of a network and seek to infer new links among nodes that are likely to occur in the near future. Link prediction can be beneficial in various fields such as social network analysis, recommendation systems and bio-informatics. For example, it can be used to predict new friendships in social networks (e.g., in Facebook) [2], or recommend

© Springer International Publishing Switzerland 2016
K.S. Xu et al. (Eds.): SBP-BRiMS 2016, LNCS 9708, pp. 162–173, 2016.
DOI: 10.1007/978-3-319-39931-7_16

new collaborations among researchers [5], which leads to better research teams. In a protein-protein interaction network, where below 1 % links are known [16], link prediction can help find possible links instead of conducting arduous and expensive experiments.

Link Prediction Challenges. The link prediction problem is challenging at least in the following four aspects. First, besides *network information* (connections among nodes), in many cases we have *external information* such as node attributes [12]. For example, in a co-authorship network among researchers, we might have research profile information as node attributes such as affiliations or published papers. Such information can be valuable in link prediction, especially when the nodes are sparsely connected (i.e., the *cold-start* problem [8]). However, most existing studies only employ network topological structures while ignoring node attribute information. A proper combination of both topological information and external information is expected to enhance link prediction performance.

Second, links may have different semantics and a simple aggregation without differentiating link semantics may be ineffective for link prediction. Classical link prediction approaches predict links based on the "similarity" of two nodes [11], assuming all similarities are *transitive*, e.g., if A and C are both similar to B, then A is similar to C. However, in social networks such assumption does not necessarily hold especially when the similarity relationships are derived from multiple dimensions (e.g., geographical distance, research interest, etc.). As shown in Fig. 1, we have a so-called "non-transitive" relationship among three researchers {A,B,C} in a co-authorship network. The similarities among {A,B,C} are determined by two different dimensions: affiliation and research interest. On affiliation dimension, A and B are similar, and on research interest dimension, B is similar to C. A simple similarity computation may yield a result indicating A and C are similar, whereas in reality A is dissimilar to C on both dimensions. Therefore, an effective leveraging of non-transitive similarities from multiple dimensions in link prediction requires significant future explorations.

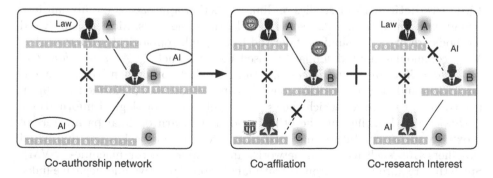

Co-authorship network   Co-affliation   Co-research Interest

Fig. 1. An illustration of non-transitive similarity in social networks. The solid line represents similarity relationship, and the dashed line with cross represents dissimilarity relationship.

The third and forth challenges are imbalance and scalability. Link prediction datasets are always extremely imbalanced: the number of positive links (e.g. links to be present) is significantly less than the number of negative links (e.g. links to be absent) [12]. Supervised classifier based methods, where a classifier is trained to discriminate between positive and negative links, use under-sampling and over-sampling to overcome imbalance [10], while leading to information loss or over-fitting. Another issue is scalability. Methods based on topological similarities between nodes like *Adamic/Adar* [1], generally scale to large graphs, by reason of only requiring simple operations on the adjacency matrix, but they ignore the node attributes. For classifiers methods, the downsampling technology can reduce computational cost but suffer from losing information.

Our Approach. We develop a semi-supervised model via a *Multi-Component Hashing* framework, in which nodes in networks are represented via multiple hashing tables. Each hash table encodes one type of similarity aspect such as previous collaboration or topical interest in co-author networks. New links are predicted based on node distance in the hashing tables.

Our main contributions include:

1. **New link prediction approach**: We propose a new link prediction model via a *Multi-Component Hashing* framework to address several challenges in link prediction problem, including effective fusion of topological and node attributes, non-similarity between nodes, data imbalance and scalability.
2. **High prediction performance**: We conduct extensive experiments on three large co-authorship networks, showing that the proposed approach outperforms all the state-of-the-art unsupervised and supervised methods.
3. **Handling Cold-start**: Experiments also show the effectiveness of our method in *cold-start* link prediction, where no or little knowledge about the nodes' network positions is given in the training phase.

2   Related Work

Link Prediction. Prior studies on link prediction approaches mainly fall into two groups: unsupervised and supervised learning methods. Most unsupervised methods are based on similarity derived from graph topological properties [11]. Most commonly-adopted similarity based methods include *Adamic-Adar*, *Jaccard Index*, *Katz Index* and *PageRank*, and etc. [9]. A comprehensive experiment for understanding the pros and cons of each method was performed by [9] on five co-author networks, which shows the usefulness of topological information.

Supervised learning, on the other hand, is to learn optimal parameters of proximity metric for future link prediction. Typical approaches are supervised classifiers [10], such as neural networks and support vector machines (SVM). Since they require a set of ground-truth labels about positive and negative links, and negative links are often several times more than positive links, data sampling approaches such as down-sampling (of negative links) or over-sampling (of positive links) strategies are adopted to overcome the data imbalance issue.

However, data re-sampling is often criticized for either causing information loss or over-fitting. Besides, feature extraction and selection are also challenging. A matrix factorization method is proposed in [12] using the linear combination of explicit and latent features for nodes or links. [2] predicts links based on supervised random walk guided by node attributes. Probabilistic models learned from the observed network could also be used in link prediction [11]. The major drawback of those supervised methods is that they do not scale well. Heterogeneous information may be combined to predict different types of links; for example [4] solves the link prediction task using a tensor decomposition model. But the present work is different and focuses on predicting new links of the same type using heterogeneous information, including different types of relations and attributes. Also, none of the existing link prediction methods consider the non-transitive similarity problem.

Hashing. Hashing is a dimension reduction approach that transforms information into a short code consisting of a sequence of bits. It has been successfully and widely applied for indexing large-scale datasets and similarity search. Various hashing methods such as locality sensitive hashing, spectral hashing and independent component analysis (ICA) hashing are proposed in current studies [21]. More recently, a multi-table hashing method is proposed in [14], to capture the latent similarity components. [15] proposed a link prediction method using locality-sensitive hashing, but the hashing is only used in searching. However, there remains insufficient research of using hashing for link prediction, which drives us to further explore this topic.

Non-transitive Similarity. Non-transitive similarity (See Sect. 1) is identified as an important problem in various fields. In social networks, many prior studies on this topic only studied in detecting overlapping communities in multiple relational networks [18]. However, they did not consider non-transitive similarity in link prediction. Our method is to use multi-hashing method to deal with non-transitive similarity in link prediction.

3   Proposed Method

In this section, we describe our semi-supervised link prediction model via a *Multi-Component Hashing* framework [14], which could be also used in unsupervised setting. In unsupervised setting, we use an undirected/ directed network at time τ to predict new links at time $\tau + 1$; in semi-supervised setting, besides network at τ, we have the information of this network at time $\tau + 1$ and to predict the remaining new links at time $\tau + 1$. We derive multiple hashing tables for nodes with each hash table corresponding to a particular type of similarity aspect. For example, in Fig. 1, we have two hash tables to capture affiliation and research interest similarity aspect in co-author network and each node is assigned with a hash code in each table. New links can be predicted based on the closeness of the nodes in the hashing tables.

In the remaining of this section, We will formulate the problem and then describe an unsupervised model. Then, we extend it to a semi-supervised one.

3.1   Problem Formulation

We are given a undirected graph $G_V^\tau = \langle V, E^\tau \rangle$ at time τ, in which each edge $e^\tau = \langle u, v, \tau \rangle \in E^\tau$ represents a link between node u and v at time τ, and $u, v \in V$. The number of nodes in this graph is N and $|V| = N$. We also have a L-dimensional feature matrix for these N nodes $\mathbf{X} \in \mathbb{R}^{L \times N}$ derived from node attributes or topological information. Our goal is to learn M hash tables with length-K hash code in each hash table (i.e., each node is assigned with a K-bit hash code in each hash table). The values in each hash tables can be any real number or binary values. The m-th hash table denoted as $\mathbf{H}_{raw}^{(m)} \in \mathbb{R}^{K \times N}$ and $\mathbf{H}_{bin}^{(m)} \in \{-1, 1\}^{K \times N}$, where each column corresponds to a hash code of one node. For simplicity, we use $\mathbf{H}^m$ to represent either $\mathbf{H}_{raw}^{(m)}$ or $\mathbf{H}_{bin}^{(m)}$. Then, the M hash tables can be represented as

$$\mathbf{H} = [\mathbf{H}^{(1)T}, \mathbf{H}^{(2)T}, \cdots, \mathbf{H}^{(M)T}]^T. \tag{1}$$

To generate the hash tables, we learn a set of linear projection matrices $\mathbf{W} = [\mathbf{W}^{(1)}, \mathbf{W}^{(2)}, \cdots, \mathbf{W}^{(M)}] \in \mathbb{R}^{L \times KM}$. The m-th hash table can be calculated by

$$\mathbf{H}_{raw}^{(m)} = \mathbf{W}^{(m)T}\mathbf{X}, \tag{2}$$

$$\mathbf{H}_{bin}^{(m)} = \mathrm{sgn}(\mathbf{H}_{raw}^{(m)}), \tag{3}$$

where the sgn function of a real number is $+1$ for a positive number, and -1 for zero or a negative number.

To encode the *non-transitive* similarity, we define three types of relationship between node pairs: *similar*, *dissimilar* and unknown. One can express the relationship information by defining a $N \times N$ matrix $\mathbf{R}^\tau \in \{-1, 0, 1\}^{N \times N}$. $\mathbf{R}_{ij}^\tau = 1$ means node pair (v_i, v_j) are similar; $\mathbf{R}_{ij}^\tau = -1$ means dissimilar and $\mathbf{R}_{ij}^\tau = 0$ means the relationship is unknown. The similarity can be approximated using hash codes in the m-th hash table similarity through inner product:

$$\mathbf{S}^{(m)} = \mathbf{H}^{(m)T}\mathbf{H}^{(m)}. \tag{4}$$

The goal is to lean a $\mathbf{W}$ so that the hashing codes satisfy the properties that S_{ij} is large when $R_{ij}^\tau = 1$ and small when $R_{ij}^\tau = -1$. As we have M hashing tables, the aggregated similarity is defined as the maximum of similarities in M hash tables [14]:

$$S_{ij} = g(S_{ij}^1, S_{ij}^2, \cdots, S_{ij}^M) = \max\{S_{ij}^m\}_{m=1}^M. \tag{5}$$

The objective is to learn matrix $\mathbf{W}$ by minimizing the difference between of $\mathbf{R}^\tau$ and $\mathbf{S}$ through a loss function $f(\cdot)$:

$$\min_{\mathbf{W}} f(\mathbf{R}^\tau, \mathbf{S}). \tag{6}$$

The loss function will be defined concretely in the next sections.

3.2   Unsupervised Model Formulation

Multi-component Hashing. In this section, we use the binary hashing code $\mathbf{H}_{bin}$ (defined in Eq. (3)) to calculate similarity among nodes. As Eq. (5) is hard to solve, we use the softmax function [14] to approximate the maximum function:

$$S_{ij} = \frac{\sum_{m=1}^{M} S_{ij}^m e^{S_{ij}^m}}{\sum_{m=1}^{M} e^{S_{ij}^m}} \tag{7}$$

We use the same objective function as in [14], which matches the assumption that similar node pairs are closer in hashing tables:

$$\mathcal{LE} = \sum_{ij} \log(1 + e^{-R_{ij}^{\tau} S_{ij}}). \tag{8}$$

To avoid overfitting, one would like to maximize entropy principle by maximizing the variance of a bit [20]. Also, we restrict the magnitude of linear projections to be small. Therefore, two regularization terms are added [14]:

$$\mathcal{LR} = \gamma_1 \sum_{i \neq j} (\mathbf{W}_i^T \mathbf{W}_j)^2 + \gamma_2 \|\mathbf{W}\|_{\mathcal{F}}^2. \tag{9}$$

The final objective function is

$$\min_{\mathbf{W}} \mathcal{L} = \mathcal{LE} + \mathcal{LR}. \tag{10}$$

However, the hashing function (Eq. 3) is not continuous, making the final objective intractable, so we use the smooth sigmoid function for approximation.

$$\mathbf{H} = \frac{2}{1 + e^{-\mathbf{W}^T \mathbf{X}}} - 1. \tag{11}$$

Then the problem could be solved by many optimization methods efficiently when relational matrix R is sparse [14] (here we use Gradient Decent method).

Multi-component Hashing Raw Value. In the above section, we encode nodes features into a sequence of binary values, which loses much information. Therefore, we can also use the raw value $\mathbf{H}_{raw}$ defined by Eq. 2, which contains more information than $\mathbf{H}_{bin}$. The linear projection function $\mathbf{W}$ could be obtained by substituting Eq. (2) into objective function (Eq. (8)) and use the same regularizers (Eq. (9)). Unlike Eqs. (2), and (3) is continuous which makes objective function tractable; therefore, we do not need the smooth sigmoid function as shown in Eq. (11).

3.3   Semi-supervised Extension

If we consider link prediction as classification problem to discriminate between positive links labeled as 1 (i.e. links that will form at time $\tau + 1$), and negative links labeled as 0 (i.e. links that will not form at $\tau + 1$), the aforementioned methods are unsupervised methods because we do not have any labeled data in training the hash codes. In this section, we extend our methods to be semi-supervised ones, where we add the information of partially labeled data (i.e., a subset of nodes and actual links among them) at $\tau + 1$ as regularizer to the original objective function (Eq. 10) in Sect. 3.2 to train the model and to predict for the remaining data at time $\tau + 1$.

In addition to the graph at time τ (G_V^τ), we also know $G_{V_L}^{\tau+1}$, which is the graph of labeled nodes V_L (where $V_L \subset V$ and $|V_L| = N_L \ll |V| = N$) at time $\tau + 1$. The task is to predict links at time $\tau + 1$ for the remaining nodes V_U (i.e., unlabeled data), where $V_L \cup V_U = V$. Similar to Sect. 3.1, we divide the labeled nodes (V_L) into categories: "similar" (i.e. node pairs having actual links) and "dissimilar" (i.e. node pairs having no links). Note that we only have two categories instead of three. A relational matrix $\mathbf{R}^{\tau+1} \in \{-1, 1\}^{N_1 \times N_1}$ is built, where $\mathbf{R}_{ij}^{\tau+1} = 1$ means node pair (v_i, v_j) are linked and $\mathbf{R}_{ij}^{\tau+1} = -1$ means not linked at time $\tau + 1$. The similarity S_{ij} between hashing codes generated by linear projection matrix $\mathbf{W}$ should be large when $R_{ij}^{\tau+1} = 1$ and small when $R_{ij}^{\tau+1} = -1$. Similar to Sect. 3.2, the regularizer could be written as

$$\mathcal{LE}_{label} = \sum_{ij} \log(1 + e^{-R_{ij}^{\tau+1} S_{ij}}). \tag{12}$$

Then the final objective function for semi-supervised method becomes:

$$\min_{\mathbf{W}} \mathcal{L}_{semi} = \mathcal{LE} + \mathcal{LR} + \mathcal{LE}_{label}. \tag{13}$$

3.4   Link Prediction with Hashing Codes

Once the linear projection $\mathbf{W}$ is learned, the M hashing tables ($\mathbf{H}_{bin}$ or $\mathbf{H}_{raw}$) for all nodes in the network could be easily computed using Eqs. (2) and (3). In this section, we describe how to perform link prediction using multiple hashing tables.

The main idea is that two nodes are more likely to form link if their hashing codes are "closer". The most common measurement for hashing code is hamming distance, but we do not use it because of two reasons: (1) The nodes will not be discriminative enough, since hamming distance only have $K + 1$ discrete values (i.e., $\{0, 1, 2, \cdots, K\}$) if the length of hashing code is K. (2) Hamming distance is not suitable for $\mathbf{H}_{raw}$, where each element is not binary. Therefore, we use cosine of similarity between two hashing codes to measure similarity between hashing codes, defined as $Sim_{cosine}^{(m)} = (h_1^{(m)} \cdot h_2^{(m)})/\|h_1^{(m)}\|\|h_2^{(m)}\|$, where $h_1^{(m)}$, $h_2^{(m)}$ belong to each hashing code in m-th hash table. For a given node v_s, we rank other nodes (with nonexistent link to v_s) by the hashing code similarity with v_s with descending order and the top nodes are most likely to form link with v_s. However, as we have multiple hash tables, each node has different cosine similarities to the given node v_s in different hash tables, which makes it hard to rank the nodes directly. To address this, we apply the similar strategy described in [14], but use cosine similarity instead of hamming distance. Particularly, given a node v_s, to determine the order of two other nodes (say v_i, v_j), we first sort the M cosine similarities of v_i and v_j with descending order and then we compare their sorted similarity list beginning from the maximum similarity until one's (say v_i) cosine similarity is larger than the other's similarity (say v_j); thus, we say that node v_i is more likely to form link with the given node v_s than node v_j.

4 Experiments

4.1 Datasets

We use the dataset scraped from Microsoft Scholar Database in [19], which contains 5 million publications from 1997 to 2013 and each publication is consisting of its authors, paper content (title and abstract) and authors' affiliations. In our link prediction experiments, we retrieve data in year Y ($Y = 2009, 2010, 2011, 2012$) and build four undirected co-authorship networks. A link between two authors is added if they co-published at least one paper. We are to predict new links in year Y ($Y = 2010, 2011, 2012$) using the network in year $Y - 1$. We focus on the "core" (or active) authors with at least 3 publication in years Y and $Y - 1$. The statistics of predicted networks are shown in Table 1. Based on the "core" authors, we also sample "cold-start" authors by defining a cold-start author as those who have at least three links with "core" authors in year Y while having no links with the "core" authors before year Y. In order to have the minimum available information, these "cold-start" authors should have at least one paper before year Y. The last column in Table 1 shows the number of "cold-start" nodes for each year.

Table 1. Statistics of three co-authorship networks

Dataset	# author	#paper	# average link	# core-nodes	# cold-nodes
2010	21,018	20,158	2.90	2,937	178
2011	24,848	18,027	3.87	3,301	352
2012	25,396	26,334	3.56	3,560	201

4.2 Feature Extraction

We extract one network feature and three node attribute features. We consider the network feature **Adamic/Adar (AA)** [1] since it is found to be the most effective approach in several prior studies. We also create content features based on all papers (titles and abstracts) published by the given author using two approaches. (a) **word2vec**: We employ a distributional representation method word2vec [13] to obtain the average of 100-dimensional vectors of all words in an aggregated papers belonging to one author. (b) **LDA**: Latent Dirichlet Allocation [3] is performed on the aggregated content to get the distribution on 100 LDA topics as feature. **Geographic distance (GD)** is the distance between each author's latest affiliation location (latitude and longitude).

4.3 Experiment Setup

1. Unsupervised Link Prediction. In the unsupervised setting, we apply our unsupervised multi-hashing methods: $\mathbf{H}_{\mathbf{Raw}}^{U}$ and $\mathbf{H}_{\mathbf{Bin}}^{U}$. We compare our methods with the following eight baselines: (1) **Adamic/Adar (AA)** [1], and (2) **Jaccard (JD)** [6] are based on common neighbors. (3) **Random Walk with Restart (RWwR)** [7] computes the probability for a random walker moving among nodes. We also evaluate pure content feature methods, and thus we use (4) **word2vec-Raw** and (5) **LDA-Raw** to present authors' research interests and then compute author similarity based on cosine. When comparing to $\mathbf{H}_{\mathbf{Bin}}^{U}$, we map word2vec and LDA feature into binary space (**word2vec-Bin** and **LDA-Bin**) for a more fair comparison. (8)**Geographic distance (GD)** measures the geo-graphic distance between authors' affiliations. The last method is (9) **Combined (CB)** simple concatenation of all node content features.

2. Supervised Link Prediction. We compare our semi-supervised framework ($\mathbf{H}_{\mathbf{Raw}}$ and $\mathbf{H}_{\mathbf{Bin}}$) with two supervised baseline methods. One is classical classifier method **Support vector machine (SVM)** [17] and one is a state-of-art approach **Supervised Random Walk (SRW)** [2], using node attributes to guide random walk on the graph.

3. Cold-Start Link Prediction. In this setting, we will compare $\mathbf{H}_{\mathbf{Raw}}$ and $\mathbf{H}_{\mathbf{Bin}}$ with CB and SVM in predicting links between "cold-start" nodes and "core" nodes.

Parameter Settings. For our four methods $H_{Raw}^{U}, H_{Bin}^{U}, H_{Raw}$ and H_{Bin}, we define two nodes as "similar" if they have link and as "dissimilar" if no link. To make the relational matrix sparse, we sample the "dissimilar" set to make the ratio of "similar" and "dissimilar" to be 1 : 1. We apply PCA on AA and GD, letting the dimension of all features the same (100-dimension). The feature matrix could be built by concatenating selected features (AA, GD, word2vec

Table 2. AUC (in percentage) results of all methods. Numbers in parentheses are standard deviations.

Type	Unsupervised									Semi/supervised			
Method	AA	JD	RWwR	word2vec-Raw/Bin	LDA-Raw/Bin	GD	CB-Raw/Bin	H_{Raw}^{U}	H_{Bin}^{U}	H_{Raw}	H_{Bin}	SVM	SRW
2010	77.8	75.2	87.5	86.2/72.0	85.3/68.7	67.0	85.3/71.5	90.3	81.5	**91.5(1.6)**	83.6(1.5)	72.8(2.3)	89.3(1.5)
2011	74.2	73.6	86.7	85.4/69.8	84.7/70.2	68.5	85.2/69.5	92.5	80.1	**93.1(0.4)**	83.5(2.0)	73.0(1.7)	87.8(0.3)
2012	73.2	72.3	88.6	86.2/75.4	85.3/70.6	67.0	85.5/73.1	91.1	82.3	**92.3(0.7)**	82.4(1.1)	72.7(2.2)	89.1(1.0)

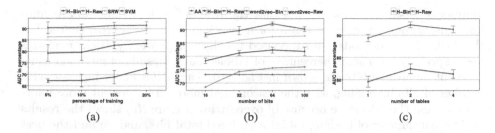

Fig. 2. (a) AUC (in percentage) with different percentage of training for supervised methods (2 hashing tables and 32bits in each table); (b) AUC (in percentage) with different bits in total (fixed 2 tables); (c) AUC (in percentage) of different number of hashing tables (fixed 64bits in total)

and LDA) together after normalization. After trying all different combinations of features, we get the optimal feature combination: AA+word2vec+GD. We set parameters by grid search and get the optimal parameters $\gamma_1 = 1.0 \times 10^{-6}$, $\gamma_1 = 1.0 \times 10^{100}$ and best setting is 2 hashing tables and 32bits in each table. The same features are used in CB, SVM and SRW. Positive and negative sample ratio for SVM is $1:1$. As recommended in [7], we adopt the Wilcoxon-Mann-Whitney (WMW) loss function and logistic edge strength in SRW. All methods are evaluated using AUC: Area under the Receiver Operating Characteristic (ROC) Curve.

4.4 Results

Table 2 shows the performances of the predictions on three networks in unsupervised setting and semi/supervised setting. We can see that the AUCs of H_{Raw}^U outperforms all unsupervised baselines. Average AUCs of 5-fold cross-validation (20 % as training) are used in semi/supervised setting. We can see that H_{Raw} achieve the best performance.

Table 3. AUC for cold-start link prediction. For SVM, H_{Bin} and H_{Raw}, the results are average AUCs (percentage) of 5-fold cross-validation with 20 % of the dataset as training and the rest of testing. The numbers in parentheses are standard deviations.

Methods	CB-Raw/Bin	SVM	H_{Bin}	H_{Raw}
2010	62.5/55.3	60.2(5.50)	61.5(3.80)	**63.5(4.20)**
2011	61.2/54.7	59.8(3.40)	60.3(2.80)	**62.7(2.00)**
2012	62.8/55.2	60.7(4.50)	61.2(3.60)	**64.2(3.70)**

Sensitivity Test. In this section, we test the sensitivity for different parameter settings by changing the size of training dataset, number of bits in total and number of hashing tables. We find that the results for three datasets are similar; thus, we only provide the result for network in 2012. Figure 2(a) shows that the AUC of our methods and supervised baselines. Apparently, H_{Raw} performs best especially for a small percentage of labeled data. Results of methods using different number of bits are shown in Fig. 2(b). The AUCs firstly increase with number of bits increasing, with the reason that more bits represent more information, and then decrease because of over-fitting. Figure 2(c) shows the results of different number of hashing tables with fixed total bits, and we find the best AUC peaks at 2, which suggested that two latent components in the feature matrix are sufficient for link prediction in our experiment datasets.

Cold-start Link Prediction. Table 3 shows the performance of the proposed method, unsupervised and supervised baseline methods in cold-start experiment. Again, H_{Raw} performs the best. The results suggest that our proposed method can better extract node similarity by fusing network topology pattern and node attributes.

5   Conclusion

In this paper, we develop a semi-supervised link prediction method via a *Multi-Component Hashing* framework. Our proposed method derives hashing tables for nodes and predicts new links based on the closeness of the hashing codes. Results on three co-authorship networks show that our approach outperforms all the state-of-the-art unsupervised and supervised methods and also show the effectiveness of the proposed method in *cold-start* setting, where we have no or few knowledge of network topology. The key contribution of our work is a *Multi-Component Hashing* framework in link prediction that addresses several challenges and achieves the best performance comparing with the existing methods. As part of the future work, we plan to further explore a weighting scheme of latent similarity components to further enhance the link prediction accuracy.

References

1. Adamic, L.A., Adar, E.: Friends and neighbors on the web. Soc. Netw. **25**(3), 211–230 (2003)
2. Backstrom, L., Leskovec, J.: Supervised random walks: predicting and recommending links in social networks. In: Proceedings of the Fourth ACM International Conference on Web Search and Data Mining, pp. 635–644. ACM (2011)
3. Blei, D.M., Ng, A.Y., Jordan, M.I.: Latent dirichlet allocation. J. Mach. Learn. Res. **3**, 993–1022 (2003)
4. Gao, S., Denoyer, L., Gallinari, P.: Tensor decomposition model for link prediction in multi-relational networks. In: 2010 2nd IEEE International Conference on Network Infrastructure and Digital Content, pp. 298–302. IEEE (2010)
5. Han, S., He, D., Brusilovsky, P., Yue, Z.: Coauthor prediction for junior researchers. In: Social Computing, Behavioral-Cultural Modeling and Prediction, pp. 274–283 (2013)

6. Jaccard, P.: Etude comparative de la distribution florale dans une portion des Alpes et du Jura. Impr. Corbaz (1901)
7. Jeh, G., Widom, J.: Simrank: a measure of structural-context similarity. In: Proceedings of the Eighth ACM SIGKDD, pp. 538–543. ACM (2002)
8. Leroy, V., Cambazoglu, B.B., Bonchi, F.: Cold start link prediction. In: Proceedings of the 16th ACM SIGKDD, pp. 393–402. ACM (2010)
9. Liben-Nowell, D., Kleinberg, J.: The link-prediction problem for social networks. J. Am. Soc. Inf. Sci. Technol. **58**(7), 1019–1031 (2007)
10. Lichtenwalter, R.N., Lussier, J.T., Chawla, N.V.: New perspectives and methods in link prediction. In: Proceedings of the 16th ACM SIGKDD International Conference on Knowledge Discovery and Data Mining, pp. 243–252. ACM (2010)
11. Lü, L., Zhou, T.: Link prediction in complex networks: A survey. Phys. A Stat. Mech. Appl. **390**(6), 1150–1170 (2011)
12. Menon, A.K., Elkan, C.: Link prediction via matrix factorization. In: Gunopulos, D., Hofmann, T., Malerba, D., Vazirgiannis, M. (eds.) ECML PKDD 2011, Part II. LNCS, vol. 6912, pp. 437–452. Springer, Heidelberg (2011)
13. Mikolov, T., Sutskever, I., Chen, K., Corrado, G.S., Dean, J.: Distributed representations of words and phrases and their compositionality. In: Advances in Neural Information Processing Systems, pp. 3111–3119 (2013)
14. Ou, M., Cui, P., Wang, F., Wang, J., Zhu, W.: Non-transitive hashing with latent similarity components. In: Proceedings of the 21th ACM SIGKDD International Conference on Knowledge Discovery and Data Mining, pp. 895–904. ACM (2015)
15. Sarkar, P., Chakrabarti, D., Jordan, M.: Nonparametric link prediction in dynamic networks, arXiv preprint arXiv:1206.6394 (2012)
16. Stumpf, M.P., Thorne, T., de Silva, E., Stewart, R., An, H.J., Lappe, M., Wiuf, C.: Estimating the size of the human interactome. PNAS **105**(19), 6959–6964 (2008)
17. Suykens, J.A., Vandewalle, J.: Least squares support vector machine classifiers. Neural Process. Lett. **9**(3), 293–300 (1999)
18. Szell, M., Lambiotte, R., Thurner, S.: Multirelational organization of large-scale social networks in an online world. PNAS **107**(31), 13636–13641 (2010)
19. Tsai, C.-H., Lin, Y.-R.: The evolution of scientific productivity of junior scholars. In: International Conference 2015 Proceedings (2015)
20. Wang, J., Kumar, S., Chang, S.-F.: Semi-supervised hashing for large-scale search. IEEE Trans. Pattern Anal. Mach. Intell. **34**(12), 2393–2406 (2012)
21. Wang, J., Shen, H.T., Song, J., Ji, J.: Hashing for similarity search: A survey, arXiv preprint arXiv:1408.2927 (2014)

TELELINK: Link Prediction in Social Network Based on Multiplex Cohesive Structures

Di Jin[1]($\boxtimes$), Mengdi Wang[2], and Yu-Ru Lin[3]

[1] School of Computer Science, Carnegie Mellon University, Pittsburgh 15213, USA
dijin@andrew.cmu.edu
[2] Intelligent System Program, University of Pittsburgh, Pittsburgh 15260, USA
[3] School of Information Sciences, University of Pittsburgh, Pittsburgh 15260, USA

Abstract. Given a network where the same set of nodes have multiple types of relationships, how do we efficiently predict potential links in the future (e.g., interactions between social actors), and how do we predict links using information from other relationships? These problems have been widely studied recently, most of the existing methods either aggregate multiple types of relationships into a single network or consider them separately and ignore the correlations across relationships, leading to information loss. In this work, we present TELELINK, a general link prediction model that works for networks with single and multiple relationships. TELELINK predicts potential links based on community detection and improves link prediction by bringing in a cohesive structure across multiple networks constructed by different relationships or node attributes. To further improve the prediction performance, we extend TELELINK to a semi-supervised scheme, incorporating partially labeled information. Our extensive experiments show that TELELINK outperforms existing methods in predicting new links. Specifically, among the various datasets that we study, TELELINK achieves a precision improvement by up to 110 % compared to the baselines.

1 Introduction and Background

Recent years have witnessed a surge of interest for understanding and characterizing the properties of *social networks*, where nodes represent people or other entities embedded in a social context and links denote relationships or interactions, such as friendship, collaboration, or influence between entities. An important problem in this context is *link prediction*, which is to predict links that will appear in the network during the interval from time T to a later time $T + 1$, given a snapshot of a network at time T [1] or before. Link prediction is useful in various areas. In social networks, link prediction algorithms can be used to predict relationships among individuals such as friendship, partnership and their future behaviors such as communications and collaborations [2]. In biological networks such as protein-protein interaction networks where over 99 % links are unknown [3], accurately predicting possible links could sharply reduce the experimental cost.

© Springer International Publishing Switzerland 2016
K.S. Xu et al. (Eds.): SBP-BRiMS 2016, LNCS 9708, pp. 174–185, 2016.
DOI: 10.1007/978-3-319-39931-7_17

A typical framework of link prediction algorithm is based on "similarity", where each pair of nodes is given a similarity measure, and node pairs with high similarity scores are assumed more likely to be connected [4]. However, most of the similarity measures mainly consider the network topological structure (e.g., the number of common neighbors [5], or the length of the shortest path between nodes [6]). These kinds of methods ignore the information provided by the community structure. A community is a densely connected group of nodes while sparsely connected to other groups. Community structure has been proved to be critical for link prediction [7]. For example, in the friendship network, a community could be a group of people in the same school, company or club and new links (i.e. friendships) are more likely to form within the group. [8] shows that group membership information can enhance the accuracy of link prediction. However, their method does not scale well to very large networks in practice.

The real world networks are always multi-relational, where links have different meanings and bring challenges to link prediction. For example, in a Twitter network, the links could be different interactions such as replying or mentioning. The problem is: how can one predict possible links representing a particular relationship using the information provided by other relationships? Two typical strategies have been employed: pre-fusion and post-fusion. Pre-fusion aggregates multiple types of relationships into a single link while post-fusion separately studies each type of links independently and ignores the correlations across types. But both approaches result in a loss of information.

Another issue in link prediction is how to develop a method that combines topological information and node attributes. Existing work has employed supervised classifiers [9], which is trained to discriminate between positive links (i.e. links that form) and negative links (i.e. links that do not form) by using multiple sources of information as features. Those methods suffer from the imbalance problem: in real networks, the number of positive links is significantly less than that of negative links. To overcome the imbalance problem, typical strategies are under-sampling and over-sampling [9], which lead to issues including overfitting.

Present Work. We present TeleLink, a general link prediction model to address the above challenges. We consider multiple layers of networks, one is target layer – the particular type of links to be predicted (e.g. Twitter follower-followee network) and others are *auxiliary* layers, which can be constructed based on other relations between nodes (e.g. reply or retweet between Twitter users) or nodes attributes (e.g. geographical information in users' profile). TeleLink predicts potential links using a probabilistic similarity measure between nodes defined by the path information of multilayer community structure revealed by *Multiplex Infomap* [10], a random-walk based community detection approach. We further extend TeleLink into a semi-supervised learning scheme to improve link prediction performance, using both networks at T and partially labeled information in the networks at time $T + 1$ in multiplex Infomap framework.

Contributions. Our main contributions include:

1. **New link prediction approach**: We extend Infomap to address several challenges in link prediction problem, including incorporating community structure, combining multiple relationships and combining topological and node attributes.
2. **Prediction performance**: We conduct extensive experiments on two different datasets. The proposed methods achieve best prediction accuracy in new link prediction, compared with existing link prediction methods.
3. **Extended analysis on dynamic networks**: We conduct experiments on real-world social networks over time and present the influence of time intervals on prediction.

2   Related Work

Single-Relational Link Prediction Methods. The seminal work of Liben-Nowell and Kleinberg [1] is the first comprehensive study on link prediction methods based on similarity measures derived from graph topology structure. Empirical results of comparison between random predictors and a variety of measures including *Jaccard's coefficient* [5], *Adamic/Adar* [11], *Katz Index* [6] and *Rooted PageRank* [12] demonstrate the usefulness of topological information. However, the weakness of these methods is that they only consider a single (topological) feature. In addition to topological information, we often have the knowledge of attributes or covariance for the nodes. Intuitively, performance is expected to be enhanced by using this extra information. A classical approach is to use supervised classifiers [2] unitizing different sources of information, including topological information and node attributes as features. [9] suggests that placing classification algorithms in an ensemble framework can benefit by reducing variance, especially for unstable algorithms like decision trees. These methods have to use down-sampling or over-sampling strategies to overcome imbalance, which lead to a loss of information or over-fitting. [13] proposes an algorithm based on *Supervised Random Walks*, which uses node attributes to guide the random walker, but it has high computational cost and does not scale well in practice.

Most existing link prediction methods do not consider community information, which is proved to be useful to link prediction. The community in a network is a densely connected group of nodes while sparsely connected to other groups. In [7], experiments on both synthetic and real-world networks unveil how the community structure affects the performance of link prediction methods: with increasing number of communities, the performance of link prediction could be improved remarkably. Recently, [14] also shows that supplement the similarity-based measures with community information could improve the accuracy of link prediction methods.

Multi-relational Link Prediction Methods. The above link prediction techniques only consider homogeneous links with the same semantic meaning, while in reality networks comprise multiple types of links or interactions among nodes. Only a few studies address link prediction problems in these heterogeneous networks. In [15], an unsupervised method extending the Admic/Adar measure and a supervised method in multi-relational networks are proposed. [16] develops machine learning approaches based on graphical models to infer new links across heterogeneous networks assuming that people will form relationships in different networks with similar principles. However, none of the multi-relational link prediction methods consider the information of multi-relational communities, which is potentially useful in improving the performance of link prediction algorithm.

TeleLink addresses the limitations of prior works in three perspectives. First, TeleLink combines the topological information and nodes attributes by extracting structures from multiple layers of networks. Second, TeleLink handles link prediction across multi-relational networks with auxiliary layers based on multiple relationships (e.g. reply or retweet of Twitter users). Last, to make use of the community information, TeleLink defines "similarity" between nodes based on the path information of multilayer community structure calculated through random walk on multiple networks.

3   Proposed Method

First we give the definitions of two fundamental concepts in our proposed method:

Definition 1 (layer). *A layer of the social network is defined as a specific relationship. A physical node n belongs to at least one layer of the social network.*

Definition 2 (link). *A link from node i to node j, l_{ij}, is defined as the behavior starting from node i to j.* For example, in the retweet layer of the Twitter network, l_{ij} denotes user i retweets user j's tweet.

The challenges discussed in previous sections can be summarized into two research questions in social networks: (a) How do we express similarity through communities, thus predicting possible new links in a single layer? (b) How do we combine the attributes and connectivity from other layers so as to improve the prediction accuracy? We present our solutions to these questions in the next two sections. The symbols used in this paper are defined in Table 1.

3.1   TeleLink

TeleLink is based on MapEquation [17], a flow-based network partitioning algorithm. MapEquation characterizes the behavior of the network through the system-wide flow of information and considers community detection as solving a coding problem.

Table 1. Table of symbols

Symbol	Definition
$\mathcal{L_C}$	a collection of all links between communities
$L_{\mathcal{C}_i\mathcal{C}_j}$	a directed link starting from community i to community j
l_{ij}	a directed link starting from node i to node j
$\mathcal{C}, \mathcal{C}_i$	a set of all communities in the network; the ith community in the set
$\mathcal{C}^{(i)}$	the community that node i belongs to
n	a physical node in the network, $n \in 1, 2, \ldots, N$
v_i	the probability that the random walker reaches node i
$w_{i,j}^\alpha$	the out-flow volume from node i to j in layer α
N, M	total number of nodes and total number of communities of the network
$\mathbf{M}$	a partition of the network with minimum Huffman coding length
α, β	specific layer of the network, represented in Greek letters
ΔT	length of time interval for community detection

These flows are described using Huffman coding, following the Random Walk algorithm to identify communities. Groups of nodes among which information flows frequently are described as well-connected communities; the paths between communities are also captured. [17] gives the community partition $\mathbf{M}$ by solving Eq. 1 to obtain the minimized expected coding length of Random Walk paths:

$$\min L(\mathbf{M}) = q_\curvearrowright H(\mathcal{L}) + \sum_{i=1}^{M} p_\circlearrowright^i H(\mathcal{P}^i), \tag{1}$$

where $L(\mathbf{M})$ denotes the Huffman coding description of random walker's path; $q_\curvearrowright$ is the probability that the random walk switches communities on any given step, and $p_\circlearrowright^i$ describes the fraction of within-community movements that occur in community i, $H(.)$ is the entropy of the community.

Given the partition $\mathbf{M}$, the network $\langle \mathcal{V}, \mathcal{E} \rangle$ could be described as $\langle \mathcal{C}_\mathcal{V}, \mathcal{L_C} \rangle$. $\mathcal{C}_\mathcal{V}$ describes the set of communities that compose the network: $\mathcal{C}_\mathcal{V} = \{\mathcal{C}_1, \ldots, \mathcal{C}_M\}$. $\mathcal{C}_i$ is the ith community. $\mathcal{L_C}$ is the collection of links between communities: $\mathcal{L_C} = \{L_{\mathcal{C}_1\mathcal{C}_1}, L_{\mathcal{C}_1\mathcal{C}_2}, \ldots, L_{\mathcal{C}_M\mathcal{C}_M}\}$. The links between communities can be interpreted as the flow volumes of all nodes reaching from the source community to the destination community. $P(L_{\mathcal{C}_i\mathcal{C}_j})$ denotes the transition probability from community $\mathcal{C}_i$ to $\mathcal{C}_j$.

Now the problem can be formulated as: given $P(\mathcal{C}_p)$ where $p \in \{1, \ldots, M\}$ and $P(L_{\mathcal{C}_p\mathcal{C}_q})$ where $p, q \in \{1, \ldots, M\}$, how do we compute the probability of a specific link l_{ij} between node i and j? We estimate the probability of l_{ij} through the product of (a) the transition probability from the community of node i to the community of node j, and (b) the conditional probability of the random walker reaching node j inside its community. An example is illustrated in Fig. 1a. To predict the link in dashed line, the random walker has to follow each possible path from node 8 to 6, which could be expensive in real-world networks. TELELINK calculates the probability based on the transition flows between communities (bold arrow): from community C to community B through community A, and the probability to reach node 6 inside community B, as illustrated in Fig. 1b.

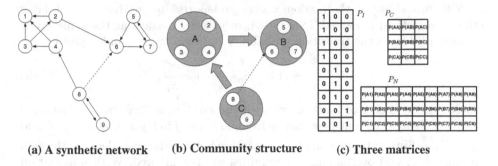

(a) A synthetic network   (b) Community structure   (c) Three matrices

Fig. 1. Single layer link prediction through Infomap

Considering nodes across the whole network, we could factorize the probability between node i and node j into three terms, as illustrated by Eq. 2:

$$P(l_{ij}) = \underbrace{\mathbb{1}_{\mathcal{C}^{(i)}}}_{P_I} \times \underbrace{P(L_{\mathcal{C}^{(i)}\mathcal{C}^{(j)}})}_{P_C} \times \underbrace{\frac{v_j}{\sum_{k \in \mathcal{C}^{(j)}}^{|\mathcal{C}^{(j)}|} v_k}}_{P_N}, \tag{2}$$

where $\mathcal{C}^{(i)}$ denotes the community that node i belongs to. The first term represents the community affiliation of node i: $\mathbb{1}_{\mathcal{C}_i}$ is the indicator matrix determining which community node i uniquely belongs to, so each row of this matrix has only one entry with value of 1; the second term describes transition probability from the community of node i to the community of node j, and the last term is the flow probability to node j normalized by the flows reaching all nodes inside its community. P_I, P_C and P_N correspond to the indicator matrix, transition matrix, and flow matrix of the example shown in Fig. 1c.

TELELINK computes the transition probability from the source community $\mathcal{C}^{(i)}$ to destination node j as the approximation to the probability of link l_{ij} used for prediction. The reason is that communities are partitioned using Random Walk algorithm, so information flows faster and more easily among nodes within the same community than those between different communities. If node i in community $\mathcal{C}^{(i)}$ has link to node j in community $\mathcal{C}^{(j)}$, i is also likely to reach other nodes in community j. In addition, by grouping nodes as a community, we don't have to compute every possible link from the source node to the destination node (which is what Random Walk algorithm does). As a result, TELELINK achieves precision with the guarantee of efficiency.

3.2 TELELINK in Multiplex Networks

Multiplex Infomap [10] extends MapEquation to multilayer networks in two ways: (1) inter-layer dynamics described in communities and (2) overlapping communities identification. The first generalization resembles PageRank algorithm to compute the flow volume between layers by introducing the relax rate r. With probability r, the random walker "teleports" to another network layer (thus the method is named TELELINK); otherwise, it stays at the same layer. Equation 3 defines the transition probability between layer α and β given r:

$$P_{ij}^{\alpha\beta}(r) = (1 - r)\delta_{\alpha\beta}\frac{w_{ij}^{\beta}}{s_i^{\beta}} + r\frac{w_{ij}^{\beta}}{S_i}, \tag{3}$$

where δ is the indicator function seeing if the random walker stays at the same layer; w_{ij}^{β} denotes the out-flow volume from node i to j in layer β; $s_i^{\beta} = \Sigma_j w_{ij}^{\beta}$ and $S_i = \Sigma_{\beta} s_i^{\beta}$. Similar to Infomap, the interaction between layers is denoted through communities, and overlapping communities are not allowed. With information from multiple layers of networks, TELELINK could predict links that cannot be predicted in single layer networks. Consider the example illustrated in Fig. 2a where network layer β consists of two disjoint communities A and B. Due to the isolation between community A and B, the random walk algorithm could not predict links such as l_{56}. However in Fig. 2b, TELELINK overcomes this limitation through flows between layers α and β with a "teleporting" relax rate r.

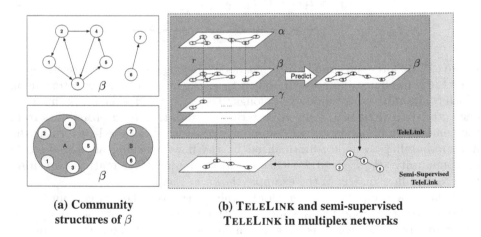

(a) Community
structures of β

(b) TELELINK and semi-supervised
TELELINK in multiplex networks

Fig. 2. Single layer link prediction through Infomap

3.3 Semi-supervised TELELINK

Semi-supervised TELELINK extends the above method by employing partially labeled data. When performing link prediction on real-world networks, historical data could be the double-edged sword: on the one hand, historical data provides valuable structural information about the network; on the other hand, it could bring in noise, impeding the prediction precision especially to rapidly changing networks. Semi-supervised TeleLink overcomes this problem by randomly selecting small parts of the structural information from the target network as an extra auxiliary layer in community detection. An example is illustrated in Fig. 2b where we use the multiplex network information at T to predict possible links in layer β at time $T + 1$. In a specific attempt, a small part of the β network at time $T + 1$ is selected: the connectivity between node 3, 4, 5 and 6. It then serves as an extra auxiliary layer for community detection at the stage of T. In this example, such information is the key to the flows between nodes in two disjoint communities and provides additional information to guide proper community assignment that could help predict future links.

4   Experiments

To demonstrate that TeleLink can be applied in different contexts, we perform experiments on two datasets to address three perspectives of TeleLink: (1) How well does TeleLink perform, comparing to existing methods? (2) How are information flows from other layers influencing the prediction in the specific layer? (3) Given different time intervals, how would the prediction results be different?

4.1   Datasets

Primary School. This dataset contains 125,773 contact records among 236 students in a primary school in Lyon, France during two days in Oct. 2009 [18,19]. We build a weighted undirected contact network with students as nodes, contacts between two students as links and contact frequency as weights. There are two features of students: class and gender, which are used as two unweighted undirected auxiliary networks. The link between two students is built if they are in the same class or of the same gender.

Twitter. We use the Twitter 2012 election dataset [20], which contains approximately 48.7 million politically active users and approximately 0.2 billion tweets during 8 weeks starting from Sept. 2012. The dataset is divided into 8 sub-datasets according to post time (one per week). In each sub-dataset, one weighted directed network is established for the mention, reply and retweet respectively. In the mention (or reply, retweet) network, a directed link with weight k is built from user A to user B if user A mentions (or replies, retweets) user B exactly k times. Given the three networks in each week, we aim to predict new links of a specific network in week $T + 1$ from the networks during the period $[T - \Delta T + 1, T]$, where $\Delta T \in \{1, 2, \ldots, 7\}$. We also ignore inactive users (with two or fewer links) because for users only posting one or two tweets during eight weeks, their impacts on the analysis is trivial. The pre-processed dataset contains 24 networks with identical 2,073 nodes. The numbers of links in week 1 for the mention, reply and retweet are 14,248, 8,707 and 8,944. As the numbers of links increase from week 1 to week 8, the numbers of new links added every week for the mention, reply and retweet are 1,990, 652 and 1,775 on average.

4.2 Experiment Setup

Baseline. We consider 2 classic approaches: *Jaccard's coefficient* and *Adamic/Adar*.

Our Methods. We apply our methods on the above datasets in three scenarios:

1. **TELELINK** TELELINK is applied to the target single layer network.
2. **TELELINK in multiplex networks (different r)** TELELINK is applied to the multiplex network, flows moving between layers with different relax rate r.
3. **Semi-supervised TELELINK** TELELINK is applied to the multiplex network, flows moving between layers with $r = 0.15$. 1/3 nodes of the target layer network are selected for training, and the result is evaluated with 3-fold cross validation.

Evaluation Metrics. The experiments are designed to predict new links appearing in $T + 1$ from networks during $[T - \Delta T + 1, T]$, $\Delta T \in \{1, 2, \ldots, 7\}$. We consider the prediction precision at Top k nodes according to the probability to which the source node is not already connected, i.e., how many of top k nodes suggested by our algorithm during $T + 1$ actually receive links not exist in $[T - \Delta T + 1, T]$. We set k equal to the total number of links in $T + 1$. In addition, we measure the improvement over baselines.

4.3 Results

Interaction Sensitivity. We first explore the sensitivity of TELELINK to the interaction between layers in the multiplex networks, which is controlled by relax rate r. Higher values of r indicate that the random walker is more likely to "teleport" to other layers while lower values of r indicate more isolation between layers. The result is shown in Fig. 3a. The relatively flat pattern in each of the diagrams indicates that when we are considering information from different relationships, the interaction between them does not lead to the significant difference in prediction precision. Therefore in the following experiments, we set $r = 0.5$ and use it for evaluation.

Prediction Precision. We measure the prediction precision on two datasets. For the primary school dataset, there are two intervals so $\Delta T = 1$. The prediction result is illustrated in Table 2. In the Twitter dataset, we also set $\Delta T = 1$ for consistency. Since there are prediction results for seven weeks, we only show the precision improvement of TeleLink over baselines in Fig. 3b. The results show that TeleLink under all settings outperform the baselines. The mention relationship is a loose way of communication between users, so TeleLink provides a relatively small improvement over the baseline AA. However, for relationships such as reply and retweet where interactions between users are stronger, TeleLink gives significant improvement over both baselines. In the reply relationship, TeleLink achieves 40 % ∼ 50 % improvement through semi-supervised learning with auxiliary layers; Multiplex TeleLink also makes great improvements, indicating the relationship of mention and retweet is helpful when we are predicting replies between users. In the retweet relationship, TeleLink performs

Table 2. Table of elementary school prediction precision

Method	InfoMap	Multiplex	Multiplex(i)	Semi-Supervised	Jaccard	AA
Precision	0.191	0.198	0.193	**0.205**	0.103	0.079

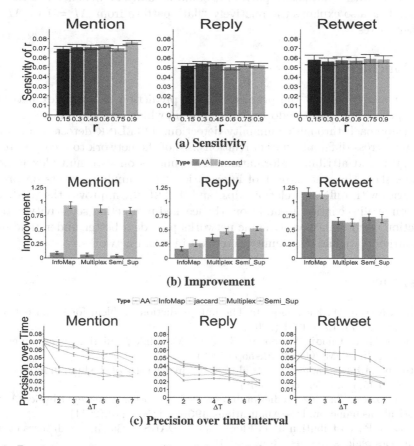

(a) Sensitivity

(b) Improvement

(c) Precision over time interval

Fig. 3. Precision performance on (from left to right) mention, reply and retweet networks

on single layer achieves approximately 110 % improvement over both baselines, which implies that for this relationship, adding other types of interactions limits or even impedes the prediction performance.

Perdiction Performance with Dynamic Networks. In this section, we focus on the impacts of different time intervals (ΔT) used for community detection to prediction precision. We apply TELELINK to the Twitter dataset partitioned into different time intervals, ranging from 1 to 7. The result is shown in Fig. 3c. Intuitively, with longer time intervals to perform community detection, more information about the network would be included, rendering higher prediction precision. However, based on the overall result curves, the prediction precision drops as ΔT increases. This means that with long-time intervals, longer-term historical data becomes outdated, which impedes the prediction. One interesting observation is the increase in retweet curve from $\Delta T = 1$ to $\Delta T = 2$. This indicates that although too much long-term historical data could be harmful, with appropriate portion it could actually be beneficial to the prediction. This also explains the relatively "flat" pattern from $\Delta T = 1$ to $\Delta T = 2$ in other curves.

5   Conclusions

In this paper, we study the problem of link prediction in multiplex networks. We propose TELELINK to address this problem, which provides a novel link prediction approach through community detection. TELELINK detects community structure across different types of relationships of the network to incorporate the topological and attribute information. Experiments on two multiplex network datasets show that the impact of information from auxiliary layers on prediction varies with different relationships, and TELELINK improves the prediction precision in all scenarios. In addition, the learned interaction sensitivity and the prediction performance on dynamic networks provide a better understanding of information flows and community detection in social networks.

References

1. Liben-Nowell, D., Kleinberg, J.: The link-prediction problem for social networks. JASIST **58**(7), 1019–1031 (2007)
2. Hasan, M.A., Chaoji, V., Salem, S., Zaki, M.: Link prediction using supervised learning. In: SDM 2006: Workshop (2006)
3. Stumpf, M.P.H., Thomas, T., et al.: Estimating the size of the human interactome. PNAS **105**(19), 6959–6964 (2008)
4. Lü, L., Zhou, T.: Link prediction in complex networks: A survey. Physica A: Statistical Mechanics and its Applications **390**(6), 1150–1170 (2011)
5. Jaccard, P.: Distribution de la Flore Alpine: DANS le Bassin des dranses et dans quelques régions voisines. Rouge (1901)
6. Katz, L.: A new status index derived from sociometric analysis. Psychometrika **18**(1), 39–43 (1953)

7. Feng, X., Zhao, J.C., Xu, K.: Link prediction in complex networks: a clustering perspective. Eur. Phys. J. B **85**(1), 1–9 (2012)
8. Clauset, A., Moore, C., Newman, M.E.J.: Hierarchical structure and the prediction of missing links in networks. Nature **453**(7191), 98–101 (2008)
9. Lichtenwalter, R.N., Lussier, J.T., Chawla, N.V.: New perspectives and methods in link prediction. In: Proceedings of the 16th ACM SIGKDD International Conference on Knowledge Discovery and Data Mining, pp. 243–252. ACM (2010)
10. De Domenico, M., Lancichinetti, A., Arenas, A., Rosvall, M.: Identifying modular flows on multilayer networks reveals highly overlapping organization in interconnected systems. Phys. Rev. X **5**(1), 011027 (2015)
11. Adamic, L.A., Adar, E.: Friends and neighbors on the web. Soc. Netw. **25**(3), 211–230 (2003)
12. Brin, S., Page, L.: Reprint of: The anatomy of a large-scale hypertextual web search engine. Comput. Netw. **56**(18), 3825–3833 (2012)
13. Backstrom, L., et al.: Supervised random walks: predicting and recommending links in social networks. In: Proceedings of the fourth ACM WSDM, pp. 635–644. ACM (2011)
14. Soundarajan, S., Hopcroft, J.: Using community info. to improve the precision of link prediction methods. In: Proceedings of the WWW, pp. 607–608. ACM (2012)
15. Davis, D., Lichtenwalter, R., Chawla, N.V.: Supervised methods for multi-relational link prediction. Soc. Netw. Anal. Min. **3**(2), 127–141 (2013)
16. Sun, Y., et al.: Co-author relationship prediction in heterogeneous bibliographic networks. In: 2011 International Conference on ASONAM, pp. 121–128. IEEE (2011)
17. Rosvall, M., Bergstrom, C.T.: Maps of random walks on complex networks reveal community structure. PNAS **105**(4), 1118–1123 (2008)
18. Stehlé, J., et al.: High-resolution measurements of face-to-face contact patterns in a primary school (2011)
19. Gemmetto, V., Barrat, A., Cattuto, C.: Mitigation of infectious disease at school: targeted class closure vs school closure. BMC Infect. Dis. **14**(1), 695 (2014)
20. Lin, Y.-R., Keegan, B., Margolin, D., Lazer, D.: Rising tides or rising stars?: Dynamics of shared attention on twitter during media events (2014)

Game-Specific and Player-Specific Knowledge Combine to Drive Transfer of Learning Between Games of Strategic Interaction

Michael G. Collins[1,2(✉)], Ion Juvina[1], and Kevin A. Gluck[2]

[1] Department of Psychology, Wright State University,
3640 Colonel Glenn Hwy, Dayton, OH 45435, USA
Michael.Collins74.ctr@us.af.mil
[2] Air Force Research Laboratory, 2620 Q Street, Building, 852,
Wright-Patterson Air Force Base, Dayton, OH 45433, USA

Abstract. Trust in others transfers between games of strategic interaction (e.g., iterated Prisoner's Dilemma– PD and Chicken Game – CG). This transfer of trust represents knowledge acquired about the other player (player-specific knowledge), carrying over from one situation to another, which is separate from what was learned about the previous game (game-specific knowledge). We examine how the transfer of both player-specific and game-specific knowledge informs one's decisions when interacting with a new player. In this paper, we present the experimental design of an upcoming study, where participants will sequentially play two games of strategic interaction (PD & CG) with the same or a different computerized confederate agent. In addition to the experimental design, we present model predictions, using a previously published computational cognitive model of trust dynamics. The model predicts transfer of learning effects in both conditions and larger effects when interacting with the same agent.

Keywords: Trust dynamics · Strategic interaction · Model predictions · Multiple agent interaction · Behavioral game theory

1 Introduction

Trust plays an important role in many social interactions by simplifying choices in complex and changing environments. Indeed, trust informs the decisions made at the start of an interaction between two people. Many are willing to place trust in another, in absence of any information about the other person [1]. It is an individual's trait trust (also referred to as trust propensity), the general disposition toward trusting others, that is thought to inform the choices made early in an interaction and in absence of any experience with another [2]. However, one's trust in another is not static, but changes over time based on the behavior of the other person. State trust (also referred to as learned trust), one's trust for a specific person in a specific situation, increases over the course of repeated interactions [3–5] and is sensitive to particular characteristics of another's behavior [6].

© Springer International Publishing Switzerland 2016
K.S. Xu et al. (Eds.): SBP-BRiMS 2016, LNCS 9708, pp. 186–195, 2016.
DOI: 10.1007/978-3-319-39931-7_18

Previous research has examined the role of trust under many different circumstances, such as interacting with a single person [4, 5] or multiple people sequentially during games of strategic interaction [7]. Often in these experimental designs, human participants interact with each other, acquiring information about both the game (game-specific knowledge –the best choices in specific contexts and their associated outcomes) and the other player (player-specific knowledge – one's knowledge about how another person is likely to behave in a particular context), one component of the latter being their trust in the other player, at the same time. A limitation of this design is that it is difficult to examine how individuals react to certain characteristics of others. For example, individuals repeatedly interact with different people in many situations and their trustworthiness may not always remain consistent, requiring one's trust to change.

A previously published computational cognitive model of trust dynamics [8] uses a trust mechanism, which monitors both the trustworthiness and the need to establish trust with the other player, to establish, repair, or recalibrate trust during games of strategic interaction. During situations where trust has been lost the model attempts to maximize the payoff of the other player, signaling the desire to reestablish trust. The model has previously been found to explain transfer of learning effects (i.e., a game's optimal outcome occurring at a different proportion when played after another game compared to before) between games of strategic interaction [8] and the behavior of individuals when interacting with different computerized confederate agents [6]. In each of these experiments, both components of the trust mechanism were necessary for the model to explain the behavior of participants. However, the model has yet to explain the reaction of individuals in sequential interactions with multiple people.

In this paper, we present predictions generated by the model of the human behavior when playing two games of strategic interaction sequentially with a computerized confederate agent who uses a different strategy during each game, whom participants will play with as well. Additionally, participants will be told that they will play games with either a single individual during both games (one-agent condition) or with a different individual during each game (two-agent condition), allowing us to investigate the transfer of trust between games. This experiment gives us the opportunity to examine human behavior during conditions in which the confederate agent changes as compared to conditions in which only the strategy of the confederate agent and the game change. In our previous work, we proposed that transfer of learning effects are driven by both game-specific and player-specific knowledge [6]. In the current work, the model makes specific predictions for conditions in which humans switch and do not switch players, allowing dissociation of the role of game-specific and player-specific knowledge in driving transfer of learning between games.

1.1 Games

A game represents an abstraction of a scenario and the outcome that occurs depends on the choices made by both players. One of the simplest types of games of strategic interaction are 2×2 games, which consist of two players simultaneously choosing one of two options (A or B), leading to one of four possible outcomes. Two common 2×2

games, and the ones which are used in the research reported here, are iterated Prisoners Dilemma (PD) and iterated Chicken Game (CG) (Fig. 1).

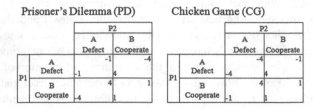

Fig. 1. The payoff matrices for the game Prisoner's Dilemma (left) and Chicken Game (right) show the two players (P1 and P2), the two possible choices they can make during each round (A or B) and the four possible outcomes that can occur during a single round (P1's payoff located in lower left hand corner, P2's payoff in upper right hand corner) depending on the choices made by both players during a round.

In PD, if both players choose to cooperate (B) (mutual cooperation), each earns one point. If both players choose to defect (A), then each loses one point. If one player chooses to defect (A) and the other chooses to cooperate (B), then the player who chose to defect (A) earns four points, while the other player loses four points. When PD is played repeatedly, the optimal strategy is for both players to choose to repeatedly cooperate (B), earning one point during each round.

In CG, if both players choose to cooperate (B), then each player earns one point, as in PD. If both players choose to defect (A), then each loses four points. If one player chooses to defect (A) and the other player chooses to cooperate (B), the player who chose to cooperate (B) loses one point and the player who chose to defect (A) earns four points. When CG is played repeatedly, the optimal strategy is for players to repeatedly asymmetrically alternate between choosing to cooperate and defect (B and A), to earn 1.5 points each during each round.

1.2   Confederate Agents

Preprogrammed agents are used in games of strategic interaction to allow for experimental control and a large number of experimental conditions. Two deterministic strategies that have been used in prior research are the Tit-for-Tat strategy (T4T) [8] and the Pavlov-Tit-for-Tat (PT4T) [9]. Each of these strategies uses the players' previous choice (N-1) to determine its move on the current round (N). T4T repeats on round N the other player's previous choice from round N-1. PT4T reciprocates the other player's choice after mutual cooperation, mutual defection, and unilateral cooperation (i.e., the strategy chose B – cooperate and the other player chose A – defect). After instances of unilateral defection (i.e., the strategy chose A – defect and the other player chose B – cooperate), the PT4T strategy will again choose to defect (choice A).

1.3   The Model

We present here only a brief summary of the model used to generate predictions for this study. A full description of the model and how it was implemented into the ACT-R architecture can be found in [6][1].

ACT-R is both a cognitive architecture and a formal theory of human cognition [10]. To account for the behavior of individuals within and between games of strategic interaction, the model uses several declarative (know-what) and procedural (know-how) architectural mechanisms as well as a novel trust mechanism. When playing a game of strategic interaction (PD and CG) the model needs to be aware of the interdependence between itself and the other player and be able to predict what choice the other player will make next, giving the model a context to base its own decision on.

The awareness of interdependence between the model and the other player is accomplished using instance-based learning (IBL) [11] a strategy that uses previous events stored in declarative memory to inform choices when in a similar context. The model predicts what choice the other player will make next using IBL and sequence learning. During each round, once the model predicts what the other player's next move will be, it chooses to either cooperate or defect depending on which choice has the greatest utility in that context. Utilities of choices in various contexts are updated through reinforcement learning.

Over the course of the game, the model also develops trust in the other player, using its trust mechanism, based on the previous outcomes that occurred in a game. The model's trust mechanism consists of two different accumulators called trust and trust-invest, each monitoring a different aspect of the interaction with the other player. Each accumulator starts at zero when interacting with a new player and increases or decreases based on the outcome in each round. The trust accumulator monitors the trustworthiness of the other player. It increases after instances when the other player has shown to be trustworthy (i.e., mutual cooperation or unilateral cooperation) and decreases after instances where they have shown to be untrustworthy (i.e., instances of mutual defection and unilateral defection). The trust-invest accumulator is used to track trust necessity, that is, the need to establish trust with the other player, increasing after instances of mutual defection and decreasing after instances of unilateral defection. Throughout the game, based on the current level of the trust and trust-invest accumulator the model uses one of three reward functions, maximizing the joint payoff of both players minus the previous payoff of the other player when the trust accumulator is greater than zero, attempting to maximize the other player's payoff when the trust accumulator is less than or equal to zero and the trust invest is greater than zero and maximize its own payoff and minimize the payoff the other player when each accumulator is less than or equal to zero. These three reward functions allow the model to learn different strategies based on the current game dynamics (see [6] for more details).

We expect to learn from this study how the model must be changed to account for circumstances when one sequentially interacts with multiple people. We currently assume that the accumulators in the model's trust mechanism, representing its state

---

[1] The model has also been made publicly available and can be viewed on http://psych-scholar.wright. edu/astecca/software.

trust, begin at zero when playing a game with a new player and are not influenced by its prior history. However, the extent that the model resets its trust accumulators when interacting with another individual may be challenged based on the findings of future research, such as the data to be collected for this study.

2   Experiment

2.1   Experimental Design

All participants will sequentially play two games of strategic interaction for 50 rounds in one of four possible game orders, playing either PD or CG twice (PDPD or CGCG order) or playing each game once (PDCG or CGPD order). During each game, participants will play with a computerized confederate agent which uses a particular strategy, playing the first game of a condition using the T4T strategy and using the PT4T strategy during the second game (T4T_PT4T order) or vice versa (PT4T_T4T order). The trustworthiness of the confederate agent will remain constant over the course of both games, manipulated to be either high (HT) or low (LT) trustworthiness. Finally, the information that participants receive about confederate agents will be manipulated. Participants will be told that they will play both games with either the same worker (one-agent condition) or that they would be randomly paired with another worker during the second game after the first game (two-agent condition).

2.2   Confederate Agents

The computerized confederate agents that will be used in this study to play games with participants are the same agents used in a previous study [5]. A confederate agent used either the T4T or the PT4T strategy (discussed previously) and its trustworthiness was manipulated by manipulating the frequency the confederate agent cooperated (HT) or defected (LT) during the game in the same way as reported in [5].

2.3   Model Simulation

All of the model's predictions were generated by placing the model in each of the 32 experimental conditions and running it 100 times, playing each game with one of ten versions of the confederate agent, mimicking the procedure that participants followed. To ensure pseudo-random variability of the confederate agent across participants and model, ten different versions of each combination of the confederate agent's strategy and trustworthiness manipulation were created[2]. Once assigned to a condition, the model was randomly assigned to play each game with one of the ten possible versions of the confederate agent, as will the participants.

---

[2] Each version of a confederate agent used the same strategy (T4T or PT4T) but each implemented its trustworthiness manipulations, cooperations or defections, on different rounds in a game.

3   Model Predictions

The model's predictions were obtained by computing the proportion of each outcome during each round across all of the model runs in each of the 32 experimental conditions. Alternation was computed by identifying instances where the model cooperated and the confederate agent defected or vice versa on round N and each chose the opposite choice during N-1. After instances of alternation were identified, the round by round proportion of alternation was computed like all of the other outcomes. All of the model's round by round predictions are available online[3].

Due to the size limitation for this article, we present only the model's predictions of transfer effects across the different game and strategy order conditions. Model predictions are aggregated across the HT and LT conditions because the trustworthiness manipulation has been investigated elsewhere [5]. Assessing the predicted transfer effects between the different conditions allows us to evaluate how both prior experience with a particular game (PD or CG), played with an agent using a particular strategy (T4T or PT4T), affects the model's predictions.

Transfer of learning effects were assessed in both the one-agent and two-agent conditions. A transfer effect was identified in either the one-agent or two-agent condition if the proportion of the outcome was significantly different than the proportion of the outcome it was compared to, being either higher or lower during the second game using a paired t-test. The overall mean and standard deviation of the proportion of the outcome during the two games being compared is also reported to show the difference between the two conditions.

3.1   Predicted Transfer Effects During Iterated Prisoner's Dilemma

When playing PD, the game's optimal outcome (i.e., mutual cooperation) can be obtained with a confederate agent that uses either the T4T or PT4T strategy. The ability to repeatedly achieve mutual cooperation during PD with either strategy means that the game's optimal outcome can always be learned during the first game in conditions with the PDPD game order. This experience can then transfer to the second game. Indeed, the model predicts that in the two-agent condition when PD is played with the confederate agent using the T4T strategy (PD T4T) ($M = .36$, $SD = .07$) after playing PD with a confederate agent that uses the PT4T strategy (PD PT4T) or when PD PT4T ($M = .30$, $SD = .01$) is played after PD T4T, that mutual cooperation will occur at a higher proportion compared to when these games are played first (PD T4T: $M = .30$, $SD = .05$: $t(49) = -8.75$, $p < .001$, PD PT4T: $M = .18$, $SD = .03$; $t(49) = -27.31$, $p < .001$) (Fig. 2A). In these conditions, the model's prior experience (game-specific knowledge) from the first game transfers to the second game, leading to mutual cooperation at a higher proportion during the second game.

Transfer effects with the mutual cooperation outcome are also predicted to occur in the one agent conditions during both PD T4T and PD PT4T. Mutual Cooperation is

---

[3] All of the model predictions are available on http://psych-scholar.wright.edu/astecca/publications.

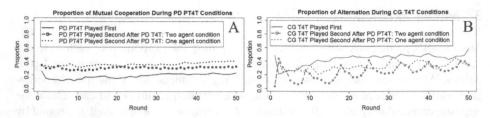

Fig. 2. The model's predicted transfer effect with the mutual cooperation outcome when PD is played with the PT4T strategy after PD with the T4T (left – A) and for CG with the alternation outcome when played with the PT4T strategy after playing PD with the T4T strategy (right – B).

predicted to occur at a higher proportion in the one-agent condition compared to the two-agent condition, when PD T4T ($M = .38$, $SD = .09$) is played after PD PT4T and when PD PT4T ($M = .35$, $SD = .02$) (Fig. 2A) is played after PD T4T ($t(49) = -4.60$, $p < .001$; $t(49) = -15.25$, p $< .001$). Here we see that in addition to the model's prior experience, the trust developed during the first game in the confederate agent (i.e., player-specific knowledge) increases the proportion of mutual cooperation during the second game compared to the two-agent condition.

In contrast to conditions with the PDPD game order, where PD's optimal outcome can be obtained with either strategy, during the conditions with the CGPD game order, whether or not the optimal outcome can be obtained during the first game (CG) depends on the strategy used by the confederate agent. The T4T strategy can repeatedly asymmetrically alternate during CG, but the PT4T strategy cannot. These differences in behavior of the confederate agent will lead to different experiences during the first game in conditions with the CGPD game order. During the two-agent condition, the model predicts a transfer effect when PD T4T is played after CG with the PT4T agent (CG PT4T) and when PD PT4T is played after CG with the T4T agent (CG T4T) (t (49) = 36.06, $p < .001$; t (49) = 18.47, $p < .01$). Instead of mutual cooperation occurring at higher proportion as in the PDPD game order conditions, mutual cooperation is predicted to occur at a lower proportion during PD T4T ($M = .10$, $SD = .03$) and PD PT4T ($M = .09$, $SD = .02$) when played after CG PT4T or CG T4T. Here the difference in the model's prior experience during the first game leads to mutual cooperation to be predicted to occur at a lower proportion during the second game in the two-agent condition.

During the one-agent condition, transfer effects are predicted to occur during PD PT4T when played after CG T4T and PD T4T when played after CG PT4T. Mutual cooperation during both PD T4T ($M = .12$, $SD = .03$) and PD PT4T ($M = .13$, $SD = .02$) is predicted to occur at a higher proportion in the one-agent compared to the two-agent condition ($t(49) = -4.72$, p $< .001$; $t(49) < -9.59$, $p < .001$). Again, as seen in conditions with the PDPD order, the trust established during the first game (CG T4T or CG PT4T) in either condition transferring to the second game (PD PT4T or PD T4T), leads to mutual cooperation being predicted to occur at a higher proportion in the one-agent compared to the two-agent condition.

3.2 Predicted Transfer Effects During Iterated Chicken Game

As mentioned previously, whether the optimal outcome can be obtained when playing CG (i.e., asymmetric alternation) depends on the strategy used by the confederate agent. For this reason, the outcome used to assess the transfer effects during CG depended on the strategy the model played with. When playing CG with the T4T strategy transfer effects were assessed using the proportion of alternation and when playing with the PT4T, transfer effects were assessed using the proportion of mutual cooperation.

As seen when examining the transfer effects predicted in PD, in conditions with the CGPD game order during the two-agent condition, playing CG first with either the T4T or PT4T strategy led to mutual cooperation occurring at a lower proportion during PD. Here no transfer effect is predicted to occur with the alternation outcome when CG T4T was played after CG PT4T ($p > .05$). However, a transfer effect is predicted to occur during CG PT4T with the mutual cooperation outcome. The proportion of mutual cooperation is predicted to occur at a lower proportion during CG PT4T ($M = .08$, $SD = .02$), when CG T4T is played compared to when played first ($M = .10$, $SD = .02$) ($t (49) = 2.90$, $p < .05$). In this situation, the learned strategy of alternation during the first game (CG T4T) inhibited learning the mutual cooperation outcome during the second game (CG PT4T).

During the one-agent condition as in the two-agent condition the strategy during the first game used by the confederate agent affected the predicted transfer effects. When CG T4T was played after CG PT4T, alternation is predicted to occur at a lower proportion in the one-agent ($M = .37$, $SD = .07$) than the two-agent ($M = .42$, $SD = .07$) condition ($t (49) = 3.30$, $p < .001$). Here we see the same predicted transfer effect with the alternation outcome when CG PT4T is played before CG T4T, just as was seen during PD T4T when played after CG PT4T. The model predicts that trust will be too low when the first game is CG PT4T to allow for trust repair during the second game. However, the opposite effect is predicted during the one-agent condition with the predicted proportion of the mutual cooperation outcome. When CG PT4T was played after CG T4T, mutual cooperation is predicted to occur at a higher proportion in the one-agent ($M = .10$, $SD = .02$) compared to the two-agent ($t(49) = -5.67$, $p < .001$) condition. Under these conditions the increase in trust during the first game (CG T4T), lead to the mutual cooperation outcome during the second game (CG PT4T) to be learned faster in the one-agent compared to the two-agent condition.

Compared to conditions with the CGCG game order, transfer effects in the PDCG game order should be prominent because PD's optimal outcome can be obtained in either game. In the two-agent condition, when CG T4T is played after PD PT4T the model predicts a transfer effect with the alternation outcome ($t(49) = 12.19$, $p < .001$) (Fig. 2B). Alternation is predicted to occur at a lower proportion in the two-agent condition when CG T4T ($M = .24$, $SD = .10$) is played second compared to when played first ($M = .43$, $SD = .08$). A transfer effect is also predicted when CG PT4T is played after PD T4T the model predicts a transfer effect, mutual cooperation being predicted to occur at a higher proportion when CG PT4T ($M = .25$, $SD = .04$) was played after PD T4T, compared to when CG PT4T was played first ($M = .09$, $SD = .02$; $t(49) = -31.46$, $p < .001$). Again, during the two-agent conditions whether an outcome occurs at a higher or lower proportion during the second game in the two-agent condition depends on model's prior experience during the first game.

In the one-agent condition, the ability to establish trust during the first game in conditions with the PDCG order allows trust to transfer to the second game and leads to predictions of transfer effects during both CG T4T and CG PT4T. During CG T4T when played after PD PT4T, alternation is predicted to occur at a higher frequency in the one-agent ($M = .31$, $SD = .07$) than the two-agent condition ($t(49) = -6.06$, $p < .001$) (Fig. 2B). The same prediction is made with the mutual cooperation outcome when CG PT4T is played after PD T4T, occurring at a higher proportion in the one-agent ($M = .27$, $SD = .04$) than the two-agent ($M = .25$, $SD = .04$) condition ($t(49) = -8.56$, $p < .001$). In each condition, as seen in the other one-agent conditions, the combination of both trust and experience during the first game transferring to the second game leads to mutual cooperation predicted to occur at a higher proportion in the one-agent compared to the two-agent condition.

4   Discussion and Conclusion

In summary, the model predicts specific transfer effects in the one-agent and two-agent conditions depending on the type of knowledge that transfers between games. Across the two-agent conditions, where the model's game-specific knowledge from the first game carries over to the second game, the model predicts that transfer effects will depend on whether the optimal outcome (i.e., mutual cooperation or alternation) learned during the first game can be applied during the second game with a particular confederate agent. If the outcome learned during the first game is applicable during the second game, then it is predicted to occur at a higher proportion during the second compared to the first game; if not, either no transfer effect is predicted or the outcome is predicted to occur at a lower proportion during the second game.

In the one-agent condition, when both game-specific and player-specific knowledge from the first game transfers over to the second game, the optimal outcome during the second game is predicted to be significantly different than during the two–agent condition. This difference due to the additional transfer of trust, a component of player-specific knowledge, leading the model to adjust its behavior in reaction to the confederate agent's change in strategy, by being more or less willing to take the risk of attempting a game's optimal strategy, because of its trust or distrust in the confederate agent.

In conclusion, the model's predictions of transfer effects in both the one-agent and two-agent conditions serve as hypotheses for the upcoming study to be run in the Spring of 2016. The comparison of the model's predictions to the human data will allow us to further test the generality of the model and compare the observed to the predicted dissociation between game-specific and player-specific knowledge as well as offer insights into how people sequentially interact with others.

Acknowledgements.   This research was supported by The Air Force Office of Scientific Research grant number FA9550-14-1-0206 to Ion Juvina. The authors also would like to thank the Oak Ridge Institute for Science and Education who supported this research by appointing Michael Collins to the Student Research Participant Program at the U.S. Air Force Research Laboratory's 711th Human Performance Wing, Cognitive Models and Agents Branch.

References

1. Berg, J., Dickhaut, J., McCabe, K.: Trust, reciprocity, and social history. Games Econ. Behav. **10**(1), 122–142 (1995)
2. Yamagishi, T., Kikuchi, M., Kosugi, M.: Trust, gullibility, and social intelligence. Asian J. Soc. Psychol. **2**(1), 145–161 (1999)
3. Yamagishi, T., Kanazawa, S., Mashima, R., Terai, S.: Separating trust from cooperation in a dynamic relationship prisoner's dilemma with variable dependence. Rationality Soc. **17**(3), 275–308 (2005)
4. Juvina, I., Saleem, M., Martin, J.M., Gonzalez, C., Lebiere, C.: Reciprocal trust mediates deep transfer of learning between games of strategic interaction. Organ. Behav. Hum. Decis. Process. **120**(2), 206–215 (2013)
5. Collins, M.G., Juvina, I., Gluck, K.A.: Cognitive model of trust dynamics predicts human behavior within and between two games of strategic interaction with computerized confederate cgents. Front. Psychol. **7**(49), 361–370 (2016). doi:10.3389/fpsyg.2016.00049
6. Juvina, I., Lebiere, C., Gonzalez, C.: Modeling trust dynamics in strategic interaction. J. Appl. Res. Memory Cogn. **4**(3), 197–211 (2015)
7. Cook, K.S., Yamagishi, T., Cheshire, C., Cooper, R., Matsuda, M., Mashima, R.: Trust building via risk taking: A cross-societal experiment. Soc. Psychol. Q. **68**(2), 121–142 (2005)
8. Axelrod, R.: The evolution of cooperation. Basic Books, New York (1984)
9. Juvina, I., Lebiere, C., Gonzalez, C., Saleem, M.: Intergroup prisoner's dilemma with intragroup power dynamics and individual power drive. In: Proceeding of Social Computing, Behavioral-Cultural Modeling and Prediction, pp. 290–297 (2012)
10. Anderson, J.R.: How can the human mind occur in the physical universe?. Oxford University Press, New York (2007)
11. Gonzalez, C., Lerch, F.J., Lebiere, C.: Instance-based learning in real-time dynamic decision making. Cogn. Sci. **27**(4), 591–635 (2003)

Electricity Demand and Population Dynamics Prediction from Mobile Phone Metadata

Brian Wheatman[2], Alejandro Noriega[1(✉)], and Alex Pentland[1]

[1] Media Laboratories, MIT, Cambridge, MA, USA
noriega@mit.edu
[2] Computer Science and Electrical Engineering, MIT, Cambridge, MA, USA

Abstract. Energy efficiency is a key challenge for building modern sus
tainable societies. World's energy consumption is expected to grow annually by 1.6 %, increasing pressure for utilities and governments to fulfill
demand and raising significant challenges in generation, distribution, and
storage of electricity. In this context, accurate predictions and understanding of population dynamics and their relation to electricity demand
dynamics is of high relevance.

We introduce a simple machine learning (ML) method for day-ahead
predictions of hourly energy consumption, based on population and electricity demand dynamics. We use anonymized mobile phone records
(CDRs) and historical energy records from a small European country.
CDRs are large-scale data that is collected passively and on a regular
basis by mobile phone carriers, including time and location of calls and
text messages, as well as phones' countries of origin. We show that simple support vector machine (SVM) autoregressive models are capable of
baseline energy demand predictions with accuracies below 3 % percentage error and active population predictions below 10 % percentage error.
Moreover, we show that population dynamics from mobile phone records
contain information additional to that of electricity demand records,
which can be exploited to improve prediction performance. Finally, we
illustrate how the joint analysis of population and electricity dynamics elicits insights into the relation between population and electricity
demand segments, allowing for potential demand management interventions and policies beyond reactive supply-side operations.

1 Introduction

In today's developing world, efficient energy procurement is a key challenge for
building sustainable societies. The world's energy consumption is expected to
grow annually by 1.6 %, increasing pressure for utilities and governments to fulfill demand and raising significant challenges in generation, distribution, and
storage of electricity. In this context, accurate predictions of electricity demand
are of salient relevance for supply-side operations by allowing efficient use of the
installed capacity for generation, distribution and storage, as well as efficient
electricity purchasing and trading. Moreover, reliable electricity demand predictions can enable the incorporation of low-carbon technologies into electricity
grids [9].

© Springer International Publishing Switzerland 2016
K.S. Xu et al. (Eds.): SBP-BRiMS 2016, LNCS 9708, pp. 196–205, 2016.
DOI: 10.1007/978-3-319-39931-7_19

Recent research has developed various methodologies for electricity demand prediction. Most prediction methodologies involve the use of diverse datasets, such as regional weather forecasts [15], calendar data, building construction materials [10], and building occupancy rates estimated from WiFi connections [11]. These data sources are often only partially available over large regions and are subject to human and institutional boundaries as well as error in their generation. Moreover, only a few of these – such as research on the use of WiFi data that estimates dynamic building occupancy rates – capture the human dynamics element underlying energy consumption.

Anonymous mobile phone records – or CDRs (Call Detail Records) – have become one of the most salient sources of information that elicit large-scale patterns of human activity. CDRs contain metadata on the social and mobility patterns of users and are generated by mobile network infrastructure on a regular basis. In recent years, researchers have developed applications of CDRs in domains relevant and diverse as crime prediction [3], population modeling for disaster response in earthquakes and floodings [1,7], modeling of epidemic outbreaks [6,14], inferring local socio-economic statistics in both the developed and developing worlds [5,12], and urban transportation systems development [2]. In addition, handsets and airtime are becoming cheaper, leading to ubiquitous mobile phone penetration, which by 2013 approached 90 % in developing countries and 96 % globally [13].

Today, few studies have explored the intersection of electricity demand and population dynamics elicited from large-scale mobile phone records datasets (CDRs). The novel intersection of these two perspectives on our built systems can prove valuable in (1) increasing performance of electricity demand predictive models, useful for efficient supply-side management of generation, distribution and storage and (2) uncovering insights on population to grid dynamic relationships, which can allow for policies that go beyond reacting to demand into shaping it.

In this work, we jointly examine population dynamics extracted from anonymous CDRs and electricity demand dynamics from hourly electric grid records. The datasets encompass all call, texts and data connections and all electricity consumptions for a small European country over an overlapping period of nine months. Section 2 describes the datasets used.

Section 3 builds the baseline purely autoregressive models – those in which feature variables are exclusively previous realizations of the predicted variable – for (1) prediction of the daily amount of active population in the country, segmented by country of origin and (2) prediction of hourly electricity demand, segmented by region within the country[1]. We show that these basic benchmark models are capable of predicting hourly energy demand at percentage errors below 3 % and daily population activity at percentage errors below 10 % (Tables 1 and 2). Section 4 explores the joint information carried by mobile phone and energy records. We show that population dynamics extracted from CDRs significantly

---

[1] There is a large amount of tourism in the country. Over the time analyzed, on roughly one in four people connecting to a cell tower were not from the local country.

correlate with the errors yielded by the energy autoregressive models (Table 3) and that these correlations can be exploited towards energy prediction performance improvements. Finally, Sect. 5 illustrates how the joint analysis of population and electricity dynamics can elicit insights on the relation between population and electricity demand segments (Table 4), potentially allowing for demand management interventions and policies beyond reactive supply-side operations.

2 Datasets Description

2.1 Mobile Phone Records

Mobile phone records (CDRs) consist of metadata about call, short message service (SMS) and data communications, such as location and time of call, but provide no information on the content of the communication. Figure 1 shows an example of the call detail record of a phone call.

CALLER ID	CALLER CELL TOWER LOCATION	RECIPIENT PHONE NUMBER	RECIPIENT CELL TOWER LOCATION	CALL TIME	CALL DURATION
X76VG588RLPQ	2°24' 22.14", 35°49' 56.54"	A81UTC93KK52	3°26' 30.47", 31°12' 18.01"	2013-11-07T15:15:00	01:12:02

Fig. 1. Example of a CDR record

Relevant characteristics of CDRs are (1) the caller and receiver identities are pseudonymized, i.e., names and phone numbers are replaced by anonymous codes and (2) the geographic location of towers used for each communication provide an approximation of users' location. Additionally, CDRs may contain mobile country codes (MCC) and type allocation codes (TAC), which encode the home country of the mobile subscription and the mobile phone model respectively.

The CDR dataset used in this study comprises metadata, including MCC and TAC codes, of all calls, SMS, and data communications of a small European country over a period of nine months in 2015. From raw CDRs, we compute daily active population, defined as the number of users that engaged in at least one communication within the country on a given day. Figure 2 shows population dynamics for locals and visitors over the period studied[2]. Segmentation by country of origin is germane as the tourism industry plays a central role in the country's economy.

2.2 Energy Records

Energy records used in this study are registries of all electric current flowing through the national grid. This and the CDRs dataset are available for an overlapping period of nine months in 2015. Temporal resolution for aggregate current volumes is hourly. Geographic segmentation is naturally defined by four main high voltage lines. Figure 3 shows the daily, weekly and seasonal dynamics of energy demand, segmented by region.

---

[2] Shown are countries with highest amount of visitors out of more than 50 countries.

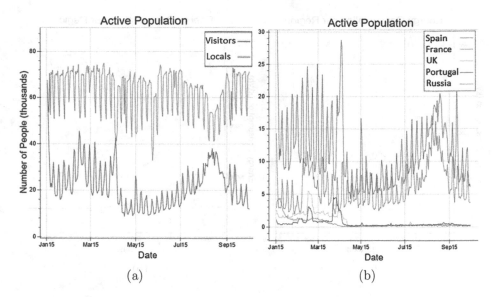

Fig. 2. (a) Active population of locals vs. visitors. (b) Active population of top five countries of visitors.

3   Baseline Autoregressive Models

We build baseline purely autoregressive models – i.e., models in which feature variables are exclusively previous realizations of the predicted variable [4] – for (1) prediction of the daily amount of active population in the country, segmented by country of origin and (2) prediction of hourly electricity demand, segmented by region within the country. All prediction tasks are solved for a horizon (h) 24 h in the future (also called day-ahead predictions).

These models are meant as modern baselines for more sophisticated prediction methodologies in terms of information inputs and statistical learning methods. We use standard statistical and machine learning (ML) methodologies:

- Feature vectors used are composed of autoregressive values of the 14 days prior to the prediction, which allows us to capture daily and weekly patterns shown in Fig. 3.
- Standard ML methodology is used for training, cross-validating, model selection, and testing[3].
- Regression models used are a linear Lasso and a support vector machine (SVM) regression, with a radial basis function (RBF) kernel [8].

We assess predictive performance in terms of percentage errors (PE), defined in Eq. 1 and normalized mean square errors (NMSE) defined in Eq. 2. PEs provide an intuitive interpretation, which is easily translatable to value metrics such

---

[3] We train on 150 days and test on subsequent 30 days. We optimize regularization parameters on a sequence of sequential 180 day blocks and assess prediction on a final set of 30 unseen days.

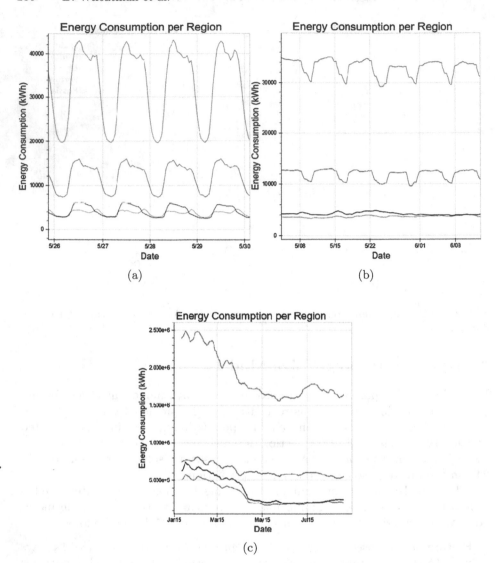

Fig. 3. (a) Daily cycle of energy demand. (b) Weekly cycle of energy demand. (c) Seasonal pattern of energy demand.

as trading cost and energy waste. Complementarily, NMSEs provide a natural benchmark for predictive models, where NMSE < 1 entails that a model performs better than the sample mean of the predicted variable [4]. Finally, we benchmark energy predictions against those yielded by the commercial predictive tool currently used by the national utility company, developed by a strong player on the European market for energy forecasts.

$$PE = \frac{1}{n} \sum_{\forall i} \frac{y_i - \hat{y}_i}{y_i} \tag{1}$$

$$NMSE = \frac{\sum_{\forall i}(y_i - \hat{y}_i)^2}{\sum_{\forall i}(y_i - \mu)^2} \tag{2}$$

3.1 Population Dynamics

We use an autoregressive linear Lasso for population dynamics prediction. This standard model uses a square loss function and an $L1$ norm, performing thus regularization and variable selection, as specified by Eq. 3.

$$\min_{A \in \mathbf{R}^{15}} \left\{ \frac{1}{N} \left(y_i - a_0 - \sum_{t=-15}^{-1} a_t y_{it} \right)^2 + \lambda \|A\|_1 \right\} \tag{3}$$

Table 1 shows that the baseline autoregressive Lasso can predict active population at percentage errors near 5 % for Total and Locals population segments. It also shows that the smaller the group, the harder it is to predict with pure autoregressive models, yielding higher PEs for the Spanish and French population predictions. All three population segments had NMSE < 1 where, for example, the Locals predictor yielded errors of only 34 % of errors yielded by the basic benchmark of using y's sample mean as a constant predictor.

Table 1. Baseline autoregression predictions of active population per country of origin

Population	Percentage error	Normalized mean square error
Total	3.61 %	0.270
Locals	5.85 %	0.336
Spain	11.81 %	0.687
France	23.89 %	0.653

3.2 Energy Demand

We predict hourly electricity demand for each region using an analogous autoregressive approach. Here we implement a linear Lasso and a SVM with RBF kernel as regression models.

Table 2 shows that baseline autoregressive models are able to predict electricity demand at percentage errors of 2.3 % for total demand and around 3 % for regional demands. All models' performances were one or two orders of magnitude better than benchmark performance of y's sample mean as predictor (NMSE $\ll$ 1). Moreover, the baseline autoregressive models yielded comparable results to the commercial tool currently used by the country's national utility and performed substantially better for Full Country and Region 3 predictions.

Table 2. Baseline autoregression predictions of energy demand per region

Energy	Commercial tool PE	Lasso PE	Lasso NMSE	SVM PE	SVM NMSE
Full Country	2.80 %	2.29 %	0.0141	2.32 %	0.0117
Region 1	2.69 %	3.39 %	0.0302	3.06 %	0.0333
Region 2	2.68 %	2.80 %	0.0192	3.17 %	0.0178
Region 3	6.41 %	3.48 %	0.0774	3.45 %	0.0755
Region 4	No Data	3.26 %	0.0205	3.26 %	0.0177

4   Energy Demand Predictions Using Population Dynamics

In this section, we explore the joint information carried by mobile phone and energy records.

We evaluate the correlation of the segmented population predictions of Sect. 3.1 with the errors yielded by the energy prediction autoregressive SVM model in Sect. 3.2. We show that population dynamics extracted from CDRs significantly correlate with the errors yielded by the energy autoregressive models and that these correlations can be exploited towards energy prediction performance improvements.

Table 3 shows that three out of four electricity demand regions present significant correlations between the errors of their autoregressive models and segmented population predictions, based on mobile phone data.

The additional information contained in mobile phone data can be leveraged towards improved prediction performance. To illustrate this application, we implemented a standard SVM regression model where the predicted variable is the hourly energy prediction error from Sect. 3.2's autoregression models and features are population predictions from Sect. 3.1's autoregression models. This simple sequential approach was able to reduce percentage errors of Sect. 3.2's baseline models in 4.2 % and 20.5 % for Region 1 and Region 4 respectively. No improvement was achieved for Region 2 and Region 3.

5   Relating Segments of Population and Energy Demand

Lastly, we illustrate in this section how the joint analysis of population and electricity dynamics can elicit insights on the relation between population and electricity demand segments. The motivation for this analysis is that such relations can, in addition to improving predictive power, potentially allow for demand management interventions and policies beyond reactive supply-side operations.

We solve simple linear regressions of electricity demand for each region against segmented population activity, as shown by Eq. 4, where explanatory variables x_{ij} denote active population of segment j for day i. Coefficients of these regressions elicit relationships between energy and population segments.

Table 3. Correlations between electricity demand prediction errors and segmented population predictions

Region 1	Correlation coefficient	P-Value
Locals	0.070	0.308
Spain	−0.213	0.0015***
France	−0.170	0.0118**
UK	−0.0516	0.447
Region 2	Correlation coefficient	P-Value
Locals	0.105	0.123
Spain	0.060	0.378
France	−0.0081	0.904
UK	−0.010	0.881
Region 3	Correlation coefficient	P-Value
Locals	−0.086	0.206
Spain	0.169	0.018**
France	0.083	0.223
UK	0.016	0.816
Region 4	Correlation coefficient	P-Value
Locals	−0.139	0.0385**
Spain	0.116	0.0862*
France	0.154	0.0227**
UK	−0.0288	0.671

* 90 %, ** 95 %, and *** 99 % confidence levels.

These coefficients capture the effect that a singular increase in the active population of a country has on the energy consumption of a region, also known as the unique or *ceteris paribus* effect.

$$\min_{\beta} \sum_{\forall i} \epsilon_i \qquad \text{S.T.} \ \ y_i = \beta_0 + \sum_{\forall j} \beta_j x_{ij} + \epsilon_i \tag{4}$$

Table 4 shows how different energy regions are affected by population segments. For example, we see how a unit increase in UK visitors tends to have a larger impact on Region 2's electricity demand than a unit increase of Spanish visitors does, and that the region most affected by an increase in visitors of any nationality is Region 2, the capital. Future research paths may explore the network of relationships among more specific energy and population segments.

6   Concluding Remarks

We proposed standard Lasso and SVM autoregressive models as basic benchmarks for segmented predictions of large-scale population and electricity demand

Table 4. Population segments on electricity segments regression coefficients

Coefficients	Locals	Spain	France	UK
Full Country	0.385**	0.567**	−0.644	6.98**
Region 1	0.0710**	0.0644*	0.0146	0.436*
Region 2	0.197**	0.240**	−0.244	3.04**
Region 3	0.0495**	0.118**	−0.164*	1.54**
Region 4	0.0682**	0.145**	−0.2509*	1.96**

* 95 % confidence level
** 99 % confidence level

dynamics. We showed that these baseline models can yield accuracies below 3 % and 10 % percentage errors for electricity demand and active population predictions respectively, and that segmented predictions are possible at similar accuracy levels. Moreover, we showed that population dynamics extracted from CDRs contain information additional to that of electricity dynamics, and that this information can be leveraged toward improved accuracy of electricity demand predictions. Finally, we illustrate how the joint analysis of these datasets can elicit the underlying network of relations among population and electricity segments.

Future work may develop more sophisticated models that incorporate richer information such as weather, calendar and building construction data and implement more sophisticated feature engineering, feature selection, and regression models. Pure autoregressive models proposed here are meant as modern benchmarks for more sophisticated models and commercial tools.

A promising research path forward consists of using finer segmentation in both predictive and explanatory models. Possible paths include disaggregated segmentation of the electric grid per town, industry or building complex and additional population segmentations such as social and spatial behavior characterizations and disposable income proxies inferred from users' mobile phone models. Finally, as pointed out in Sect. 5, further eliciting the underlying relations among energy and population segments can, in addition to improving predictive power, allow for demand management interventions and policies beyond reactive supply-side operations.

References

1. Bengtsson, L., Lu, X., Thorson, A., Garfield, R., Von Schreeb, J.: Improved response to disasters and outbreaks by tracking population movements with mobile phone network data: a post-earthquake geospatial study in Haiti. PLoS Med **8**(8), e1001083 (2011)
2. Berlingerio, M., Calabrese, F., Di Lorenzo, G., Nair, R., Pinelli, F., Sbodio, M.L.: AllAboard: a system for exploring urban mobility and optimizing public transport using cellphone data. In: Blockeel, H., Kersting, K., Nijssen, S., Železný, F. (eds.) ECML PKDD 2013, Part III. LNCS, vol. 8190, pp. 663–666. Springer, Heidelberg (2013)

3. Bogomolov, A., Lepri, B., Staiano, J., Oliver, N., Pianesi, F., Pentland, A.: Once upon a crime: towards crime prediction from demographics and mobile data. In: Proceedings of the 16th International Conference on Multimodal Interaction, pp. 427–434. ACM (2014)
4. Bontempi, G.: Machine Learning Strategies for Time Series Prediction, Hammamet (2013)
5. Eagle, N., Macy, M., Claxton, R.: Network diversity and economic development. Science **328**(5981), 1029–1031 (2010)
6. Frias-Martinez, E., Williamson, G., Frias-Martinez, V.: An agent-based model of epidemic spread using human mobility and social network information. In: 3rd International Conference on Social Computing, SocialCom (2011)
7. Global Pulse, U.N.: Using mobile phone activity for disaster management during floods. Technical report, UN Global Pulse (2013)
8. Hastie, T., Tibshirani, R., Friedman, J.: The Elements of Statistical Learning. Springer, New York (2013)
9. Kalogirou, S.A.: Artificial neural networks in renewable energy systems applications: a review. Renew. Sustain. Energy Rev. **5**(4), 373–401 (2001)
10. Kolter, Z., Ferreira, J.: A large-scale study on predicting and contextualizing building energy usage. In: Proceedings of the Conference on Artificial Intelligence (AAAI) (2011)
11. Martani, C., Lee, D., Robinson, P., Britter, R., Ratti, C.: Enernet: Studying the dynamic relationship between building occupancy and energy consumption. Energ. Build. **47**, 584–591 (2012)
12. Smith, C., Mashhadi, A., Capra, L.: Ubiquitous sensing for mapping poverty in developing countries. In: Proceedings of the Third Conference on the Analysis of Mobile Phone Datasets (2013)
13. United Nations, I.T.U.: Facts and figures. Technical report, UN International Communications Union (2013)
14. Wesolowski, A., Eagle, N., Tatem, A.J., Smith, D.L., Noor, A.M., Snow, R.W., Buckee, C.O.: Quantifying the impact of human mobility on malaria. Science **338**(6104), 267–270 (2012)
15. Zhao, Hx, Magoulès, F.: A review on the prediction of building energy consumption. Renew. Sustain. Energy Rev. **16**(6), 3586–3592 (2012)

Detecting Communities by Sentiment Analysis of Controversial Topics

Kangwon Seo[1], Rong Pan[1(✉)], and Aleksey Panasyuk[2]

[1] School of Computing, Informatics, and Decision Systems Engineering,
Arizona State University, Tempe, AZ, USA
{kangwon.seo,rong.pan}@asu.edu
[2] Air Force Research Lab, Rome, NY, USA
aleksey.panasyuk@us.af.mil

Abstract. Controversial topics, particularly political topics, often provoke very different emotions among different communities. By detecting and analyzing communities formed around these controversial topics we can paint a picture of how polarized a country is and how these communities evolved over time. In this research, we made use of Internet data from Twitter, one of the most popular online social media sites, to identify a controversial topic of interest and the emotions expressed towards the topic. Communities were formed based on Twitter users' sentiments towards the topic. In addition, the network structure of these communities was utilized to reveal those Twitter users that played important roles in their respective communities.

Keywords: Topic modeling · Sentiment analysis · Twitter · Social network analysis

1 Introduction

Using social media to detect and monitor emerging phenomena has attracted more and more attention from social scientists, political scientists, military strategists, etc. With the ever-growing Internet connectivity, including mobile connectivity, it is possible to collect millions of expressions from social media users on a current event in an almost uninterruptable fashion. Twitter, for example, is an ideal source for collecting information of public opinions and public sentiments. With less than 140 characters, each tweet is more likely to directly express a Twitter user's feeling towards a particular event. Some recent research efforts have been carried out on statistical modeling of Twitter messages (e.g., [3–5,13]); however, there is less research of using sentiment analysis for online community detection and monitoring.

In this research we demonstrated a practical approach to social network analysis that aimed at identifying distinct online communities and prominent nodes in these communities. We analyzed Twitter data during the period of the Egyptian Revolution of 2011 to identify communities around a controversial

© Springer International Publishing Switzerland 2016
K.S. Xu et al. (Eds.): SBP-BRiMS 2016, LNCS 9708, pp. 206–215, 2016.
DOI: 10.1007/978-3-319-39931-7_20

topic of interest. On January 25 of 2011 there was a large scale of demonstration in Egypt whose aim was of overthrowing the former Egyptian President, Hosni Mubarak. The Twitter messages posted during this time period showed people's expressions and feelings without any censorship. Our data analysis was conducted with the following procedure: First, we extracted several popular topics in this Twitter dataset, and then, the most interesting topic was chosen by historical or political knowledge. Second, sentiment analysis was applied on the tweets that were closely related to the selected topic and communities were constructed using the sentiment score. Lastly, we examined the network structure of these communities to deduce the characteristics of each community. These characteristics can reflect the polarization on the topic within a society and may shed light on some early signs of potential physical conflicts.

1.1   Related Work

Traditional research on community detection in a network was based on the graph theory (e.g., [8,16]). [25] defined finding community in a social network as "to identify a set of nodes such that they interact with each other more frequently than with those nodes outside the group". These types of research often involved solving optimization problems to discover subgraphs or clusters that represent regions with higher density of edges than other portions of the graph. [18] provided a survey on the performance characteristics of several community detection methods for social media. Some other recent work on this aspect included [6,20]. Meanwhile, some researchers considered not only network structure but also social media contents. The analysis of these contents may provide more insights about the polarized opinions expressed on specific topics in the online society. [19] analyzed both opinions and social interactions to find similar people and see how this similarity relates to their social interactions. They built graphs based on the available opinions and social data and then used the Infomap community detection algorithm [23] to identify community structures. [5] described methods for predicting the political alignment of Twitter users based on the content and structure of their political communication in the run-up to the 2010 U.S. midterm elections. They used manually annotated data set and trained it with several classification features such as TF-IDF and hashtags obtained by content analysis, and cluster structures obtained by network analysis.

The social media data related to the Arab's revolutions had been of great interest, as it played an important role in shaping the events in the region and had been studied by several researchers. [3], for example, analyzed content and network of Twitter to detect users who switched their polarity during the turmoil in Egypt in 2013. They used a supervised method to classify the tweets into two groups, pro-military intervention and anti-military intervention, then examined if a user polarity has been swapped during some time period. [4] examined reply and retweet networks of tweets related to the revolution in Egypt and the civil war in Libya in order to identify interactions between the English and Arabic language groups. [13] analyzed information flows across different types of users (activists, bloggers, journalists, mainstream media outlets, and other engaged participants). They described the symbiotic relationship between media outlets and individuals and the distinct roles between different user types.

Theories in the social-psychological science indicate that sentiments and emotions of an individual often directly influence the individual's behavior [1]. Political scientists have postulated hypotheses on how the emotion of a political group may escalate or deescalate conflicts with opponents and how a political actor may manipulate the general populace sentiment towards an event or an institution to their own advantage [17]. From this research it is seen that the sentiments expressed in tweets could be an important factor in predicting possible trends of a political event. In addition, by analyzing opposing sentiments along with different ways of expression of emotions towards a political topic, we can group tweeter users to different communities. The characteristics of these communities, such as its size, emotion intensity, etc., will provide a quantitative measure of the political environment in a country.

1.2   Overview of Dataset

The Twitter dataset used in this research comes from a collection of tweets about Egypt during 19 days, from February 1 to 19, 2011, provided by Army Research Lab (ARL). It consists of over 950,000 tweets. Each data entry has 14 fields, which includes identification number, language code, text message, user name of who posted the tweet("from user"), user name of whom the tweet replies to("to user"), date and time the tweet is posted("created at") and some other metadata of each tweet. We extracted only tweets in English, which consist of 619,635 tweets. The type of a Twitter message can be categorized by Original tweet, Retweet and Reply. Retweet messages include a string pattern "RT @username" in their texts and Reply messages start from "@username" and they also have a user name of whom the tweet replies to in the "to user" fields. Our dataset consists of 75 % of Original tweets, 20 % of Replies and 5 % of Retweets, approximately. The total number of users who posted at least one tweet is 144,648. Meanwhile, the total number of users who received at least one reply from the other user is 36,149. We found that majority of users posted only one or two tweets. The number of tweets for each day varies between 20,000 to 40,000, and peaked around February 2, corresponds to Mubarak's refusal to resign and on February 11, corresponds to Mubarak's resignation.

2   Methodology

A flowchart of our data analysis process is shown in Fig. 1. It has two phases: In Phase I, a controversial topic of interest is identified and irrelevant tweets to this topic are filtered out; in Phase II, we perform sentiment analysis and identify communities.

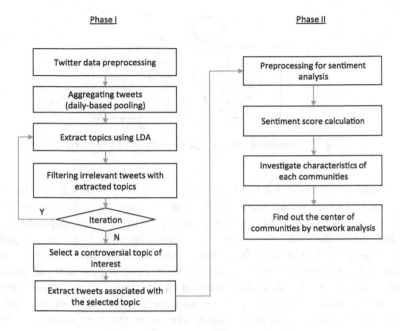

Fig. 1. Summary of data analysis

2.1   Phase-I: Topic Extraction

The first step of making sense of online community formation is to understand what people are talking about. This requires topic modeling, a special area of text mining. With a compiled document of online text messages, called corpus, we are interested in finding multiple sets of keywords (topics) in the corpus that are closely related to each other and the association of each message to any topic. Latent Dirichlet Allocation (LDA) is a flexible probabilistic generative model for this topic-modeling problem [2]. It treats a document (or a message) as a random mixture of latent topics and these latent topics have word distributions. The parameter estimation of this model is typically carried out by the variational Expectation Maximization (EM) method. After extracting topics from all tweets, we are able to filter out irrelevant tweets based on the strength of association with the topic of interest and conduct further analysis on the remaining tweets.

Figure 2 shows the graphical model representation of LDA, where M is the number of documents in a corpus and N is the number of words in a document. The boxes represent replicates. The generative model of LDA consist of the following steps: (1) The term (word) distribution β for each topic is determined; (2) The topic proportion θ for a document is chosen from a Dirichlet distribution, $Dirichlet(\alpha)$; (3) Choose a topic Z from $Multinomial(\theta)$ for each word in a document; (4) Choose a word W from a multinomial probability distribution conditioned on the topic Z. The LDA result shows which words are most likely to appear for a topic and the proportions of topics for each document.

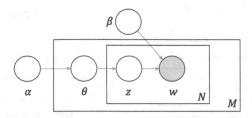

Fig. 2. Graphical model representation of LDA [2]

One challenge of applying LDA on twitter messages comes from the reality that a single twitter message is too short to be used as a document [27]. In order to overcome this difficulty, a few suggestions have been proposed that we categorize as two distinct approaches. The first approach is to use aggregated Twitter messages without modifying the standard LDA model. For examples, [10] obtained a higher quality of learned model by training a topic model on aggregated messages; [15] described various tweet aggregating schemes such as author-based pooling, temporal-pooling and hashtag-based pooling, and they suggested the hashtag-based aggregating scheme as the best one evaluated by the ability of topics to reconstruct clusters and topic coherence. The second approach is to modify the standard LDA model so that it can be adapted to short documents. [27] assumed a single tweet is associated with a single topic and proposed the Twitter-LDA model. The labeled LDA proposed by [21] is another variation of LDA for microblog data. In this research, we used the first approach to aggregate messages and analyzed the data using an LDA implementation available in R package: "topicmodels" [9]. Specifically, we randomly selected 50 % of all tweets to constitute the training dataset, and then aggregated them by the date that they were posted (daily-based pooling) so that aggregated documents have relatively similar lengths compared to cases of author-based or hashtag-based aggregating schemes.

2.2 Phase-II: Sentiment-Based Community Detection

A basic task in sentiment analysis is to quantify the polarity in a given text - whether the expressed opinion in the text is positive, negative, or neutral [12]. A commonly used method is to compare the text with known lexicons of positive or negative words. We used Bing Liu's sentiment lexicon [11] as a dictionary of positive and negative words, which includes around 2,000 positive words and 4,800 negative words. This lexicon is useful for social media text because it includes mis-spelling, morphological variants, slang and social media mark-up [26].

With the selected topics from twitter corpus, we assess the twitter user's attitude to these topics based on the words used and the way of expression. Subsequently, a similarity measurement based on the sentiment score between two tweets can be obtained, which represent the likeness of the minds of two Twitter users in terms of their attitudes to a common topic of interest. We will partition these users to different communities based on similarity measures. It needs to be mentioned that our proposed community detection task is different from the other graph based community detection problems since it is utilizing sentiment analysis unlike other approaches introduced in Sect. 1.1. We also exploit tweet-retweet and tweet-reply networks to identify the users who play a role of opinion leader in the community.

3   Data Analysis and Results

It is reasonable to assume that the topics of reply-tweets are closely related to those of original tweets they are replying to and hence reply-tweets should be considered with their original tweets. In our dataset, about 20 % of tweets are reply-tweets and unfortunately the original tweet of a reply-tweet is unknown in the dataset we have been working with. Instead, they only provide the user names who posted the original tweets through "to user" attribute(or @username in text). For these reasons we assume that a reply-tweet has been written for relatively recent post of an original tweet, so we add at-most 5 recent tweets, if any exist, posted by @username before the posting of a reply-tweet. In our dataset approximately 30 % of reply-tweets have at least one original tweet posted by @username before the reply was posted and these original messages are added to the reply-tweet. Meanwhile, URL, @username, punctuations and English stop-words are removed from all tweet messages. We also removed words with less than 3 letters and words that appeared less than 10 times in the whole dataset.

We chose to fit 30 topics based on perplexity [2] from 10-fold cross validation. Table 1 shows the most heavily weighted words from some selected topics. It is observed that most topics are dominated by a few similar terms such as "egypt", "tahrir", "jan25" and "cairo"; while some terms are relatively unique, such as "cbs", "logan", "lara" in topic 15. We also found that the document of each day involves only one or two topics, and hence we could determine the topical subjects of each day by looking at the highly ranked words in corresponding topics.

Table 1. Extracted topics(part) by LDA

1	5	8	12	15	19	23	28	30
egypt	egypt	egypt	egypt	egypt	egypt	cairo	egypt	egypt
cairo	cairo	cairo	cairo	cbs	libya	egypt	cairo	cairo
bahrain	tahrir	jan25	tahrir	logan	jan25	jan25	tahrir	mubarak
revolution	can	google	will	lara	tahrir	revolution	jan25	tahrir
tahrir	video	protests	jan25	cairo	news	like	square	people
people	will	tahrir	like	news	square	news	news	square

The next step is to filter Twitter messages which may not be related to any topic extracted from LDA. For this we fit the individual tweets to LDA model once again given the term distributions of β for each topics provided by the previously fitted model, and obtain the topic proportions θ for each tweet. We calculate the following measure to determine relevance to the topics from LDA of each tweet.

$$relevance(T) = \frac{\sum_{i=1}^{n} \log p(w_i)}{n} \tag{1}$$

where $T = \{w_1, w_2, \ldots, w_n\}$ is a tweet with n words, $p(w_i)$ is the likelihood of the word w_i. If $relevance(T)$ of a tweet is too small, then it probably be a message that can be ignored. We removed tweets which have $relevance$ smaller than the 5 % quantile of the normal distribution fitted to $relevance$ values.

With the reduced dataset, topic modeling can be repeated. This time, a smaller number of topics can be given to LDA in order to zoom in the topics which have more exposure to the Twittersphere. We repeated the whole process of topic modeling on the remaining Twitter data with 20 topics. We then removed more irrelevant tweets using $relevance$. The topics obtained in the second iteration are similar to the topics in the first iteration. Therefore, this iterative modeling/filtering process is stopped.

Now we can choose a controversial topic that we are interested in. During this time period in Egypt, the most important historical event was the resignation of Mubarak. Among the 20 topics, Topic 7 and Topic 15 were thought to be highly related to the resignation of Mubarak since those two topics allocate relatively high probability on the term "resignation", comparing to other topics. Then, we extracted the tweets that had the largest weights on Topic 7 or Topic 15. In addition, based on the fact that the resignation of Mubarak was announced at 4 pm on Feb. 11, we removed all tweets posted before this time point, resulting in 27,494 tweets of interest.

The sentiment scores of each tweet were calculated by a simple voting between the number of positive words and of negative words in the tweet [14]. Treating the sentiment scores of $\{-1, 0, 1\}$ as neutral, we used the tweets with scores higher than 1 or lower than -1 to construct two opposite communities. We also investigated some categorized emotional items in each community using the R package "sentiment". It classifies the emotion of a set of texts using a naive Bayes classifier trained on Strapparava and Valitutti's emotions lexicon [24]. It shows the magnitudes of 6 emotional items - anger, surprise, sadness, joy, fear and disgust. Table 2 summarizes the results of sentiment and emotion analysis. From the sentiment analysis result, it shows that the size of positive community (for Mubarak's resignation) is much larger than the size of negative community (against Mubarak's resignation). In addition, the average sentiment scores of positive and negative communities show that the positive community has stronger sentiment intensity. From the emotion analysis, the positive community has a much higher score for joy and the negative community has a relatively higher score for anger than the opposite community.

The network structure of tweets provides useful information of how Twitter users communicate with each other. Using the R package "igraph" we construct

Table 2. Summary of sentiment & emotion analysis

	Sentiment			Emotion					
	users	tweets	score	anger	disgust	fear	joy	sadness	surprise
Pos	2,978	3,831	2.36	1.74	3.12	2.11	6.82	2.15	3.46
Neg	525	635	−1.98	2.27	3.32	2.47	2.40	2.64	3.04

a network from all pairs of users in both communities who were linked by tweet-retweet or tweet-reply. We ignore independent pairs of users which do not have a link with other part of the network to focus on a dense region. Figure 3 shows the graph, in which the network is dominated by users of the positive community and we observe some sub-networks of radial shape which have one user playing the role of center. It shows that the user "ghonim" is in the center of the network and has the highest degrees of connection with 35, indicates that "ghonim" is likely to be the opinion leader in this network. In fact we found that it was the username of Wael Ghonim, who had led anti-government protesters during Egypt revolution. The username of the U.S. president, "BarackObama", ranks the second highest degrees with 16, and it is no surprise based on a fact that he expressed welcome at Mubarak's decision to step down. Lastly "samiyusuf", which has 12°, is the username of Sami Yusuf, who is a British Muslim singer-songwriter and released a song prompted by Egypt revolution.

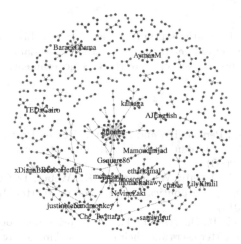

Fig. 3. Tweet-retweet & tweet-reply network visualization: *Blue dots* represent users in the positive community, *red dots* are users in the negative community, and *gray dots* are neutral. User names (*vertex label*) of Twitter users who have more than 5° of connection are presented. (Color figure online)

4   Conclusion

Although Twitter users do not represent the whole range of people in a country, it is still useful for understanding the popular topics of the day and how people express their feelings online. This knowledge is valuable for predicting some future events such as violent civil conflicts. In this research we proposed a two-phased approach that integrated statistical topic modeling, sentiment analysis and network structure analysis for social network. We analyzed the Twitter data in the period of Egypt revolution in 2011. We are able to group Twitter users to different communities based on their sentiment expressions on a topic of interest. It is found that for this historical event, the resignation of Mubarak, online society is dominated by sentiments supporting the resignation. In addition, we provide the network structures of these communities and identify the center of information flow.

Some limitations of our study could be addressed in future research. First, in this study we used a simple unsupervised way to evaluate the sentiment of a message by the pre-specified lexicon. In recent years, however, more advanced supervised learning methods such as SVM or maximum entropy are actively used in the task (e.g., the SemEval shared task on Sentiment Analysis in Twitter [22]). Secondly, other languages (e.g., Arabic) could be included in the analysis by translating them into English. Finally, more elaborate sentiment analysis requires several natural language processing techniques; e.g., part of speech. By utilizing these techniques, we expect to achieve more reliable results of sentiment orientation.

Acknowledgments. The authors thank Sue E. Kase and Liz Bowman at the Army Research Lab for their help in getting access to the Egypt data. The views expressed in this paper do not represent the views of the U.S. government.

References

1. Ajzen, I.: Attitudes, Personality, and Behavior. Dorsey Press, Chicago (1988)
2. Blei, D.M., Ng, A.Y., Jordan, M.I.: Latent dirichlet allocation. J. Mach. Learn. Res. **3**, 993–1022 (2003)
3. Borge-Holthoefer, J., Magdy, W., Darwish, K., Weber, I.: Content and network dynamics behind egyptian political polarization on twitter. In: CSCW 2015, pp. 700–711. ACM (2015)
4. Bruns, A., Highfield, T., Burgess, J.: The arab spring and social media audiences english and arabic twitter users and their networks. Am. Behav. Sci. **57**(7), 871–898 (2013)
5. Conover, M.D., Gonalves, B., Ratkiewicz, J., Flammini, A., Menczer, F.: Predicting the political alignment of twitter users. In: PASSAT/SocialCom 2011, pp. 192–199. IEEE (2011)
6. Du, N., Wu, B., Pei, X., Wang, B., Xu, L.: Community detection in large-scale social networks. In: WebKDD/SNA-KDD 2007, pp. 16–25. ACM (2007)
7. Feinerer, I., Meyer, D., Hornik, K.: Text mining infrastructure in R. J. Stat. Softw. **25**(5), 1–54 (2008)

8. Fortunato, S.: Community detection in graphs. Phys. Rep. **485**, 75 (2010)
9. Grün, B., Hornik, K.: topicmodels: An R package for fitting topic models. J. Stat. Softw. **40**(13), 1–30 (2011)
10. Hong, L., Davison, B.D.: Empirical study of topic modeling in twitter. In: SOMA 2010, pp. 80–88. ACM (2010)
11. Hu, M., Liu, B.: Mining and summarizing customer reviews. In: KDD 2004, pp. 168–177. ACM (2004)
12. Liu, B.: Sentiment Analysis and Opinion Mining. Morgan & Claypool Publishers, San Rafael (2012)
13. Lotan, G., Graeff, E., Ananny, M., Gaffney, D., Pearce, I.: The arab spring—the revolutions were tweeted: information flows during the 2011 tunisian and egyptian revolutions. Int. J. Commun. **5**, 31 (2011)
14. Marwick, B.: Discovery of emergent issues and controversies in anthropology using text mining, topic modeling, and social network analysis of microblog content. In: Data Mining Applications with R, p. 514. Academic Press, New York (2014)
15. Mehrotra, R., Sanner, S., Buntine, W., Xie, L.: Improving LDA topic models for microblogs via tweet pooling and automatic labeling. In: SIGIR 2013, pp. 889–892. ACM (2013)
16. Newman, M.E.J., Girvan, M.: Finding and evaluating community structure in networks. Phys. Rev. **E69**(2), 026113 (2004)
17. O'Brien, S., Shellman, S.: Effects of Emotions on Dissident and Government Behavior. Strategic Analysis Enterprise (SAE) Inc., Williamsburg, VA (2013)
18. Papadopoulos, S., Kompatsiaris, Y., Vakali, A., Spyridonos, P.: Community detection in social media. Data Min. Knowl. Disc. **24**(3), 515–554 (2012)
19. Parau, P., Stef, A., Lemnaru, C., Dinsoreanu, M., Potolea, R.: Using community detection for sentiment analysis. In: ICCP 2013, pp. 51–54. IEEE (2013)
20. Qi, G.J., Aggarwal, C.C., Huang, T.: Community detection with edge content in social media networks. In: ICDE 2012, pp. 534–545. IEEE (2012)
21. Ramage, D., Dumais, S.T., Liebling, D.J.: Characterizing microblogs with topic models. In: ICWSM 2010, p. 1 (2010)
22. Rosenthal, S., Nakov, P., Kiritchenko, S., Mohammad, S.M., Ritter, A., Stoyanov, V.: SemEval-2015 Task 10: Sentiment analysis in twitter. In: SemEval 2015, pp. 451–463 (2015)
23. Rosvall, M., Bergstrom, C.T.: Maps of random walks on complex networks reveal community structure. PNAS **105**(4), 1118–1123 (2008)
24. Strapparava, C., Valitutti, A.: WordNet affect: an affective extension of wordnet. In: LREC, vol. 4, pp. 1083–1086 (2004)
25. Tang, L., Liu, H.: Community detection and mining in social media. Synth. Lect. Data Min. Knowl. Discovery **2**(1), 1–137 (2010)
26. Xie, R., Li, C.: Lexicon construction: A topic model approach. In: ICSAI 2012, pp. 2299–2303. IEEE (2012)
27. Zhao, W.X., Jiang, J., Weng, J., He, J., Lim, E.-P., Yan, H., Li, X.: Comparing twitter and traditional media using topic models. In: Clough, P., Foley, C., Gurrin, C., Jones, G.J.F., Kraaij, W., Lee, H., Mudoch, V. (eds.) ECIR 2011. LNCS, vol. 6611, pp. 338–349. Springer, Heidelberg (2011)

Event Detection from Blogs Using Large Scale Analysis of Metaphorical Usage

Brian J. Goode[1,2(✉)], Juan Ignacio Reyes M.[3], Daniela R. Pardo-Yepez[3],
Gabriel L. Canale[3], Richard M. Tong[4], David Mares[3], Michael Roan[2],
and Naren Ramakrishnan[1]

[1] Discovery Analytics Center (Department of Computer Science),
Virginia Tech, 900 N. Glebe Rd., Arlington, VA, USA
{bjgoode,naren}@vt.edu
[2] Department of Mechanical Engineering, Virginia Tech, 445 Goodwin Hall,
Blacksburg, VA, USA
mroan@vt.edu
[3] Center for Iberian and Latin American Studies (CILAS), UC San Diego,
9500 Gilman Dr., La Jolla, CA, USA
{jireyes,dpardoye,gcanale,dmares}@ucsd.edu
[4] Tarragon Consulting Corporation, 1563 Solano Avenue, #350, Berkeley, CA, USA
rtong@tgncorp.com
http://dac.cs.vt.edu
http://www.me.vt.edu
https://cilas.ucsd.edu
http://www.tgncorp.com

Abstract. Metaphors shape the way people think, decide, and act. We hypothesize that large-scale variations in metaphor usage in blogs can be used as an indicator of societal events. To this end, we use metaphor analysis on a massive scale to study blogs in Latin America over a period ranging from 2000–2015, with most of our data occurring within a nine-year period. Using co-clustering, we form groups of similar behaving metaphors for Argentina, Ecuador, Mexico, and Venezuela and characterize overrepresented as well as underrepresented metaphors for specific locations. We then focus on the metaphor's potential relation to events by studying the tobacco tax increase in Mexico from 2009–2011. We study correspondences between changes in metaphor frequency with event occurrences, as well as the effect of temporal scaling of data windows on the frequency relationship between metaphors and events.

Keywords: Metaphors · Blogs · Open source indicators · Event detection

1 Introduction and Related Work

Conceptual metaphors associate an abstract target concept with a concrete source concept. So, we say things like "life is a journey" or "poverty is a disease". Individual mappings of source-target concept pairs in language are linguistic metaphors. Such constructions facilitate knowledge dissemination, bridge

© Springer International Publishing Switzerland 2016
K.S. Xu et al. (Eds.): SBP-BRiMS 2016, LNCS 9708, pp. 216–225, 2016.
DOI: 10.1007/978-3-319-39931-7_21

concepts over a variety of subject areas, and unwittingly affect human behavior through intrinsic associations and world-views [4,7,10]. There have been a number of attempts to quantify these effects in an experimental framework [14–16]. Despite the confirming evidence, the effects of metaphors in everyday interactions and mental constructs remains unclear [5]. Metaphor exposure and adoption alone is not necessarily responsible for action, because anchoring and time also dictate decision outcomes [14]. The impact of metaphor on behavior may largely depend on its age or relative conventionality (i.e., becomes a definition) in society [6,8]. Applied to complex social interactions and events in everyday scenarios, capturing changes in metaphor usage can become an untenable undertaking. At present, we adopt the assumption that conceptual mappings governing action within a population are present to some degree in language, and can be used for event detection.

Prior effort has focused on the use of Open Source Indicators (OSI) as a tool for improving the forecasting of events [13]. In such analyses, models with keywords relevant to events such as protests and disease epidemics steer algorithms to predict the probability of occurrence [3,11]. However, we are not blind to the operational modeling challenges that exist when incorporating metaphors [2]. The view taken in this paper is that we can re-purpose the idea of the keyword dictionary to extract relevant text that now consists of a tuple, $M_L := (source, target)$. This tuple is one of many linguistic metaphors mapped to a given conceptual metaphor. Extracting these linguistic metaphors in documents is a subject of continued research (e.g., [1,12]).

This is the first study to present a method of large scale analysis of metaphorical usage in blogs. Our data derives from 327 Latin American blogs spanning the years 2000–2015. We extracted 589,089 documents from Argentina, Ecuador, Mexico, and Venezuela. Blogs from these countries were chosen for their rich use of metaphorical language and discussion of significant political events. We performed spectral co-clustering over the time-series correlation matrix to identify groups of metaphors that appeared in the text in close temporal proximity and linear proportionality. Through the course of the analysis, we investigate:

1. What do concepts mapped to linguistic metaphors reveal about differences in blogger populations?
2. Do prevailing conceptual references of linguistic metaphors change over the course of specific events?

In general, identifying equilibrium behavior is highly subjective. We can detect clusters of different slants on particular target concepts like taxation and government, but locally prevailing source concepts can be subject to variations that are dependent on the size of the data window (scaling effects).

2   Methods

Blog Selection. Blogs are studied, because initial investigations revealed more metaphorical language than similarly timed news articles containing more factual description of events. Candidate blogs were required to be located in the country of interest, express political opinion, and reference events published in a major news outlet. These were found using either a directory (e.g., http:// blogsdemexico.com.mx/) or through Google searches. The main search format was: [Political blog, name of the country/topic]. Examples of searches include: "Blog politica, Argentina, Venezuela", "blogs problemas sociales", "blogs politicos", etc. Specific references to hosting sites were also used such as "blogspot Pena Nieto". Topics of the blogs include societal issues, the economy, and criticism of current or former government and politicians. Most of the blogs were written in Spanish; however, some of the Venezuelan blogs were written in English for a number of possible reasons. This is an unavoidable source of bias resulting from inconsistencies across languages in metaphor usage. Therefore, we only consider blogs written in Spanish. Summary details of the corpus are given in Table 1. The ratio of documents containing linguistic metaphors (LMs) remains fairly consistent across countries. The LMs included are only those mapped to seven target concepts discussed below, not all LMs present in the data.

Table 1. Summary details of the blog corpus by country.

Country	# blogs	# docs	# LMs	LM-doc ratio
Argentina	89	111437	53283	0.48
Ecuador	59	13899	5278	0.38
Mexico	97	306101	157553	0.51
Venezuela	82	157652	75869	0.48
Total	327	589089	291983	0.49

Metaphor Extraction. Linguistic metaphors were extracted using software, the Metaphor Detection System (MDS), developed on the IARPA Metaphor program[1]. The goal of this program was to develop automated systems to identify the beliefs and world-views of different cultures by examining their use of metaphorical language. The MDS automatically detects linguistic metaphors in the original Spanish text and then maps them to pairs of pre-defined conceptual constructs. The MDS uses a mixture of techniques, including lexico-syntactic pattern matching and classification using semantic and psycholinguistic features. Internal program testing showed the MDS capable of achieving an LM detection F-score of 0.74 (precision 0.82, recall 0.68) in Spanish. As stated earlier, the goal of this paper is not to focus on the mechanics of MDS itself (covered elsewhere) but to present the first large-scale analysis of metaphors in blogs.

Both linguistic metaphors and their mappings to source-target concept pairs are used in this analysis[2]. The conceptual targets we consider in this analysis are $m_t \in$ {BUREAUCRACY, DEMOCRACY, ELECTIONS, GOVERNMENT,

---

[1] See: http://www.iarpa.gov/index.php/research-programs/metaphor.
[2] Target and source concepts will be represented in all caps: e.g., ELECTIONS.

POVERTY, TAXATION, and WEALTH}. We analyze 84 conceptual sources $m_s \in$ {CRIME, MACHINE, MOVEMENT, ...}. The linguistic sources and targets are a dictionary of political keywords, such as *"government"*, that map to a conceptual source or target domain. Given the translated text from our corpus, *"the usurper government"*, *usurper* is the linguistic source and *government* is the linguistic target. Similarly, *usurper* maps to the source concept CRIME, and *government* maps to the target concept GOVERNMENT. These source and target concepts form the tuple, $M := (m_s, m_t) \in \mathcal{M}$. The set $\mathcal{M}$ consists of all source-target pairs, and has cardinality ($|\mathcal{M}| = 672$). Our dataset is $X \subseteq T \times \mathcal{M} \times \mathcal{B}$ for daily timestamps $t \in T$ and blogs $B \in \mathcal{B}$. Time-series are generated for each blog and concept pair, $f_{M,B}(t) = |\{x \in X | x = (t, M, B)\}|$, that indicate the frequency of the metaphors appearing at time, t.

3   Results and Discussion

Blogs are heterogeneous in nature, because they can range in publication frequency, content length, and production effort. The prototypical blog post may not exist on all dimensions, because they display characteristics of both academic articles and Twitter posts. Although germane to the discussion, we do not directly address the qualities composing the blogs, and focus the analysis to aggregating a variety of different blogs by target concept through co-clustering.

Clustering of Concept Pairs. Clusters of related metaphors are formed over summed times series of metaphor frequencies,

$$F_M(t) = \sum_{\mathcal{B}} w_B f_{B,M}(t) \tag{1}$$

where we assume $w_B = 1$. Setting $w_B = 1$ assumes that the blogs contributing to the proportional response are of equal weighting. This is perhaps sub-optimal, but it is a maximum-entropy approach for reducing noise fluctuations in individual blogs given no prior information. We could reduce the argument further by considering the relative contribution per writer (multiple per blog) given certain types of events, but then lose the information inherent in the collection. For now, w_B is a parameter for investigation in future work.

The spectral co-clustering algorithm [9] forms clusters of similar source concepts, m_s for a given target concept m_t. Co-clustering is performed over the Pearson correlation matrix C, to find evidence of similar linear proportional responses. The correlation matrix is $N \times N$, where N is the number of source concepts m_s for a given target concept,

$$c_{ij} = \frac{\sum_1^K \hat{F}_{M_i}(t_k)\hat{F}_{M_j}(t_k) - K\bar{\hat{F}}_{M_i}\bar{\hat{F}}_{M_j}}{\sqrt{\sum_1^K \hat{F}^2_{M_i}(t_k) - K\bar{\hat{F}}^2_{M_i}}\sqrt{\sum_1^K \hat{F}^2_{M_j}(t_k) - K\bar{\hat{F}}^2_{M_j}}} \tag{2}$$

where $\hat{F}_M = (F_M * W)(t)$ is a windowed time-series from $t \in \{1, K\}$ for metaphor M_i and M_j, and $\bar{\hat{F}}$ is the mean of the time-series. The Hamming window is applied in the convolution of F_M,

$$W(\gamma) = 0.54 - 0.46 \cos\left(\frac{2\pi\gamma}{\Gamma - 1}\right), 0 \leq \gamma \leq \Gamma - 1 \tag{3}$$

where Γ is a parameter that controls the length of the window. The Hamming window places more weight on the current time step while the future and past extrema are minimized. The convolution is phase centered to preserve temporal accuracy and assume a symmetric lead-lag duration between blog and event dates. This choice of filter (as opposed to cross-correlation) reduces the effects of inconsistent lead-lag responses of publication dates around events for small Γ. For large Γ, this window compensates for unknown lead-lag relationships to frequencies across multiple events.

The number of clusters were chosen to minimize the objective function,

$$\beta^* = \arg\min_{\beta} \phi(C(\beta)), \ \phi(C) = \sum_{i=1, j>i} ijc_{ij} \tag{4}$$

where $\phi(C)$ minimizes entries appearing on the off-diagonal, and β is the optimal number of clusters. Minimizing this objective discourages highly correlated time-series from appearing in separate clusters. This increases discrimination of clusters by forcing smaller highly correlated clusters, when possible. The co-clustering algorithm was initialized randomly for 20 trials at each choice of β.

The source concept clusters grouped by target for $\Gamma = 365$ are shown for Mexico[3] in Fig. 1. Source concepts with higher frequency counts are in the center of the cluster along the y-axis. For each target concept, different countries have different groupings of source concepts. However, there are some common trends in all of our datasets. ELECTIONS, as a target concept, tend to have a thin cluster having COMPETITION as the most frequent source concept. In our data for both Mexico and Argentina, we tend to see increased metaphor usage around the time of elections. The target concepts, POVERTY and GOVERNMENT, tend to have one large cluster and one small cluster of source concepts. The large cluster tends to be the source BUILDING, and the smaller cluster varies. POVERTY, however, has a different set of source clusters for each country indicating the possibility of regional variations. Lastly, BUREAUCRACY tends to have many smaller source clusters, which indicates more variety in the descriptions of this target.

Quantitatively, with this amount of data, it is likely that the distributions of source concepts associated with targets are going to be different[4], if not by chance alone. Therefore, we consider the table of standard residuals to find the relative target and source pairs that are markedly different between countries

---

[3] Argentina, Ecuador, and Venezuela are not shown to conserve space. Qualitative results from these countries are discussed in terms of their similarities to our Mexico dataset. Differences are explicitly highlighted in Table 2.

[4] χ^2 test ($\Gamma = 365$), $p < 0.001$ for all targets in all countries.

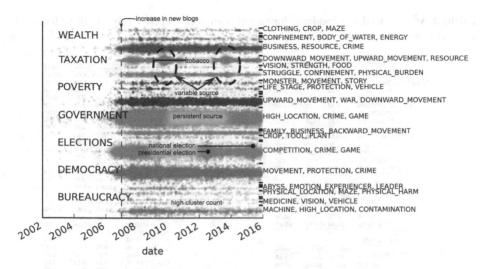

Fig. 1. Source clusters of linguistic metaphors are shown for blogs in Mexico, $\Gamma =$ 365. Each horizontal position corresponds to a particular source concept, M. Clusters are identified by the partitions on the right y-axis. Each time-series is color-coded to a target, and the opaqueness shows increased metaphor frequency per blog at that timestamp, and marker size indicates the frequency of usage on a given date by a particular blog. Larger clusters show source names.

as shown in Table 2. This table shows results for summed blog data, $F_M(t)$ over the entire window of time from 2000–2015. The subtables are grouped by target, and the source concepts showing the most differences are shown in bold. Ecuador is not included for having too few data relative to Argentina, Mexico, and Venezuela to make a proper comparison. Of the highest relative standard residuals, highlighted in bold, Mexico tends to deviate from the distributions of Argentina and Venezuela for BUREAUCRACY as a HIGH_LOCATION and for the increased usage of ELECTIONS as both a CRIME and BUSINESS, without as much ELECTIONS as a GAME (i.e., something won). Mexican blogs also the DEMOCRACY as BUILDING tends to be underrepresented as well. However, target concept WEALTH shows more diverse differences where compared to Argentina, Mexico shows increased usage of BUSINESS, CRIME, RESOURCE, and LOW_LOCATION sources as opposed to MOVEMENT and BUILDING.

Cluster Variation Around Events. One of the challenges of using blogs and metaphors to detect events is that there is no one specific type of event that corresponds to a particular conceptual metaphor. For example, POVERTY as a MONSTER can refer to a number of different events in the realm of government, economics, civil unrest, and politics. Modeling an individual interpretation of an event through metaphor is likely not to converge to a particular expected outcome. However, given enough data points of correlated trends in a cluster surrounding a particular event may have insight into this complex relationship.

Table 2. χ^2 standardized residuals for target-source concepts; larger (± 10) differences in source concepts relative to sample countries are shown in bold.

BUREAUCRACY

	Argentina	Mexico	Venezuela
CRIME	-2.19	-5.05	7.70
UPWARD_MOVEMENT	3.20	-4.65	2.56
RESOURCE	4.00	-4.17	1.31
BARRIER	6.56	-5.33	0.42
PHYSICAL_BURDEN	-0.52	-1.02	1.62
PROTECTION	1.03	2.16	-3.37
MORAL_DUTY	-2.57	5.10	-3.63
HUMAN_BODY	3.98	-4.04	1.18
HIGH_LOCATION	**-10.35**	**18.89**	**-12.69**
MACHINE	-1.26	-0.81	2.03
WAR	4.75	**-10.07**	7.43
BUILDING	3.08	-1.48	-0.97
CONTAMINATION	-2.37	1.59	0.24

DEMOCRACY

	Argentina	Mexico	Venezuela
CRIME	-2.47	3.81	-2.37
MOVEMENT	-0.07	6.00	-6.39
PROTECTION	-7.70	5.08	-0.08
BUILDING	**15.70**	**-19.52**	**10.01**

ELECTIONS

	Argentina	Mexico	Venezuela
BUSINESS	**-11.30**	**21.26**	**-14.85**
CRIME	**-40.19**	**50.06**	**-23.98**
GAME	**36.45**	**-52.60**	**29.88**
COMPETITION	8.87	-7.41	1.19
STRUGGLE	-3.94	7.34	-5.10

POVERTY

	Argentina	Mexico	Venezuela
FORCEFUL_EXTRACTION	-3.94	-6.02	**10.41**
CRIME	-1.64	3.45	-2.46
RESOURCE	1.25	1.06	-2.33
CONFINEMENT	4.25	-7.11	4.28
PHYSICAL_LOCATION	1.19	-1.38	0.51
DISEASE	-2.97	3.15	-0.92
STRUGGLE	-1.54	-2.24	3.94
WAR	**-12.04**	**15.33**	-6.66
SIZE	0.77	3.93	-5.18
BUILDING	**22.49**	**-15.62**	-2.41
DOWNWARD_MOVEMENT	8.64	**-11.48**	5.32
PHYSICAL_HARM	-4.39	6.35	-3.29
UPWARD_MOVEMENT	-5.40	6.27	-2.30

GOVERNMENT

	Argentina	Mexico	Venezuela
FURNISHINGS	**-11.85**	**18.04**	**-10.34**
BUSINESS	-5.11	**11.73**	-8.93
CRIME	**-18.78**	**27.48**	**-15.12**
STRENGTH	-0.14	3.61	-3.97
GAME	-4.50	1.72	1.89
PROTECTION	2.13	0.03	-1.84
RESOURCE	6.66	-5.40	0.45
PHYSICAL_BURDEN	0.11	2.90	-3.37
SHAPE	4.72	-9.12	6.30
COMPETITION	1.23	7.86	-9.95
MORAL_DUTY	-0.85	3.78	-3.55
CONFINEMENT	6.80	-4.79	-0.37
ENSLAVEMENT	1.00	-5.62	5.51
THEFT	**-13.79**	**27.67**	**-19.57**
HUMAN_BODY	4.34	-9.88	7.49
HIGH_LOCATION	**-20.94**	**19.83**	-4.62
LEADER	-4.36	6.93	-4.13
MOVEMENT	3.86	**-11.09**	9.26
STORY	**12.49**	**-13.67**	4.85
PHYSICAL_LOCATION	-1.39	0.04	1.14
MACHINE	5.40	**-15.14**	**12.54**
STRUGGLE	**13.01**	-6.38	-3.85
WAR	6.02	**-25.65**	**23.91**
BUILDING	6.68	-9.49	5.06
WEAKNESS	2.36	-0.66	-1.25
PHYSICAL_HARM	6.72	-6.82	2.00
CONTAMINATION	-4.86	7.95	-4.86
UPWARD_MOVEMENT	8.75	-4.70	-2.13

TAXATION

	Argentina	Mexico	Venezuela
RESOURCE	1.91	-2.98	2.15
BUILDING	-1.56	2.94	-2.60
DOWNWARD_MOVEMENT	**13.47**	**-16.59**	7.68
CONFINEMENT	1.81	-1.53	-0.15
UPWARD_MOVEMENT	**-10.51**	**12.30**	-4.89

WEALTH

	Argentina	Mexico	Venezuela
BUSINESS	**-15.72**	**21.21**	-7.93
CRIME	**-18.70**	**20.35**	-2.79
GAME	-6.67	7.07	-0.75
RESOURCE	**-19.83**	**16.99**	3.28
MOVEMENT	**40.14**	**-39.03**	-0.34
LOW_LOCATION	-6.82	**11.28**	-6.28
BUILDING	**17.91**	**-23.49**	8.12

As an example, we study the metaphors associated with the target concept, TAXATION in Mexico. In Fig. 1, we see that there a number of intervals of time, particularly between 2009–2011, and just after 2014 where there is an increase in both the number of blogs and source concepts referenced in our data. These are noted by the more opaque patches and increased path width, respectively. During the 2009–2011 interval, Mexico had an increase in the tobacco tax. Figure 2 shows the time-series produced by the top three sources of both time-varying TAXATION metaphor clusters shown in Fig. 1. The clusters are labeled "1" and "2", respectively. The black time-series is the frequency of occurrence of the words "tabaco" and "cigarrillo" in our Mexican blog corpus. These metaphors tend to vary proportionally with respect to the appearance of the tobacco keywords. However, the CONFINEMENT, STRUGGLE, and PHYSICAL_BURDEN source concepts tend to focus around the same interval as the tobacco keywords peak. In these time-series 82 % of the tobacco references and taxation metaphors occurred within the same sentence, indicating a strong relationship to the metaphors referring to the tax increase during this time period.

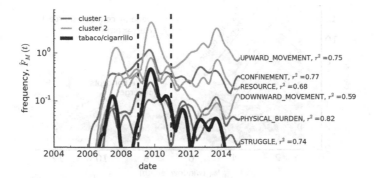

Fig. 2. $\hat{F}_M(t)$ ($\Gamma = 365$) for the TAXATION target concept in Mexico. The vertical lines indicate the years during the period of increased tobacco tax.

The association of concept pairs to events is limited by the mapping of linguistic metaphors. In this example, the TAXATION as an UPWARD_MOVEMENT is commonly referring to a tax increase or "impuesto" and "elevar". The linguistic metaphors supporting TAXATION as a CONFINEMENT conceptual metaphor are referring to evasion or being captive. There is no guarantee that any of these metaphors is strictly referring to the tobacco tax, but in aggregate we see mutual linearly correlated fluctuations in the times series. The least correlated is TAXATION as a DOWNWARD_MOVEMENT, and the most correlated is PHYS-ICAL_BURDEN. Therefore, we can extract a linearly proportional relationship between metaphors and events, but is currently on a case-by-case basis.

Effects of Varying the Data Window, Γ. Inconsistencies in publication times, metaphor usage and events affect the temporal relationship between metaphors and events. Our results up to this point have focused on clustering using a data window of size, $\Gamma = 365$. We assume that applying a Hamming window of length Γ over the data helps to disperse signal energy symmetrically in time to smooth variations in blog publications around relevant events. However, in larger data windows, we expect a decrease in the number of clusters, and certain metaphors may exhibit more similar behaviors over different temporal windows.

If the metaphors had similar temporal characteristics at every windowing parameter, Γ, we would expect to see the clusters merging as the window length increases. However, the network in Fig. 3 shows a different path of cluster formation. Each edge weight corresponds to the number of same sources in each node (cluster). Initially, there are many small clusters containing few sources. As the window length increases, many of these smaller clusters are absorbed as would be expected. These are identified by the downward arrow showing similar temporal trends. There are also a number of crossings between clusters, and often these crossings contain groups of 10 metaphors. As the window approaches the length of the time-series (4096 time steps), the clusters converge to two nodes with

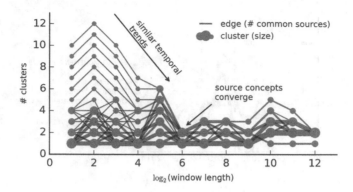

Fig. 3. The path of source concept clusters for TAXATION in the Mexico dataset reveals different cluster configurations, and the "right" time-scale is less obvious.

one containing most of the sources. Although there is a trajectory of thick edges connecting the windows of increasing size, the cross-overs indicate the possibility of different temporal similarities between source-target concept pairs, and the possibility of no "characteristic" time window for analysis of events.

4   Conclusions

This is the first large scale analysis of metaphors in Latin American blogs for event detection. We discuss qualitative similarities of cluster formation when grouped according to correlation, but source concepts will vary when aggregated by country. Using a co-clustering approach, we identify times of interest and show an example of a linear proportional relationship of metaphor usage regarding taxation around a period of time when Mexico enacted a tobacco tax. However, when generalizing the approach to multiple scales, the clusters do not necessarily show a characteristic time-scale. This indicates that different relationships could be realized over different data windows. Future work will consist of a two-pronged approach. First, we aim to better understand what the different window lengths may come to represent. Second, aggregating heterogeneous blogs is a non-trivial process. To refine our approach, we aim to better understand the uncertainty of blog temporal behavior to better estimate prior probabilities of publication style and metaphor usage parameters in this diverse data source.

Acknowledgments. Supported by the Intelligence Advanced Research Projects Activity (IARPA) via DoI/NBC contract number D12PC000337, the US Government is authorized to reproduce and distribute reprints of this work for Governmental purposes notwithstanding any copyright annotation thereon. Disclaimer: The views and conclusions contained herein are those of the authors and should not be interpreted as necessarily representing the official policies or endorsements, either expressed or implied, of IARPA, DoI/NBC, or the US Government.

References

1. Bracewell, D.B., Tomlinson, M.T., Mohler, M., Rink, B.: A tiered approach to the recognition of metaphor. In: Gelbukh, A. (ed.) CICLing 2014, Part I. LNCS, vol. 8403, pp. 403–414. Springer, Heidelberg (2014)
2. Carroll, J.M., Mack, R.L.: Metaphor, computing systems, and active learning. Int. J. Hum.-Comput. Stud. **51**(2), 385–403 (1999)
3. Chakraborty, P., et al.: Forecasting a moving target: Ensemble models for ILI case count predictions. In: Proceedings of the 2014 SIAM International Conference on Data Mining, pp. 262–270 (2014)
4. Geary, J.: I Is an Other: The Secret Life of Metaphor and How it Shapes the Way We See the World. HarperCollins Publishers, New York (2011)
5. Gibbs Jr., R.W.: The real complexities of psycholinguistic research on metaphor. Lang. Sci. **40**, 45–52 (2013)
6. Golshaie, R., Golfam, A.: Processing conventional conceptual metaphors in persian: A corpus-based psycholinguistic study. J. Psycholinguist. Res. **44**(5), 495–518 (2014)
7. Johnson, M.: Metaphorical reasoning. South. J. Philos. **21**(3), 371–389 (1983)
8. Keysar, B., Shen, Y., Glucksberg, S., Horton, W.S.: Conventional language: How metaphorical is it? J. Mem. Lang. **43**(4), 576–593 (2000)
9. Kluger, Y., Basri, R., Chang, J.T., Gerstein, M.: Spectral biclustering of microarray data: Coclustering genes and conditions. Genome Res. **13**(4), 703–716 (2003)
10. Lakoff, G., Johnson, M.: Metaphors We Live By. University of Chicago Press, Chicago (2008)
11. Muthiah, S., et al.: Planned protest modeling in news and social media. In: Innovative Applications of Artificial Intelligence (2015)
12. Neuman, Y., et al.: Metaphor identification in large texts corpora. PLoS ONE **8**(4), e62343 (2013)
13. Ramakrishnan, N., et al.: 'Beating the News' with EMBERS: Forecasting civil unrest using open source indicators. In: Proceedings of the 20th ACM SIGKDD International Conference on Knowledge Discovery and Data Mining, KDD 2014, pp. 1799–1808. ACM, New York (2014)
14. Thibodeau, P.H., Boroditsky, L.: Metaphors we think with: The role of metaphor in reasoning. PLoS ONE **6**(2), e16782 (2011)
15. Thibodeau, P.H., Boroditsky, L.: Natural language metaphors covertly influence reasoning. PLoS ONE **8**(1), e52961 (2013)
16. Thibodeau, P.H., Boroditsky, L.: Measuring effects of metaphor in a dynamic opinion landscape. PLoS ONE **10**(7), e0133939 (2015)

"With Your Help... We Begin to Heal": Social Media Expressions of Gratitude in the Aftermath of Disaster

Kimberly Glasgow[1,2]([✉]), Jessika Vitak[1,2], Yla Tausczik[1,2], and Clay Fink[1,2]

[1] Applied Physics Laboratory, Johns Hopkins University, Baltimore, USA
{kimberly.glasgow,clayton.fink}@jhuapl.edu
[2] College of Information Studies, University of Maryland College Park,
College Park, USA
{jvitak,ylatau}@umd.edu

Abstract. In the aftermath of disasters, communities struggle to recover from the physical and emotional tolls of the event, often without needed social support. Social media may serve to bridge the distance between the affected community and those outside who are willing to offer support. This exploratory study uses Twitter as a lens for examining gratitude for support provisions in the aftermath of disasters. Gratitude for support is examined in the context of two significant U.S. disasters, a tornado that devastated Alabama and the mass shooting at Sandy Hook Elementary School. Observed expressions of gratitude for social support from each community differed from what would be expected based on established factors relating to social support in the aftermath of the disaster. These findings offer ways for social media – as a window into real-time community behaviors relating to response and healing after disaster – to contribute to the provision of mental health resources and monitoring community resilience and recovery.

Keywords: Gratitude · Social support · Social media · Disaster · Machine learning · Twitter

1 Introduction

Disasters are serious disruptions of the functioning of a community or a society. They involve widespread human, material, economic, and/or environmental losses and impacts that exceed the coping capacity of the affected community or society. Their impacts are wide-ranging and may involve loss of life, injury, and other negative effects on human physical, mental, and social well-being; damage to property; destruction of assets; loss of services; social and economic disruption and environmental degradation [2]. Improved understanding of the processes of disaster response and recovery, including the role of social support, could contribute to improving the health and resiliency of communities coping with disaster.

© Springer International Publishing Switzerland 2016
K.S. Xu et al. (Eds.): SBP-BRiMS 2016, LNCS 9708, pp. 226–236, 2016.
DOI: 10.1007/978-3-319-39931-7_22

Social support is critical to individual and community-level well-being and happiness, and it becomes more important when coping with disaster. Gratitude has been conceptualized as an emotion triggered in response to the supportive, helpful actions of others. Experiencing gratitude when coping with adversity, such as disasters, can contribute to healing, resilience, and growth [14].

There is a need for improvements to existing tools and procedures to support disaster behavioral health response and for additional research in the areas of risk, resilience, and recovery [29]. Current instruments rely on retrospective self-reports [11,16]. Social media such as Twitter could supplement this perspective by capturing the spontaneous expressiveness of a community as it experiences the event. Sharing emotional responses to a disaster, as well as offering and requesting support in networked publics, have become part of our collective response to crisis [5].

Relatively little research has explored the ongoing process of coping with a traumatic event as it progresses [26]. Successfully identifying social media discussions acknowledging social support could provide insight into how well a community is coping. This could help focus disaster mental health efforts or help monitor community resilience and recovery.

The present work explores expressions of gratitude for support on Twitter in the aftermath of two disasters. We demonstrate an automated method for detecting this important behavior which, while an imperfect proxy for actual social support received, is independently an important factor in resilience and healing after disaster [10]. We then measure this behavior in the context of two significant U.S. disasters, a tornado that devastated part of Alabama and a mass shooting at an elementary school in Connecticut.

2   Related Work

2.1   Characterizing Disasters

A traumatic event can be defined as one that involves exposure to death, threat of death, actual or threatened serious injury, or actual or threatened sexual violence [3]. Disasters are traumatic events that are collectively experienced, in contrast to personal tragedies.

While multiple typologies of disasters exist, we adopted a broadly applied approach that considers whether the disaster was natural or human-induced/technological [25], and if human-induced, whether it was intentional or accidental. These aspects of a disaster have been observed to shape offers of support, and the health, and influence recovery and resilience of affected individuals and communities [8,21].

In general, high levels of community distress after disaster are most likely when two or more of the following features exist: human perpetrators; intentional violence; high prevalence of injuries; threat to life; loss of life; severe, extensive property damage; and significant, ongoing financial difficulties for the community [16]. In our work, both disasters possess three or more of these features.

2.2 Disasters, Social Support, and Gratitude

Social support has been demonstrated to be critical throughout one's life course, and is particularly valuable as a moderator of life stress [6]. Victims need social support to cope with and recover from disaster, and may experience a need for support that exceeds what their immediate environs and network can supply [12].

A number of factors influence the amount of support that disaster victims are offered. A primary factor is the severity of the event [9]. Psychological factors also play a role in the behavior of potential providers of support. Both *locus of causality* and *situational controllability* have been observed experimentally to influence provision of support in disaster contexts [15,18]. Locus of causality involves the degree to which a person in need of help personally caused the negative event, while situational controllability considers if the negative event could have been foreseen, prevented or avoided.

Those who receive support or other benefits, particularly if the benefit was not anticipated, are likely to experience gratitude [4]. Gratitude has been found to be psychologically protective, notably in the context of human-induced disasters, such as the September 11th attacks [8] and in other types of events such as accidents, natural disasters, and personal experiences of violence [14].

2.3 Social Media and Disasters

Social media platforms such as Twitter have demonstrated utility in helping individuals cope with disasters [19], and enable collective response to crisis. The phenomenon of thanking others, or expressing gratitude via social media in the aftermath of a disaster has been identified as a significant portion of discourse, occurring across a wide range of disasters and spanning multiple continents, but this phenomenon is imperfectly understood [17]. This imperfect understanding has led to gaps in understanding disaster behavior and related communications [23]. Our work contributes new knowledge to help address this gap.

Olteanu [17] observed Twitter to more generally be a medium through which the "nuance of disaster events" is magnified, and a mechanism through which a disaster event is socially constructed. The scale and granularity of Twitter data provide a unique lens into the nature and dynamics of response to disaster, including social support and gratitude.

3 Current Work: Comparing Expressions of Gratitude on Twitter Following Two Disasters

The full spectrum of emotions, intentions, and outcomes relating to social support following a disaster has provided fertile ground for theorizing and study. The current work draws from the literature on social support and gratitude, disaster response and recovery, and social media to examine aspects of the recovery of two communities struck by two typologically distinct disasters. It uses machine learning to help address the challenges posed by the scale of social media data, uncovering relevant tweets in datasets far too large to be easily annotated by humans. In this work, we focus on acknowledgments and expressions of gratitude for support after a disaster. We use social media as a lens through which to view the organic responses of the community.

This work contributes to an acknowledged gap in the literature regarding understanding expressions of gratitude in social media in the aftermath of disasters [23]. It extends work using Twitter to evaluate discussion of disasters using data from random 1 % samples of Twitter data based on matching disaster key terms [17] with analysis based upon a much more complete and geographically relevant sample of tweets from members of the disaster stricken community.

In this work, we examined two disasters. One was a massive tornado, a natural disaster that tore apart Tuscaloosa and Birmingham, Alabama in 2011. The second was a brutal human-induced disaster and the deadliest elementary school shooting in US history, the Sandy Hook Elementary School shooting in Newtown, Connecticut. The current work examines disasters that differ with respect to severity, measured in terms of associated deaths, injuries, and damage. Previous work has identified severity of a disaster as increasing both willingness to help, and the amount of support offered to victims of a disaster in experimental settings [15,18]. Severity of loss has been associated with greater support from non-kin after devastating flooding [13].

4 Background on Disasters Studied

Tuscaloosa Alabama tornado: The tornado that struck Jefferson and Tuscaloosa counties in Alabama on April 27, 2011 was the most devastating tornado in terms of damage that the United States had experienced, causing an estimated $2.4 billion in damage. The tornado sustained wind speeds up to 190mph, and left an 80-mile long swath of destruction. It killed 64 people and injured an additional 1500 [1]. Afterwards, the President visited the region and a federal state of emergency was declared.

Newtown school shootings: On December 14, 2012, 20 school children and six faculty members were shot and killed at Sandy Hook Elementary School in Newtown, Connecticut. This was the deadliest primary school shooting in US history. Most victims were six years old. In the days following this disaster, the President visited Newtown, spoke at a vigil, and met with families and first responders.

Assessed by typical quantitative measures, the tornado was the far more destructive event (see Table 1). Over twice as many people were killed, and orders of magnitude more were injured. The Alabama communities sustained over 50 times as much property damage. Thousands were left homeless by this natural disaster, and many saw their workplaces destroyed as well. The most fundamental difference between these two events – and one much harder to quantify – is the type of disaster. The tornado that devastated Alabama is a *natural disaster*. The school shooting was a *human-induced disaster*, a deliberate act of commission, not a consequence of error or negligence.

Table 1. Scale and destructiveness of Alabama tornado and Newtown school shooting

	Alabama tornado	Newtown school shooting
Number of victims killed	64	26
Number of children killed	19	20
Number of persons injured	1500	2
Estimated property damage	$2.4 billion	$42 million (replace school)
Displaced households	6 k homes destroyed, 15 k damaged	0

5 Methods and Results

One challenge posed in working with Twitter content is identifying relevant tweets in large corpora. Social media discussions relating to the disasters and their aftermath are varied – providing information, describing volunteering opportunities, soliciting contributions, etc. Tweets acknowledging or expressing gratitude for support are a fraction of the general discussion relating to the disasters, which itself is a small fraction of the total data.

For both events, we used Twitter's geo API to collect tweets from the affected counties. We analyzed data covering an 11-day span, beginning the day after the event. This helped eliminate potential effects of time of day variance. We retrieve 2.8 million tweets from 74,819 user from April 28-May 9 from Jefferson and Tuscaloosa counties, and 119,651 tweets from 3648 users from 15–26 December for Fairfield county. We collected many more tweets from Alabama than Connecticut. This may be attributable to changes in Twitter's rate limit policy enacted in 2012, before the Newtown school shooting, though we cannot be certain.

Machine learning methods were used to help identify relevant tweets. Only a small percentage of the tweets were relevant to the topic of acknowledgement of social support following the tornado or the shooting. We estimated that less than about 1 % of tweets were relevant to our task, based on random samples from each dataset. Table 2 provides examples.

Thus, we face a common challenge seen in real-world classification tasks. The classes we are interested in are only a small percentage of the actual data. To address this, we queried the dataset for tweets containing terms likely to indicate expressions acknowledging social support relating to each respective disaster. For Alabama, these included terms relating to thanks or support (*thank, help, support, aid*), and terms relating to the tornado (*tornado, twister, storm*). For Newtown, we queried the Twitter data for tweets containing terms likely to indicate expressions acknowledging social support relating to the school shooting. These included the previous terms relating to thanks or support, and terms relating to the shooting (*shoot, loss, kill, grief*). For each event, we further queried for tweets containing terms from both classes, support-related and event-related.

We randomly sampled from each of these sets of tweets, in addition to a random sample of tweets from the full dataset, to ensure sufficient positive tweets in the dataset and address class imbalance [28]. The dataset for the Alabama tornado contained 705 tweets annotated for ground truth, drawn from the sets described above. The equivalent dataset for Newtown consisted of 670 annotated tweets. Each dataset contained 15 % gratitude tweets. We used a semi-supervised active learning framework that incorporates Twitter-specific features for building classifiers [22]. We used two classes to distinguish between tweets expressing gratitude for social support relating to disaster, and all others. The highest-scoring class (>.50) can be considered the label generated by the model for that tweet. We created a separate model for each dataset. Both models performed well at classifying tweets in the relevant class, achieving 96 % recall on the Alabama test data and 94 % for Newtown. However they produced a substantial proportion of false positives. One technique demonstrated to help reduce false positives rates is cascading classifiers, or using the output of an initial classifier as input for a second classifier [24]. The cascade can alternately be considered a focus-of-attention mechanism that discards regions unlikely to contain objects of interest. We developed a second classifier employing Linguistic Inquiry and Word Count (LIWC).

Table 2. Example positive and negative tweets for expressions of gratitude for social support

	Alabama tornado	Newtown school shooting
Positive	@A we are thankful for your help in our time of need from the #tornado #BhamSalvArmy	@B i know.thank you, a lot of people are making me feel better by just sending condolences. it's hard, but we're #203strong
Negative	Locked my keys in the house for the 2nd time in maybe 12 hours...?	@C baby your the best girlfriend ever I would never trade you for anything your perfect

LIWC is a widely used text analysis program [20,27]. Its central premise is that the words people use can reveal their mental, social, or emotional state. LIWC incorporates a number of psychologically relevant variables that might pertain to this research, such as positive emotion, first person plural (we), money, and work. This final model incorporates LIWC features and the previous classifier score.

We created our training, validation, and testing data sets from an additional 1000 annotated tweets from the 34,600 Alabama tweets and 506 tweets from the 5,400 Newtown tweets output as positive by their respective classifiers. We used Rattle [30], which implements a number of R's machine learning libraries, to build our final classifiers.

For the Alabama data, the best performing final model, an SVM, achieved 82 % accuracy on test data. For the Newtown data, the final SVM had 87 % accuracy on test data. A confusion matrix for both models is provided in Table 3.

Table 3. Confusion matrices for Newtown and Alabama classifiers

Actual class	Predicted class			
	Newtown		Alabama	
	Not Gratitude	Gratitude	Not Gratitude	Gratitude
Not Gratitude	0.74	0.07	0.58	0.09
Gratitude	0.07	0.13	0.09	0.24

Our task involves comparing relative percentages of gratitude tweeting for each disaster. The total gratitude tweets predicted by the models mirror the number of true positives in the data.

Findings from previous work suggest we should observe a higher percentage of tweets from Alabama in the positive class, since that was the more severe event [9,13]. The tornado killed more than twice as many people, injured thousands more, caused over \$2.3 billion more in damage, and physically devastated a far greater geographic area. Both events are similar in having an external locus of causality and minimal situational controllability.

However, when we compare data gathered from Newtown and Alabama in the days following their respective disasters, a different picture emerges. The community that experienced the quantitatively less severe disaster, Newtown, generated proportionally more support acknowledgments (1.1 % vs. 0.15 % of tweets, $p < .001$). A far larger percentage of their population expressed gratitude for support received (13.33 % vs. 3.43 %, also $p < .001$). and per capita gratitude tweeting is 6.33 times higher (Table 4).

Table 4. Tweets and users expressing gratitude in the Tuscaloosa tornado and Newtown school shooting

	Events	
	Alabama tornado	Newtown shooting
Population of affected counties	852 k	917 k
Unique Twitter users	74,819	3542
Gratitude tweets (final model)	4289	1330
Percent of gratitude tweets	0.15	1.11
Users with gratitude tweets	2564	472
Percent of users with gratitude tweets	3.43	13.33
Per capita gratitude tweets	0.06	0.38

Because rates of tweeting show daily variability, and Alabama suffered interruptions to power and Internet access immediately after the disaster, it is preferably to normalize daily rates of gratitude tweets as a percentage of total tweets per day. The two communities have the highest percentage of gratitude tweets in the days following the disaster. Both communities show about a 60 % drop in the first week. Over the entire timeframe, the minimum rate for expressing gratitude in Newtown exceeds the maximum measured in Alabama, as shown in Fig. 1.

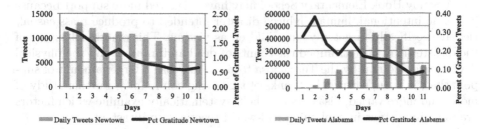

Fig. 1. Tweet activity and expressions of gratitude in the 11 days following the Newtown school shooting and Alabama tornado

6   Discussion

This study examines social media expressions of gratitude for social support received in the wake of two large-scale, traumatic events. Both events had deep impacts on their respective communities. The support provided to these communities after their disasters would be essential to any recovery process.

Yet, judging from patterns observed in Twitter data from these communities, their experiences and response to support differed markedly, and in ways not always anticipated by theory. In Alabama, the community that suffered the more quantitatively severe disaster, there were proportionally fewer expressions of gratitude for support received. This stands in contrast to theoretical predictions that support should be related to the severity of the event [13].

This work makes contributions in two primary areas: (1) it presents an approach and methodology for studying gratitude in response to support, (2) it suggests that additional social and psychological factors should be considered in theorizing about social support and disaster response.

Leveraging affordances of social media to focus on affected communities, rather than global discussion of disaster, provides unique insight into community-level phenomenon. Data collection over an extended time period has advantages over snapshots taken at one or a few points in time, and helps cope with disruptions of utility service or access to social media platforms.

Given the quantity of data that may be produced by a community after a disaster, automated text processing and machine learning approaches may be essential. While identifying expressions of gratitude across many types of disasters will not be amenable to a single cookie-cutter approach, bootstrapping off the language commonly used to express thanks and gratitude, and to describe the disaster should aid development of classifiers for gratitude for social support received after disasters.

Beyond the severity of a disaster, how bystanders view a disaster and the victims is known to affect how much support given. While the two disasters were similar in many of the ways disasters can engender different psychological responses, such as locus of causality and situational controllability, there must be other factors that explain the dramatic differences seen our analyses. The shooting at Sandy Hook Elementary School may have triggered more support because it was an intentional, human-induced disaster intended to produce mass casualties. It specifically targeted young children as victims. This emotional aspect of the disaster may have trumped the counts of dead and injured, and the physical devastation wrought by the Tuscaloosa tornado to evoke a larger amount of support that, in turn, earned the thanks of a grieving community. Further study of more, and more varied, disasters may help systematically examine which factors are most important in explaining real-world reactions to disasters.

6.1 Limitations and Future Work

We observe only acknowledgments of social support transmitted in Twitter by users who had provided geolocational data indicating they lived in the studied counties. Because we chose to focus on the response of community members to the disasters, and must rely on Twitter's affordances for representing location, the social media discussions of users who had associated themselves with different geographic locations, or no location, but actually were members of the community, are not captured.

Further, social media usage is not universal or representative within the US population. Thus we may be privileging the voices and responses of those who are active on social media, and underrepresenting those who are not.

This work compares social media data from two disasters, a natural and human-induced disaster, and finds the human-induced disaster produced expressions of gratitude more than an order of magnitude more frequently than the natural disaster. But human-induced disasters are not a unitary phenomenon. Comparison of the Newtown school shooting data expressing gratitude with expressions of gratitude in other, more similar events would likely be informative.

7   Conclusions

We presented findings of an exploratory study that treated social media as lens for examining gratitude for social support in the aftermath of disaster. Gratitude is an important factor in resilience and healing after disaster, and because it was triggered by receipt of support, may provide some insight into the dynamics of support itself. We then measured this behavior in the context of two significant U.S. disasters, a tornado and a mass shooting. The health and resilience of our communities after traumatic events is not easy to measure and monitor. Analyses of user activity in the aftermath of a disaster may prove an important adjunct to existing survey methods, providing more data and more temporal detail than would otherwise be obtainable [7], and enabling triangulation of data sources. The organic expressions and responses to events by community members, if closely examined, may advance our understanding of the complex linkages between the experience of disaster, support, gratitude, and post-traumatic growth.

References

1. Tornado Fatality Information (2011). http://www.spc.noaa.gov/climo/torn/2011deadlytorn.html
2. Terminology - UNISDR (2015). http://www.unisdr.org/we/inform/terminology
3. Association, A.P.: DSM 5. American Psychiatric Association (2013)
4. Bono, G., Emmons, R.A., McCullough, M.E.: Gratitude in practice and the practice of gratitude. In: Positive Psychology in Practice, pp. 464–481 (2004)
5. Bruns, A., Burgess, J., Highfield, T., Kirchhoff, L., Nicolai, T.: Mapping the australian networked public sphere. Soc. Sci. Comput. Rev. 0894439310382507 (2010)
6. Cobb, S.: Social support as a moderator of life stress. Psychosom. Med. **38**(5), 300–314 (1976)
7. Dredze, M.: How social media will change public health. IEEE Intell. Syst. **27**(4), 81–84 (2012)
8. Fredrickson, B.L., Tugade, M.M., Waugh, C.E., Larkin, G.R.: What good are positive emotions in crisis? a prospective study of resilience and emotions following the terrorist attacks on the United States on September 11th, 2001. J. Pers. Soc. Psychol. **84**(2), 365–376 (2003)
9. Greitemeyer, T., Rudolph, U., Weiner, B.: Whom would you rather help: an acquaintance not responsible for her plight or a responsible sibling? J. Soc. Psychol. **143**(3), 331–340 (2003)
10. Joseph, S., Butler, L.: Positive changes following adversity. PTSD Res. Q. **21**(3), 1–8 (2010)
11. Joseph, S., Linley, P.A.: Psychological assessment of growth following adversity: a review. In: Trauma, Recovery, and Growth: Positive Psychological Perspectives on Posttraumatic Stress, pp. 21–38 (2008)
12. Kaniasty, K., Norris, F.H.: Social support in the aftermath of disasters, catastrophes, and acts of terrorism: altruistic, overwhelmed, uncertain, antagonistic, and patriotic communities. In: Bioterrorism: Psychological and Public Health Interventions, pp. 200–229 (2004)
13. Kaniasty, K.Z., Norms, F.H., Murrell, S.A.: Received and perceived social support following natural disaster1. J. Appl. Soc. Psychol. **20**(2), 85–114 (1990)
14. Linley, P.A., Joseph, S.: Positive change following trauma and adversity: a review. J. Trauma. Stress **17**(1), 11–21 (2004)
15. Marjanovic, Z., Greenglass, E.R., Struthers, C.W., Faye, C.: Helping following natural disasters: a social-motivational analysis1. J. Appl. Soc. Psychol. **39**(11), 2604–2625 (2009)
16. Norris, F.H., Friedman, M.J., Watson, P.J., Byrne, C.M., Diaz, E., Kaniasty, K.: 60,000 disaster victims speak: Part I. An empirical review of the empirical literature, 1981–2001. Psychiatry Interpersonal Biol. Process. **65**(3), 207–239 (2002)
17. Olteanu, A., Vieweg, S., Castillo, C.: What to expect when the unexpected happens: social media communications across crises. In: Proceedings of the 18th ACM Computer Supported Cooperative Work and Social Computing (CSCW 2015) (2015)

18. Ouden, D.: Russell: sympathy and altruism in response to disasters: a Dutch and Canadian comparison. Soc. Behav. Pers. Int. J. **25**(3), 241 (1997)
19. Palen, L., Vieweg, S., Liu, S.B., Hughes, A.L.: Crisis in a networked world: features of computer-mediated communication in the April 16, 2007, Virginia Tech event. Soc. Sci. Comput. Rev. **27**(4), 467–480 (2009)
20. Pennebaker, J.W., Francis, M.E., Booth, R.J.: Linguistic inquiry and word count: LIWC 2001, p. 71. Lawrence Erlbaum Associates, Mahway (2001)
21. Russell, G.W., Mentzel, R.K.: Sympathy and Altruism in Response to Disasters. J. Soc. Psychol. **130**(3), 309–316 (1990)
22. Settles, B.: Closing the loop: fast, interactive semi-supervised annotation with queries on features and instances. In: Proceedings of the Conference on Empirical Methods in Natural Language Processing, pp. 1467–1478 (2011)
23. Shaw, F., et al.: Sharing news, making sense, saying thanks: patterns of talk on Twitter during the Queensland floods. Aust. J. Commun. **40**(1), 23 (2013)
24. Shen, J., Li, L., Dietterich, T.G., Herlocker, J.L.: A hybrid learning system for recognizing user tasks from desktop activities and email messages. In: Proceedings of the 11th International Conference on Intelligent User Interfaces, IUI 2006, pp. 86–92. ACM, New York (2006)
25. Smith, E., Wasiak, J., Sen, A., Archer, F., Burkle, F.M.: Three decades of disasters: a review of disaster-specific literature from 1977–2009. Prehospital Disaster Med. **24**(04), 306–311 (2009). http://journals.cambridge.org/abstract_S1049023X00007020
26. Stone, L.D., Pennebaker, J.W.: Trauma in real time: talking and avoiding online conversations about the death of Princess Diana. Basic Appl. Soc. Psychol. **24**(3), 173–183 (2002)
27. Tausczik, Y.R., Pennebaker, J.W.: The psychological meaning of words: LIWC and computerized text analysis methods. J. Lang. Soc. Psychol. **29**(1), 24–54 (2009)
28. Van Hulse, J., Khoshgoftaar, T.M., Napolitano, A.: Experimental perspectives on learning from imbalanced data. In: Proceedings of the 24th International Conference on Machine Learning, pp. 935–942 (2007). http://dl.acm.org/citation.cfm?id=1273614
29. Watson, P.J., Brymer, M.J., Bonanno, G.A.: Postdisaster psychological intervention since 9/11. Am. Psychol. **66**(6), 482–494 (2011)
30. Williams, G.J.: Data Mining with Rattle and R: The Art of Excavating Data for Knowledge Discovery. Use R. Springer, New York (2011)

Methodology

Spot the Hotspot: Wi-Fi Hotspot Classification from Internet Traffic

Andrey Finkelshtein[✉], Rami Puzis, Asaf Shabtai,
and Bronislav Sidik

Ben Gurion University of the Negev, Beersheba, Israel
{andreyfi, sidik}@post.bgu.ac.il,
{puzis, shabtaia}@bgu.ac.il

Abstract. The meteoric progress of Internet technologies and PDA (personal digital assistant) devices has made public Wi-Fi hotspots very popular. Nowadays, hotspots can be found almost anywhere: organizations, home networks, public transport systems, restaurants, etc. The Internet usage patterns (e.g. browsing) differ with the hotspot venue. This insight introduces new traffic profiling opportunities. Using machine learning techniques we show that it is possible to infer types of venues that provide Wi-Fi access (e.g., organizations and hangout places) by analyzing the Internet traffic of connected mobile phones. We show that it is possible to infer the user's current venue type disclosing his/her current context. This information can be used for improving personalized and context aware services such as web search engines or online shops, without the presence on user's device. In this paper we evaluate venue type inference based on mobile phone traffic collected from 115 college students and analyze their Internet behavior across the different venues types.

Keywords: Smartphone · Machine learning · Classification · Wi-Fi · Hotspot

1 Introduction

The widespread use of ubiquitous devices with Internet access such as laptops, smartphones, and tablets is on the rise. According to Ofcom [2], in 2014, 62 % of adults in the UK used smartphones, and 52 % considered smartphones (22 %) and tablets (30 %) the most important devices for Internet access. Browsing the Internet is one of the most common activities (performed daily) of smartphone users. [11], usually via a Wi-Fi hotspot connection. According to "iPass Wi-Fi growth map" (http://www.ipass.com/wifi-growth-map), at the start of 2015, there were 50 million worldwide Wi-Fi hotspots, and their number is expected to hit 340 million by 2018.

Wi-Fi traffic analysis can provide interesting insights about users and their interests in various hotspot venues. Previous works have studied users' behavior and Wi-Fi hotspot properties in order to cluster users according to their engagement and length of stay [5, 10] or to improve network quality of service [3, 4].

We investigate the network usage patterns of mobile devices connected to hotspots located across different types of venues, such as home, work, or public transportation. These patterns can be used to infer user context and the type of venue a user is visiting

© Springer International Publishing Switzerland 2016
K.S. Xu et al. (Eds.): SBP-BRiMS 2016, LNCS 9708, pp. 239–249, 2016.
DOI: 10.1007/978-3-319-39931-7_23

without the need to know the user's exact location. Deriving user context can be used to provide context-aware personalized services such as recommendation systems or targeted advertisements [8]. For example, a recommendation in push notification can be appropriate while the user is waiting for a train but not while the user is at work. The type of venue a user is located in can also support context-aware access control policies [7]. For example, an organizational VPN can restrict access to specific resources to users located in public areas. In addition, venue type inference can be used to enrich maps and Wi-Fi hotspot databases such as WiGLE (https://wigle.net) in a non-intrusive manner.

In this paper we introduce the venue classification problem. Based on the properties extracted from the traffic of a single user the solver should infer the type of the venue the user is located in. Experimental results show that the type of venue can be identified with an accuracy rate of 75 %. In addition, in order to understand the users' behavior across venues, we distinguish between two types of properties: user agnostic *hotspot properties* such as quality of service (QoS) and *user behavior properties* such as categories of browsed websites. Analysis of *user behavior properties* distributions in the collected data and feature selection results show that the Internet behavior of smartphone users changes in different types of venues. On one hand, the most popular domain categories (e.g., social, search, business) are similar across different types of venues. On the other hand, the browsing patterns for less popular website categories (e.g., blogs, travel, and news), together with the domains' popularity and security ranks, differ significantly based on the venue type.

To the best of our knowledge, no work has been done on venue type inference based on the Internet traffic traces of mobile device users.

2 Related Work

Learning users' behavior and optimizing network performance are two important tasks associated with Internet communication, specifically when connected via Wi-Fi hotspot. A great deal of valuable information can be extracted from Internet traffic dumps. Previous works have studied users' behavior and hotspot properties in wireless networks. Afanasyev *et al.* [3] studied the Google Wi-Fi network deployed in Mountain View, California. In 2008, the authors collected information for a period of 28 days. They analayzed the diversity of coverage, temporal activity, traffic demands, and mobility of the network. The users were grouped by device: smartphones, laptops, and static devices. They found significant differences between the three groups in terms of session lengths, network usage, and the diversity of application layer protocols.

Balachandran *et al.* [4] also characterized user behavior in wireless networks using Internet traffic. Traffic traces were collected from a wireless network during a computer networking conference. These traces contained aggregated packet level statistics of the link and network, transport, and application layers, together with information about users in different hotspots such as MAC addresses and signal-to-noise-ratios (SNR). The authors investigated session duration, data rates, popular protocols, and user mobility. Moreover, they studied network performance at different hours of the day. Their main findings were as follows: (1) users primarily use the Internet for browsing and SSH communication, (2) they tend to connect for short time sessions, and (3) their data usage is uneven. The authors observed specific data patterns from wireless hotspots at the conference, with patterns of use associated with specific concentrated locations and specific periods of time. This pattern is similar to that of classrooms, airport gates, etc.; however, it is not characteristic of all wireless hotspots in public areas.

Other studies employed machine learning in order to classify users' behavior based on Wi-Fi traffic. ToGo [5] is a system used to predict the length of a user's stay at a Wi-Fi hotspot. The authors presented SVM models for predicting this, starting with a basic model with no feedback, solely based on time and RSSI (signal strength)/bitrate, and progressing to models that use smartphones sensors such as the accelerometer. The paper also demonstrates the use of ToGo in an experiment involving 15 users. The Mo-Fi system [10] predicts the length of stay of users in a hotspot. It extracts features from three types of Wi-Fi packets: Wi-Fi probe requests, probe responses, and data messages. Using the k-means clustering algorithm, the system clusters Wi-Fi users into four different groups: outside, walkbys, bounced, or engaged. The writers evaluated their system in a real life office environment and achieved a human presence detection rate of 87.4 %.

Namiot [8] presented SpotEX – an Android application that displays ads to users based on the SSID and other publicly available information about nearby hotspots. Unfortunately, SpotEX requires an agent to be present on the user device and it relies on a-priory knowledge of the hotspot types in order to infer the user context. In contrast to SpotEX we infer the hotspot types from the Internet traffic only and do not require an agent on the user's phone.

3 Wi-Fi Traffic Dataset

In the course of the current study we collected the traffic dumps of smartphone users within a specific geographic area. In this section we describe the data collection process, cleanup, and feature extraction as presented in Fig. 1.

3.1 Data Collection

Internet traffic data was collected from the smartphone devices of 115 students from Ben-Gurion University of the Negev for a period of 30 to 60 consecutive days during

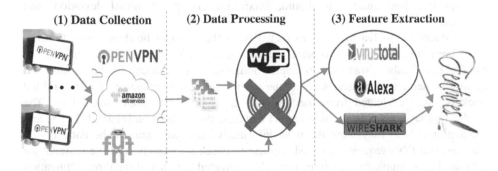

Fig. 1. Dataset processing scheme: (1) Client application redirects Internet traffic to a server and records WiFi connection/disconnection events. (2) Cellular traffic is filtered out. (3) Features are extracted for each session.

2014 and 2015. The traffic includes both Wi-Fi and cellular data that was collected using a dedicated VPN service. As such, traffic dumps contain IP packets but no data-link layer frame headers. In addition to the traffic dumps, we collected information about Wi-Fi hotspots using a dedicated Android application. All data was securely stored on a research server and analyzed using software tools for the purpose of Internet usage analysis in accordance with the permission of the university's ethics committee and subjects' consent.

3.2 Data Processing

We aggregated the collected IP packets into sessions (a session was either a TCP session or a UDP message and its response). First, we associated each session with a venue. Since much of the activity was performed on a university campus, an urban area with multiple venues such as offices and coffee shops located in the same buildings, we could not rely on GPS localization for reliable labeling of venues. Therefore, in the current study we used only Wi-Fi traffic. Every session was associated with a venue according to the SSID and BSSID of the hotspot which were collected by our application (a total of 978 hotspots). All sessions taking place between consecutive Wi-Fi connection and disconnection events were associated with the respective BSSID. The rest of the sessions were considered cellular traffic and were disregarded.

Next, we associated each hotspot with the venue it is located in. The hotspots were manually labeled based on SSID and BSSID. If the hotspot had very few sessions or it could not be labeled, its traffic was removed from the dataset. At the end of this process the data contained sessions for 738 different hotspots.

3.3 Feature Extraction

Following the hotspot labeling and aggregation of traffic into sessions, we extracted the following sets of features (summarized in Table 1).

Communication features focus on traffic volume, the ratio between sent and received traffic, duration of sessions, packet arrival times, and the amount of lost packets. Statistical values were extracted for each of those attributes: average, median, first quarter, third quarter, minimum, maximum, entropy, standard deviation, and variance.

Protocol-based features are extracted from the protocol headers. From the network layer (IP protocol) we extracted statistics about the TTL (time to live) values, and the GeoLite database was used to map IP addresses to their countries. Port usage and quality of service attributes, such as the number of lost segments and retransmitted packets, were evaluated from the transport layer protocols (TCP and UDP). The application layer was also used to extract features. We focused on HTTP, HTTPS, and DNS protocols as they were the most informative and prevalent in the data. HTTP cookies, bad DNS requests, and SSL certificate check are examples of the features we extracted. Communication features were also extracted for each of the three application layer protocols separately (e.g., traffic volumes of HTTP traffic).

The protocol-based features and the communication features were both extracted using T-Shark (www.wireshark.com).

Domain-based features are related to the popularity, security rank, and domain category (based on the browsing activity of users). These features were extracted using third party services. The domain names of sessions were taken from the DNS requests and HTTP/S host names. Domain categorization was based on Websense Threat Seeker and Bit Defender categories. A single category was derived by grouping similar categories and integrating the information provided by both services. The domains' security scores were based on the WoT (Web of Trust) rating tool, and the popularity rank was taken from the Alexa ranking service.

Table 1. Extracted features.

Feature Category	Sub-Category	Features
Communication	Sessions	total bytes, bytes out, bytes in, duration, in/out ratio
	Packets	Aggregation of the packets' arrival time, total bytes, bytes out, bytes in, and lost packets to sessions, computed by calculating the average, median, first quarter, third quarter, minimum, maximum, standard deviation, entropy, and variance
Protocol-based	Network Layer	TTL statistics, countries (nominal feature)
	Transport Layer	Port ratios including 80, 443, 5223 (used by Google Play), and other, lost segments, retransmitted/out of order/duplicated packets
	Application Layer	HTTP cookie count, HTTPS cipher key/certificate count, bad DNS requests. Communication features calculated separately for HTTP, HTTPS, and DNS traffic.
Domain-based	Category	Nominal features: Bit Defender + Websense category, WhatsApp/ Facebook
	Security	WoT score, Webutation score
	Popularity	Alexa Rank

Notice that the feature extraction process is automatic and does not require any manual process. The information is extracted from the traffic itself and from third party services (e.g., WoT) using their API. Therefore, the process can be deployed in real life applications.

4 Venue Type Inference

4.1 Types of Hotspots

Wi-Fi hotspots can be roughly divided into private and public hotspots. Private hotspots usually implement access control and are password protected. In contrast, public hotspots are designed to serve a wide range of casual users and are usually open or their password is readily available to the targeted population. In this study, we divide the hotspots into four main classes: home, organization, waiting, and hangout hotspots. We consider the first two as private, while the latter two are considered public.

Home hotspots (H) are used to connect multiple devices to the Internet at private residences. Usually, they are characterized by a single access point, a small number of users, and high quality service. Most home hotspots are protected with passwords.

Organization hotspots (O) represent the second class of hotspots. This class of hotspot is maintained by IT professionals employed by commercial organizations, education facilities, etc. In most cases these are restricted password protected networks which are subject to strict access controls. Organizational hotspots are characterized by high QoS and tight security policies.

Hangout hotspots (HO) are located in public venues such as bars, restaurants, shopping malls, etc. Similar to waiting hotspots, the security and QoS of hangout hotspots are not usually high.

Waiting hotspots (W) are defined as public hotspots which are located in waiting areas such as public transportation, hair salons, etc. Hotspots of this type have high user turnover during short periods of time. The Internet behavior of users in waiting hotspots is expected to be characterized by brief browsing and "time killers" such as games and social networks. The security and QoS of waiting hotspots tends to be low.

4.2   Connection Windows

We infer the type of venues based on the traffic of smartphone users connected to Wi-Fi hotspots located in the respective venues. User's Wi-Fi traffic is generally available to ISP providers, VPN, and proxy servers. These parties can use user context for commercial and security applications as described in Sect. 1, even when other information sources (such as location services) are disabled.

The general data and statistics we collect for a single session are not sufficient to determine the type of the hotspot a user is connected to. Therefore, in the following analysis we aggregate sessions in small chunks denoted as connection windows (CW) during which the smartphone is connected to the same hotspot. Since devices are disconnected and reconnected to the same hotspot within minutes (for example, due to low signal strength), we allow an idle time (up to 30 min) between consecutive sessions within the same CW. In total we identified 37,714 CWs, most of which were associated with home hotspots, as shown in Table 2.

Table 2. Distribution of CWs based on hotspot type.

Hotspot type	Number of connection windows
Home (H)	27,367
Organization (O)	7,929
Hangout (HO)	2,708
Waiting (W)	720
Total	**37,714**

We aggregated the features of the sessions within each CW in order to define CW features; the following method was used: For every numerical session feature (see Sect. 3.3), we calculated the average, median, minimal, and maximal values across all of the sessions associated with the CW. For domain category features, we created numerical features that represent the categorical value's incidence in the CW. For example, if 50 sessions occurred in a CW, of which 30 are from the "search" category and 20 are categorized as "news," the value of the "search" and "news" features for the CW is 0.6 and 0.4, respectively, and the values of other domain category features are 0. The feature aggregation process resulted in ~ 250 numerical features, after filtering out features that provided no information.

Three additional features were defined for CWs: the number of sessions in the CW, its day of the week (e.g., Sunday), and the time of day (8am-12am, 12am-16 pm, etc.). Both the "day of the week" and the "time of day" attributes were nominal, with seven possible values each. CWs that occurred in more than one "time of day" or "day of the week" were assigned based on the majority of session start times. For example, if a CW began at 7:55am and ended at 8:10am, with the majority of the sessions starting between 7:55 and 7:59, the label would be 5am-8am. Finally, each CW was labeled with the type of hotspot.

4.3   Hotspot Type Classification

Next we evaluate a model that classifies the type of hotspot a user was connected to. The model was based on the CW data defined above. Every hotspot was associated with numerous CWs creating a diverse dataset with sufficient representation of each hotspot type. However, there were significantly more CWs associated with home hotspots than those associated with other types of hotspots. Therefore, we generated three random balanced datasets (Dataset1, Dataset2, and Dataset3), each containing 3,600 CW instances. Waiting hotspots accounted for 20 % (720 instances) of the dataset, while home, organization, and hangout CWs accounted for 26.67 % each (960 instances).

For each of the datasets we selected the best features using the correlation feature selection (CFS) algorithm with GreadyStepwise search. Then, we built classification models using the rotation forest meta-classifier with random forest as the base classifier. The accuracy of the classifiers was around 57 % with the area under the ROC curve (AUC) ranging from 0.73 to 0.90 as presented in Table 3.

Table 3. Multiclass classification results.

	Dataset1	Dataset2	Dataset3
Accuracy	57.75 %	57.67 %	58.5 %
Weighted AUC	0.81	0.82	0.83
Home AUC	0.75	0.73	0.76
Hangout AUC	0.90	0.90	0.90
Org. AUC	0.80	0.81	0.83
Waiting AUC	0.80	0.81	0.82

These basic classifiers were able to distinguish well between waiting and hangout CWs; however they often confused home and organization CWs. To solve this issue, we used the "1-vs-all" (1vsA) approach. In this approach, four different classifiers are trained. Each classifier tries to classify whether the CW is within a single hotspot type or within the group consisting of the other three hotspot types, e.g., a classifier for "home" or "other." After training these classifiers, every CW has been classified by the four classifiers and the label of the CW is determined by the model with the highest confidence. This approach showed some improvement in the results. More importantly, we noticed that it was better at classifying home and organization CWs, while it often confused between hangout and waiting CWs.

In order to combine the pros of both approaches, we employed a meta-classifier that combines the multiclass and 1vsA classifiers. This model first classifies instances using the 1vsA classifier. In case the output label is "home" or "organization", we use this label. Otherwise, we classify it using the multiclass classifier. Experiments with the meta-classifier were performed on the same datasets as the previous models. Each dataset was randomly divided into train and test groups (test groups of 100, 200, 400, and 800 instances). Table 4 summarizes the performance of the meta-classifier.

Table 4. Results of combining multiclass and 1-vs-all classifiers.

	Dataset1				Dataset2				Dataset3			
	H.	O.	HO	W.	H.	O.	HO	W.	H.	O.	HO	W.
Avg. Precision	.73	.55	.55	.63	.70	.50	.46	.67	.66	.62	.51	.70
Avg. Recall	.81	.66	.48	.50	.84	.61	.34	.57	.81	.71	.46	.51
Avg. F-measure	.77	.60	.51	.56	.76	.55	.39	.62	.73	.66	.48	.59
Accuracy	**76.5 %**				**75.1 %**				**78.1 %**			

4.4 Public Vs Private Classification

The inference of hotspot type is important, but sometimes simpler classification is preferred in order to achieve higher accuracy. For example, specifying whether the hotspot is public or private may satisfy a context oriented access control application. Therefore, we decided to classify whether a user is connected to a private (home or organization) or public (waiting or hangout) hotspot. We balanced the CW dataset by randomly removing private instances until we obtained a ratio of 50:50 between the labels. This process resulted a dataset of 4,856 instances. The classifier we created used AdaBoost with resampling and random forest algorithms. In a 10-fold cross-validation, the model's accuracy rate was 78.95 %. The ROC graph and its AUC measure are presented in Fig. 2.

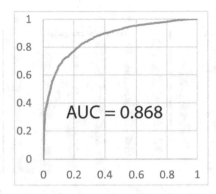

Fig. 2. ROC graph of public vs private CWs' classification.

5 Discussion

The experiments' results show that Internet behavior of users changes in different types of hotspots. Next we analyze the collected data in order to understand the nature of the differences in users' behavior in different types of venues. We define attributes of venue types similar to the CW features defined in Sect. 4.2. Specifically, we calculated the average values of all numeric session-features (e.g. WoT score) and derived the domain category incidence among all sessions associated with each type.

We depict the variability of the venue type attributes in Fig. 3. The bars in Fig. 3 present the normalized venue type attributes. For example, the incidence of search sessions is almost the same across different venue types while the incidence of entertainment sessions is larger in waiting venues than in all other venue types alto-gether. The relative incidences of popular domain categories (accounting for ~66 % of sessions) are similar across venue types in contrast to the less popular categories. We attribute this phenomenon to the fact that popular information services became a part of the everyday life and are used always, regardless location and context. In addition a large amount of the traffic to domains in popular categories is generated by the smartphone regardless the user behavior. Therefore, the difference in user behavior across different hotspots is reflected in actions associated with less popular categories, such as playing games (entertainment category) and reading news.

In order to stress the importance of unpopular categories for hotspot type classi-fication we present in Fig. 3 the results of the CFS algorithm. This algorithm selects a set of features that have high merit to the classification and low correlations between themselves. Unpopular categories were selected by the CFS algorithm because together they contribute to the classification. In contrast, the popular categories correlate to each-other and therefore, only one popular category was selected. In addition the differences in users' Internet behavior are reflected in the Alexa popularity rank[1],

---

[1] We reversed the Alexa rank such that popular domains receive high ranks.

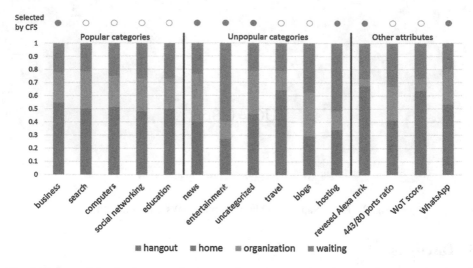

Fig. 3. Internet behavior feature selection results and comparison between hotspot types based on these features.

WoT security scores, and the use of the WhatsApp (instant messaging) application. We believe that the WoT feature was not selected by the CFS, because it correlates to the Alexa rank score, i.e., popular domains are often more secure than unpopular domains.

6   Conclusions

In this paper we study the user behavior in different types of venues and present the hotspot type classification problem. We show that venue type can be inferred from the Internet traffic of smartphone users. This type of inference can be used for advertisements, recommendations, access control and other context aware services. The prediction process can be automated as all the features.

The analysis of Internet behavior attributes shows that the majority of Internet traffic is similar in terms of domain categories and port usage. Nevertheless, users' behavior differs in access to less popular domain categories, instant messaging traffic, and in the domains' popularity and security ranks.

The subjects of the experiment were all students studying at a university in Israel; thus the data represents user behavior properties of only a segment of smartphone users. Despite the similar demographic properties of our subjects, the results show the feasibility of venue type classification based on user Internet traffic. In future we intend to expand and diversify the dataset in terms of demographics and geography.

Furthermore, we aim to classify the hotspot type using other information sources. For example, classify the hotspot type locally on the device using information that can obtained by applications (e.g., Wi-Fi connections, internet usage statistics and sensors).

References

1. Adomavicius, G., Tuzhilin, A.: Context-aware recommender systems. In: Ricci, F., Rokach, L., Shapira, B., Kantor, P.B. (eds.) Recommender Systems Handbook, pp. 217–253. Springer, US (2011)
2. Adults' Media Use and Attitudes Report. Ofcom (2014)
3. Afanasyev, M., Csirao, B.Q., Chen, T., Voelker, G., Snoeren, A.: Usage patterns in an urban WiFi network. IEEE/ACM Trans. Netw. 18(5), 1359–1372 (2010)
4. Balachandran, A., Voelker, G.M., Bahl, P., Rangan, P.V.: Characterizing user behavior and network performance in a public wireless lan. In: ACM SIGMETRICS International Conference on Measurement and Modeling of Computer Systems, pp. 195–205 (2002)
5. Manweiler, J., Santhapuri, N., Choudhury, R., Nelakuditi, S.: Predicting length of stay at WiFi hotspots. In: INFOCOM, 2013 Proceedings IEEE. Turin (2013)
6. Mark Hall, E.F.: The WEKA data mining software: an update. SIGKDD Explor. 11(1), 11 (2009)
7. Miettinen, M., Heuser, S., Kronz, W., Sadeghi, A.-R., Asokan, N.: ConXsense – context profiling and classification for context-aware access control. In: ASIACCS (2014)
8. Namiot, D.: Context-aware Browsing – a practical approach. In: 2012 6th International Conference on Next Generation Mobile Applications, Services and Technologies. Paris (2012)
9. Pentland, A.S., Aharony, N., Pan, W., Sumter, C., Gardner, A.: Funf: Open sensing framework (2013)
10. Qin, W., Zhang, J., Li, B., Zhu, H., Sun, Y.: Mo-Fi: discovering human presence activity with smartphones using Non-intrusive Wi-Fi sniffers. In: 2013 IEEE International Conference on Embedded and Ubiquitous Computing (HPCC_EUC) (2013)
11. The Infinite Dial 2013. Navigating Digital Platforms. Edison Reseach and Arbitron (2013)

I²Rec: An Iterative and Interactive Recommendation System for Event-Based Social Networks

Cailing Dong[1], Yilin Shen[2], Bin Zhou[1(✉)], and Hongxia Jin[2]

[1] University of Maryland, Baltimore County, USA
{cailing.dong,bzhou}@umbc.edu
[2] Samsung Research America, San Jose, USA
{yilin.shen,hongxia.jin}@samsung.com

Abstract. Event-based social networks (EBSNs) such as Meetup and Plancast have been emerging in recent years. In addition to the online virtual groups and connections on many existing Online Social Networks (OSNs), EBSNs also provide a platform for users to initialize and manage offline physical events in which user's activities are strongly geographically constrained. As an increasing number of users are attracted by EBSNs, it is highly desirable to provide users with accurate recommendations of both online groups and offline events, which has become an urgent need. In this paper, we propose a comprehensive study for recommending users online groups and offline events on EBSNs. To represent user's interactions via a timeline horizon, we design an *Iterative and Interactive Recommendation System (I² Rec)*, which couples both online user activities and offline social events together for providing more accurate recommendation. Our proposed *I² Rec* system infers a user's online and offline activities in turn and iteratively enriches the training information based on user's feedback. Using the large-scale real-world dataset crawled from Meetup, our recommendation system outperforms other baseline approaches significantly. More importantly, the empirical results also validate that our proposed system can continuously provide accurate recommendation over time by capturing users' changing interests.

1 Introduction

In the past decades, a large number of social networks have been emerging for different purposes. One of the most popular classes of social networks is *Event-Based Social Network* (EBSN) [8], such as Meetup (http://www.meetup.com), Eventbrite (http://www.eventbrite.com), and so on. According to Statistics of 2015, there are on average 525,716 Monthly Meetups and 3.61 million Monthly RSVPs on Meetup, and over 1 million events all around 187 countries have been planned via Eventbrite. These social sites not only act as *online* social networks, on which users can build their profiles, join different social groups to share interests and opinions on the web, but also provide a platform for people to initialize and manage face-to-face *offline* social events, such as dining

© Springer International Publishing Switzerland 2016
K.S. Xu et al. (Eds.): SBP-BRiMS 2016, LNCS 9708, pp. 250–261, 2016.
DOI: 10.1007/978-3-319-39931-7_24

out, techniques symposium, business meetings and so on. As such, it is highly desirable to provide users accurate recommendations of both online groups and offline events.

Recently, EBSNs have attracted more and more research attentions. Gomez-Rodriguez *et al.* [4] first observed the impact of offline events on the evolution and temporal dynamics of the online connectivity. Liu *et al.* [7] later discovered some significantly different aspects of EBSNs from conventional social networks, including regular temporal and spatial patterns of social events, positively correlated online and offline social interactions and the investigations of communities and flows on EBSNs. However, their approaches and derived observations are not sufficient to provide users high-quality recommendations.

On the other hand, unfortunately, most existing solutions introduced in the literature are developed either on Online Social Networks (OSNs) only for online recommendations [9,12] or using barely offline interactions for event recommendation [3,5,10], and ignored much valuable information which is coherent on EBSNs. First, a user's participation of offline events and his/her online activities affect each other substantially. For example, there is a dramatic increase in the number of connections between event attendees shortly after the date of the event [4]. In addition, the majority of users who are in the same online group indeed attend some local events frequently together [7]. Secondly, users' social interests are discovered to be varied and drifting over time [2,6]. Google trends showed that the topics discussed over time might be very different [1]. Therefore, it is extremely desirable and challenging to *continuously* provide users accurate recommendation of both online groups and offline events over time, which indeed becomes an urgent need.

In this paper, we propose the *first Iterative and Interactive Recommendation System (I²Rec) for EBSNs*. In order to tackle the challenges of combining large amount of information and lacking user future interest information, I²Rec couples online user activities and offline social events together to predict user's future behaviors, in an interactive manner by taking users' feedback into account. Such iterative inference of a user's online and offline activities enriches the training information based on user's feedback, thereby not only maintaining highly accurate recommendation over time but also capturing how users' interest changes as well.

Our contributions are summarized as follows: first, *we propose an iterative and interactive recommendation system named I²Rec for event-based social networks*. The proposed method monitors user's activities using a time horizon and iteratively provides recommendation of online groups and offline events whenever the user performs an action. Our I²Rec system supports both content-based and collaborative filtering-based recommendation algorithms. The recommendation generated by I²Rec is dynamic.

Second, *we utilize both user's online and offline interactions in EBSNs to generate recommendations*. Some existing research [7] showed that user's online and offline interactions are coherent in EBSNs. Our work provides a comprehensive study to bridge both online and offline information for more accurate recommendation in EBSNs.

Third, *we conduct an empirical study using a large real-world dataset crawled from Meetup*. The proposed recommendation methods are conducted using the iterative and interactive manner. The experimental results strongly validate the effectiveness of high-quality recommendations and users' interest tracking of our proposed method.

The remaining of the paper is organized as follows. In Sect. 2, we first discuss the properties of event-based social networks (EBSNs) and adopt Meetup, a popular EBSN, to illustrate the important characteristics. Then, in Sect. 3, we formalize the recommendation problem in EBSNs, and provide an iterative and interactive recommendation system. The detailed recommendation algorithms are presented afterwards in Sect. 4, followed by a set of results based on our empirical evaluation in Sect. 5. Finally we conclude the paper and outline some future directions in Sect. 6.

2   Event-Based Social Networks

Users' online and offline activities in EBSNs have distinct characteristics. First, *user's offline social interactions are strongly driven by their physical locations*. In [8], Liu *et al.* analyzed a large Meetup dataset and revealed that more than 80 % events that Meetup users participated in were held within 10 miles of users' home location. In addition, about 85 % Meetup users who participated in same offline events actually lived within 10 miles to each other. The results verified the locality of users' offline activities in EBSNs.

Second, *user's online social activities are also greatly affected by their geographical locations*. In many EBSNs such as Meetup, users can become a member of some online groups and communicate with others in the same group. Being a member of some online groups can help users to identify interesting events to attend. As events are strongly location-driven, user's online social activities in EBSNs are also location constrained. For example, Liu *et al.* [8] reported that more than 70 % Meetup users who are in the same online group in fact lived within 10 miles to each other.

Third, *user's online and offline social activities are coherent in EBSNs*. Users in the same online group will likely attend one or more local events together. Moreover, users that frequently participate in some offline events together tend to share common interests. Thus, the groups they join online would have large overlaps. A positive correlation between user's online and offline interactions was reported in [8].

To further explore the properties of online and offline activities in EBSNs, we adopt *Meetup* as an example. Meetup is a popular event-based social networking platform which has attracted more than 13 million active users from more than 190 countries. Meetup users can organize offline social events and disseminate

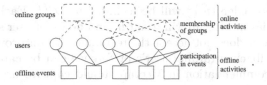

Fig. 1. The general structure of Meetup, one of the most popular event-based social networks (EBSNs) in the literature.

the event information to relevant users. Users in Meetup can also formalize online social groups with others who share similar interests. Figure 1 sketches the general structure of Meetup.

3 Recommendation in EBSNs

A recommender system can predict user's preferences on specific data items. Existing recommendations in conventional social networks are mainly generated based on user's online social activities. However, as we discussed before, due to the coherent relationships and their own properties between user's online and offline activities, they should be coupled tightly to provide accurate recommendation in Meetup.

In this paper, we model the Meetup dataset as a simple graph $D = (U, G, E, A_G, A_E)$, where U, G, and E represent a set of users, online groups, and offline events, respectively. $A_G \subseteq U \times G$ represents users' membership of those online groups, and $A_E \subseteq U \times E$ represents users' participation in some offline events. User's online social activities are captured using a bipartite graph $D^{on} = (U, G, A_G)$ and user's offline social activities are modeled using another bipartite graph $D^{off} = (U, E, A_E)$.

In Meetup, online groups and offline events are the two major items related to users. Therefore, the recommender system in Meetup needs to generate lists of both online groups and offline events as recommendation. As users' preferences may be very different, recommendation of online groups and offline events needs to be customized to meet each individual's expectation. Intuitively, a good recommender system should take into consideration rich information in the Meetup dataset, especially user's online and offline activities, to provide accurate recommendation.

Another important issue in recommendation is the timepoint when the recommendation is conducted. In EBSNs, user's online and offline activities are continuous and interleaving. The various timepoints of user's activities in Meetup in fact can be modeled as a time sequence data stream $\mathcal{T} = \{(t_0, a_0), (t_1, a_1), \ldots, (t_i, a_i), \ldots\}$, where t_0 is the timepoint when user performs the action of registration in Meetup. At timepoint t_i ($i \geq 1$), the user can conduct an action a_i, either joining a new online group or attending a new offline event. When a user $u \in U$ joins a new online group $g \in G$, a new edge (u, g) will be inserted into D^{on}. Similarly, when a user $u \in U$ attends a new offline event $e \in E$, a new edge (u, e) will be included in D^{off}. Each of user's online and offline activities in Meetup will result in an update of user's activity history. In addition, if a user does not accept the recommendation provided by Meetup, some implicit feedback regarding user's preference can be captured as well. In summary, the recommendation in Meetup should capture user's feedback and reflect the changes timely.

To meet the requirements of recommendation in EBSNs, we propose $\mathbf{I^2Rec}$, an **I**terative and **I**nteractive **Rec**ommendation system depicted in Fig. 2. The proposed system monitors user's activities in Meetup using a time horizon and iteratively provides recommendation of online groups and offline events whenever a user conducts some activities. Each time when a user joins a new online group or RSVPs a new offline event, the recommendation system will keep track of user's activities. I^2Rec will automatically trigger the recommendation algorithm to suggest to the user some interesting groups and events.

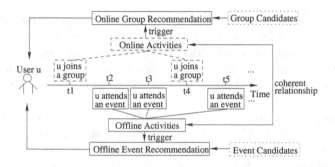

Fig. 2. $\mathbf{I^2Rec}$, an iterative and interactive recommendation system in EBSNs.

The proposed I^2Rec system has several advantages. First, the recommendation of I^2Rec is not a one-time decision. Instead, it provides recommendation of online groups and offline events dynamically. Second, I^2Rec generates the recommendations considering both online and offline activities in EBSNs. Due to coherent relationship between user's online and offline activities, our system can couple user's online and offline behaviors tightly. Third, user's preferences in EBSNs may fast evolve. Our I^2Rec system monitors user's behavior and quickly responds if user performs some online/offline activities. Thus, the recommendation of I^2Rec can be personalized in a timely manner. Last but not the least, the proposed I^2Rec framework is highly customizable. In addition to user's explicit activities in EBSNs such as joining a group or participating in an event, there are some mechanisms to capture user's implicit feedback. For example, if Meetup recommends a group "Mid-Atlantic Hiking Group" to a user; however, the user does not accept the recommendation. It indicates implicitly that the user is not a hiking fan. Whenever such implicit feedback is captured by the system, I^2Rec can also trigger the recommendation algorithms. The triggering timepoints in I^2Rec can be customized to meet domain-specific requirements.

4   Recommendation Algorithms

We propose a probabilistic model for group and event recommendation in Meetup. In order to provide accurate online group recommendation, we attempt to estimate the likelihood that a user may join an online group. Such probabilistic estimation is based on the influence of both user's online and offline social interactions. A fusing technique is introduced to bridge the influence from different social interactions. In addition, considering the intrinsic property of users' tendency to participate in those geographically nearby groups, we also provide an effective *candidate selection* process to reduce the candidate groups for recommendation.

Preliminary of the Probabilistic Method.   In EBSNs, user's social interactions in an online group are greatly influenced by other members in the same group. Intuitively, if the majority of users in a group g are closely connected to u, u has a strong relationship with the group g. Mathematically, we can model the intrinsic relationship between a user u and a group g, denoted as $R(u, g)$ using the following formula:

$$R(g, u) = \frac{\sum_{u' \in g} Sim(u, u')}{|U_g|},$$ (1)

where $U_g = \{u | u \in U \wedge (u, g) \in A_G\}$ represents the set of users joined the group g and $Sim(u, u')$ is a symmetric metric which denotes how closely the two users u, u' are connected. A_G represents the membership of groups in the online bipartite graph D^{on} in Meetup.

The metric $Sim(u, u')$ can be modeled in different ways. A simple solution is to utilize the tags assigned to each user in Meetup. Based on the given tag vocabulary, the tags of a user can be modeled as a vector using the Vector Space model. The *cosine similarly* is widely used to measure the distance between vectors. Thus, we have the *interest-based user similarity*, denoted as $Sim^i(u, u') = \frac{T_u \cdot T_{u'}}{\|T_u\| \|T_{u'}\|}$, where T_u and $T_{u'}$ are the tag vectors of users u and u', respectively.

Other than interest tags, user's social interactions such as event participation can also be used for evaluating whether users are closely connected. Suppose users u and u' have participated in $|E(u)|$ and $|E(u')|$ events, respectively. User's behaviors of event participation can be represented using two event vectors $V_{E(u)} = [e_1^u, e_2^u, \ldots, e_{|E(u)|}^u]^T$ and $V_{E(u')} = [e_1^{u'}, e_2^{u'}, \ldots, e_{|E(u')|}^{u'}]^T$. Thus, based on the *cosine similarity*, we have the *behavior-based user similarity*, denoted as $Sim^b(u, u') = \frac{V_{E(u)} \cdot V_{E(u')}}{\|V_{E(u)}\| \|V_{E(u')}\|}$.

By constituting the $Sim(u, u')$ metric in Eq. 1, we can obtain two different ways of modeling relationship between a user u and a group g, that is,

$$R^i(g, u) = \frac{\sum_{u' \in g} Sim^i(u, u')}{|U_g|},$$ (2)

which is based on user's interest tags, and

$$R^b(g, u) = \frac{\sum_{u' \in g} Sim^b(u, u')}{|U_g|}, \tag{3}$$

which is based on user's behavior.

Probabilistic Method for Group Recommendation in Meetup. In our probabilistic method, we attempt to estimate the likelihood that a user u may join a group g, based on the fact that u has been a member of a few groups $G(u)$. We use the notation $P(g|G(u), u)$ to represent such probability.

Social interactions in Meetup play an important role when user joins groups online. Thus, $P(g|G(u), u)$ is largely affected by other users that are socially related to u. We use $I(u)$ to denote the set of users who have interactions with u. For each user $u' \in I(u)$, if u' and u are closely connected, and u' has strong relationship with a group g; then, u is likely to join group g as well due to the large influence of u'. Specifically, we have:

$$P(g|G(u), u)$$
$$\propto \frac{\sum_{u' \in I(u)} \sum_{g^* \in G(u)} R(g^*, u') \times R(g, u') \times Sim(u, u')}{|I(u)|}$$
$$\propto \sum_{u' \in I(u)} \sum_{g^* \in G(u)} R(g^*, u') \times R(g, u') \times Sim(u, u'). \tag{4}$$

where $R(\cdot, \cdot)$ represents the relationship between a user and a group, and $Sim(u, u')$ denotes the connection between two users. As we discussed before, both user's interests and behaviors are useful for the metrics $Sim(u, u')$ and $R(\cdot, \cdot)$. To integrate them together, we introduce a parameter $\alpha \in [0, 1]$, which controls the weights of user's interests and behaviors in the probabilistic method. Thus, Eq. 4 can be rewritten as:

$$P(g|G(u), u)$$
$$\propto (1 - \alpha) \times \sum_{u' \in I(u)} \sum_{g^* \in G(u)} R^i(g^*, u') \times R^i(g, u') \times Sim^i(u, u')$$
$$+ \alpha \times \sum_{u' \in I(u)} \sum_{g^* \in G(u)} R^b(g^*, u') \times R^b(g, u') \times Sim^b(u, u') \tag{5}$$

Equation 5 is the core of our probabilistic-based method for recommendation. For each user u and any group $g \in \{G - G(u)\}$, the equation calculates a probabilistic score which indicates the likelihood that u will be a member of g. To generate group recommendation, the probabilistic model can first calculate the score $P(g|G(u), u)$ for all the groups $g \in \{G - G(u)\}$. The groups will be sorted based on the probabilistic score descending order and the k top-ranked groups can be returned as recommendation.

There is still one issue left unsolved in the probabilistic model: *how to determine $I(u)$, the set of users that interact with u in Meetup?* Obviously, both online social activities and offline social activities in Meetup should be considered. We propose two different ways to implement our probabilistic method for recommendation.

Online **Co-member-based Probabilistic Method (Co-mPM).** Users interact with others online via being members of the same online group. Thus, for a specific user u, other users that may interact with u online must come from those groups u has joined. We have: $I(u)^{Co-mPM} = \{u'|(u' \in U) \land (u' \neq u) \land (\exists g^* \in G, (u, g^*) \in A_G \land (u', g^*) \in A_G)\}$, where A_G represents the membership of groups in the online bipartite graph D^{on} in Meetup.

Based on $I(u)^{Co-mPM}$, Eq. 5 corresponds to our online co-member-based probabilistic method. The metric $R(\cdot, \cdot)$ can be simply calculated in the following way. Consider a user $u' \in I(u)^{Co-mPM}$ and a group $g \in G$, if u' is already a member of g, we let $R^i(g, u') = 1$ and $R^b(g, u') = 1$. If $g \notin G(u')$, we adopt Eqs. 2 and 3 to calculate $R^i(g, u')$ and $R^b(g, u')$, respectively.

Offline **Co-participant-based Probabilistic Method (Co-pPM).** Users also interact with others by attending some offline events together. Users' offline activities are coherent with their online activities. We consider offline activities to model $I(u)$. For a user u, other uses that may interact with u offline must have participated in some offline events together before. Thus, we have: $I(u)^{Co-pPM} = \{u'|(u' \in U) \land (u' \neq u) \land (\exists e^* \in E, (u, e^*) \in A_E \land (u', e^*) \in A_E)\}$, where A_E represents users' participation in offline events in the offline bipartite graph D^{off} in Meetup.

Therefore, Eq. 5 based on $I(u)^{Co-pPM}$ corresponds to our offline co-participant-based probabilistic model. The similar procedure discussed in the previous section can be adopted to calculate the values of $R^i(g, u')$ and $R^b(g, u')$.

Probabilistic Model for Event Recommendation in Meetup. Intuitively, an offline event can be treated as an "offline group" where its members interact with others in some face-to-face local gatherings. Thus, the probabilistic model for group recommendation can be extended for event recommendation. However, comparing to online groups, the attendance of offline events is even more influenced by user's physical locations.

We adopt the *candidate selection* process to facilitate the unique property of offline events for event recommendation in Meetup. In Meetup, offline events are held by specific online groups. When a user RSVPs an event, the user should be a member of the corresponding group. Thus, we focus on those events which are held by the online groups that the user has joined. In addition, most of the groups in Meetup hold the events weekly. At a specific timepoint, we only need to consider those events that will take place in the same week. As a result, the events in the candidate set will be both location and time constrained.

5 Experimental Results

In this section, we report some findings from the experiments. All the experiments were carried out on a workstation running Windows 7 operating system, with a 3.2 GHz Intel Core i5 CPU, 8.0 GB main memory, and a 500 GB hard disk. The programs were implemented in Java.

5.1 Experimental Preparation

To evaluate our proposed I^2Rec system and various recommendation methods, we crawled a large-scale Meetup dateset using Meetup API from www.meetup.com. The crawling starts with the extraction of groups for three seed users from different states, followed by the extraction of group members from those groups, and finally obtains all of the events and users' RSVPs from Meetup. The above crawling procedure was repeated three times to obtain a large Meetup dataset. The statistics of our dataset are shown in Table 1.

Table 1. The statistics of the Meetup dataset.

#Groups	#Users	#Events
86,396	6,532,384	6,088,084
#RSVPs	#Topic-tags	#Interest-tags
45,173,583	880,176	52,654,436

To simulate the recommendation in Meetup, we have to organize user's activities using a time horizon. For a testing user u who has joined $|G(u)|$ online groups and participated in $|E(u)|$ offline events, we first sort all of u's activities (both online and offline) based on the timestamps of those activities. We use t_0 to denote the timepoint when u registers with Meetup and t_i ($1 \leq i \leq |G(u)| + |E(u)|$) to denote the i-th timepoint in the sorted activity list of u.

The recommendation of groups and events is regarded as a ranking problem in this paper. We adopt *Mean Average Precision (MAP)* as our evaluation metric. MAP provides a single-figure measure in precision across all recall levels. This is ideal as our evaluation needs to be conducted on a time horizon.

5.2 Experimental Results of the Evaluation

For group recommendation, we considered the following methods described in the paper: *online* co-member-based probabilistic method (**Co-mPM**) and *offline* co-participant-based probabilistic method (**Co-pPM**). In addition, as baseline methods, we also considered the simple Group Tag Similarity method (**GTS**) based on *Jaccard similarity*, and the conventional Probabilistic Matrix Factorization (**PMF**) [11] in Recommender Systems.

Table 2. Overall performance of group recommendation and event recommendation.

Group	Co-mPM	Co-pPM	GTS	PMF
$Avg(MAP)$	0.505	0.524	0.088	0.048
Event	Co-mPM	Co-pPM	Content	PMF
$Avg(MAP)$	0.672	0.775	0.529	0.487

For event recommendation, we considered the following recommendation methods: *online* co-member-based probabilistic method (**Co-mPM**) and *offline* co-participant-based probabilistic method (**Co-pPM**). As comparison, we also considered the simple content-based event recommendation method (**Content**) based on calculating event similarity using features of event keywords, time and location, and the conventional Probabilistic Matrix Factorization (**PMF**).

All of the recommendation methods were incorporated into the iterative and interactive system, I^{2}Rec, as described in Sect. 3.

The results in Table 2 show the average recommendation performance for all the users. Specifically, for each user, we calculated the average MAP values for all the iterations of recommendation. The MAP values are then averaged based on the total number of users in the dataset. The results in Table 2 verify that the probabilistic methods achieve the best performance for both group recommendation and event recommendation.

We also examined the average performance at different timepoints. Figure 3 shows the average performance of online group recommendation in each iteration averaged over all the users. The x-axis represents the index of each iteration

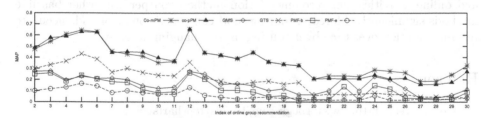

Fig. 3. Overall Performance of online group recommendation

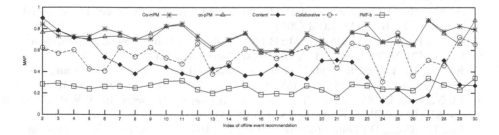

Fig. 4. Overall Performance of offline event recommendation

and the y-axis depicts the corresponding average MAP values. We have several interesting findings. First, the probabilistic methods Co-mPM and co-pPM outperform other methods significantly along with the time horizon. Second, at some iterations, the recommendation performance drops. This may be related to user's interest change. However, the MAP values can be picked up after a few iterations. This verifies that our I^2Rec system can be used to capture user's dynamic interest drift timely. Third, during the simulation, the number of positive instances decreases continuously. However, the MAP values are still at a high level, even for the last few iterations. The results clearly verify that our I^2Rec system is capable of capturing user's dynamic interest change in a timely manner.

In Fig. 4, we depict the average performance of event recommendation along with the iterations. Similar findings can be observed. The probabilistic methods generate more stable and better results in general. In addition, the MAP values are kept at a higher level at different iterations.

6 Conclusion

In this paper, we studied the problem of recommendation in EBSNs. To model user's continuous activities in EBSNs, an iterative and interactive recommendation system, I^2Rec, is developed. I^2Rec infers a user's online and offline activities in turn and iteratively enriches the training information based on user's feedback. We designed *probabilistic*-based recommendation algorithms, which is able to couple user's online and offline interactions to provide accurate recommendation in EBSNs. The experimental results verified that by coupling online and offline activities, our recommendation methods outperform other baseline methods significantly. Moreover, our method can continuously provide accurate recommendation over time by capturing users' changing interests.

References

1. Google trend. http://www.google.com/trends/topcharts
2. Cao, H., Chen, E., Yang, J., Xiong, H.: Enhancing recommender systems under volatile userinterest drifts. In: CIKM 2009, pp. 1257–1266. ACM, New York, NY, USA (2009)
3. de Macedo, A.Q., Marinho, L.B.: Event recommendation in event-based social networks. In: HT 2014, ACM (2014)
4. Gomez Rodriguez, M., Rogati, M.: Bridging offline and online social graph dynamics. In: CIKM 2012, pp. 2447–2450. ACM, New York, NY, USA (2012)
5. Khrouf, H., Troncy, R.: Hybrid event recommendation using linked data and user diversity. In: RecSys 2013, pp. 185–192. ACM, New York, NY, USA (2013)
6. Li, X., Guo, L., Zhao, Y.E.: Tag-based social interest discovery. In: WWW 2008, pp. 675–684. ACM, New York, NY, USA (2008)
7. Liu, B., Xiong, H.: Point-of-interest recommendation in location based social networks with topic and location awareness. In: SDM, pp. 396–404. SIAM (2013)

8. Liu, X., He, Q., Tian, Y., Lee, W.-C., McPherson, J., Han, J.: Event-based social networks: Linking the online and offline social worlds. In: KDD 2012, pp. 1032–1040. ACM, New York, NY, USA (2012)
9. Ma, H., Zhou, D., Liu, C., Lyu, M.R., King, I.: Recommender systems with social regularization. In: WSDM 2011, pp. 287–296. ACM, New York, NY, USA (2011)
10. Qiao, Z., Zhang, P., Zhou, C., Cao, Y., Guo, L., Zhang, Y.: Event recommendation in event-based social networks. In: AAAI 2014 (2014)
11. Salakhutdinov, R., Mnih, A.: Probabilistic matrix factorization. In: NIPS 2008, vol. 20 (2008)
12. Yang, X., Steck, H., Liu, Y.: Circle-based recommendation in online social networks. In: KDD 2012, pp. 1267–1275. ACM, New York, NY, USA (2012)

Modeling Influenza by Modulating Flu Awareness

Michael C. Smith[✉] and David A. Broniatowski

The George Washington University, Washington DC, USA
{mikesmith,broniatowski}@gwu.edu

Abstract. It is important for public health officials to follow both the incidence of disease and the public's perception of it, especially in the Internet-connected age. In the specific context of influenza, disease surveillance through social media has proven effective, but public awareness of influenza and its effects are not well understood. We build upon the existing Epstein model of coupled contagion with the aim of including modern media mechanisms for awareness transmission. Our agent-based model captures the unique effects of news media and social media on disease dynamics, and suggests potential areas for policy intervention to modulate the spread of the flu.

Keywords: Agent-based modeling · Influenza · Awareness · Coupled contagion · Surveillance

1 Introduction

Accurate and timely surveillance methods are important for monitoring potential epidemics. Influenza poses an annual threat across the US, leading much recent work to focus on observing the flu in particular [3,7]. Several authors draw on the Internet and social media data to perform this surveillance on flu and other diseases [2–4,6,12] because these media are readily available and offer a means of quickly and easily transmitting information.

Public health officials' ultimate goal is not merely to track a disease during an epidemic, but also to be able to attenuate the spread of the disease. This latter goal depends on effective communication: informing the public of appropriate behaviors to take that may so attenuate the spread; or in other words, it depends on measuredly making the public aware. Thus, we aim to better understand how the spread of this influenza awareness interacts with the spread of the disease when considering the effects of Internet-enabled media. In the Epstein model of coupled contagion (CC), the spread of a disease and awareness of the disease interact in emergent ways [8]. We motivate adding such Internet-enabled media into the CC model as potential mechanisms to modulate the spread of the disease. The resulting model will prove useful for theoretical insights. The outline of the rest of this paper is as follows: first we detail work in the area in a literature review, introducing among other things the CC model and the data from which

© Springer International Publishing Switzerland 2016
K.S. Xu et al. (Eds.): SBP-BRiMS 2016, LNCS 9708, pp. 262–271, 2016.
DOI: 10.1007/978-3-319-39931-7_25

we drew inspiration; next we provide the details of our model augmentation; third, we use our model to examine how these media might influence the spread of the flu; and finally, we conclude and identify directions for future work.

2 Literature Review

Information transmitted via the Internet and social media is not limited strictly to information about disease incidence. Hatfield described how emotions are "contagions" that spread through social interactions [9]; it follows that awareness of a disease is an instance of such a contagion. Because so many social interactions occur online, awareness of disease may also travel via the internet and social media. Along with other prominent work describing the role of concerned awareness on the spread of disease, Epstein's CC model [8] motivated us to recently extend our social media flu surveillance system [6] to measure signals of influenza awareness [14]. Using the social media microblogging site Twitter, Smith et al. specifically provided preliminary results comparing influenza awareness data, Twitter influenza tracking data (infection), and online news media data during the 2012–2013 flu season [1,14]. We showed that influenza awareness trends can be separated from influenza infection trends, and that news media correlate strongly with signals of influenza awareness [14]. While Broniatowski et al. have previously shown that influenza surveillance using Twitter is an effective predictor of actual influenza infection rates [2], little work has been done to examine influenza awareness and relate it to behavioral adaptations.

Prior work has shown that integrating behavioral choice into models of other diseases yields results that reflect empirical findings (as in the case of AIDS [11]). Furthermore, Qinling Yan and colleagues have found that media coverage, in particular, may have driven people's behavior during the H1N1 outbreak [16]. Along with the findings of our previous work, this current conversation motivates the novel contribution of this paper: augmenting the CC model with news media and social media as potential mechanisms to modulate the spread of the flu.

2.1 Epstein's Coupled Contagion Model

Classical epidemiological models use differential equations to track populations of Susceptible, Infected, and Recovered individuals [10]. Epstein et al. introduced the idea that "concerned awareness" of a disease spreads like a contagion, interacting with the dynamics of the disease itself [8]. Epstein et al. showed that incorporating behavioral adaptations that are driven by this awareness into models of epidemic prevalence have a qualitatively marked effect on how disease may spread. Considered adaptations include ignoring the disease, self-isolating or hiding from the disease, and fleeing from it.

Specifically, Epstein and colleagues showed that this coupled contagion model can be used to generate nontrivial trends such as multiple waves of infection. For example, Fig. 1 shows how large numbers of hiders can significantly reduce the infected population [8].

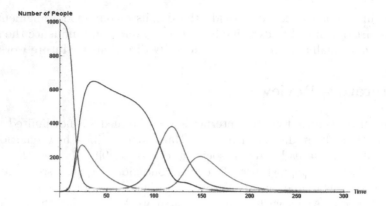

Fig. 1. "In the idealized run...susceptible individuals (blue-curve) self-isolate (black curve) through fear as the infection of disease proper grows (red curve)" Image source: [8] [Public domain, permission implicit]

CC assumed awareness required physical contact for transmission [8], but the interconnectedness of modern society provides additional social opportunities that go beyond spatial proximity. In the next section we extend the agent-based model (ABM) approach of Epstein et al. [8] by modeling media coverage and social media information exchange. Epstein showed that behavioral reactions to awareness affects disease spread [8]; we show that including the additional channels for awareness transmission changes the theoretical effects of awareness on behavior.

3   Methods

We iteratively build from the standard Susceptible-Infected-Recovered (SIR) model in the NetLogo model library [15,17] to an augmented version of CC that reflects today's interconnected society. The successive constructions are in the following subsections, with each being evaluated against the previous version of the model, holding all parameters constant between model changes where possible. We also conduct manual and automated sensitivity analyses. Steps 1–3 recreate the CC model. Steps 4–5 are our augmentations.[1]

3.1   Epstein's CC Model

1. Standard SIR. SIR models are standard in epidemiology, dividing the population into Susceptible, Infected, and Recovered populations [10]. One agent-based approach to SIR modeling (the "epiDEM basic" model in the NetLogo

---

[1] These descriptions are intended to provide insight into the motivation underlying these models. See https://bitbucket.org/mcsmith/awareness_abm for the full code that details all particularities of implementation.

library) has agents, representing people, that move in a random direction a random number of steps between 0 and their maximum movement parameter A, and after moving they then interact with a random other agent in the area if any others are there [17]. If an agent is infected and another susceptible, the infected agent has a probability B to infect the susceptible agent. Infected agents recover after being infected for C time steps. Agents are initialized in the infected state (as opposed to in the susceptible state) with density D, meaning $D\%$ of agents start infected. The model run ends when disease prevalence equals zero.

2. Contagion of Coupled Awareness. In addition to disease contagion, Epstein [8] modeled concerned awareness of the disease like a contagion, with corresponding parameters E, F, and G respectively indicating the probabilities of transmission, recovery rate, and initial density respectively. Agents can be aware of a disease, infected with the disease, or both, and agents can "catch" awareness from another agent who is aware and/or infected. Contrasting with the epiDEM model [17], CC introduced a set world size to the model to allow measuring if the disease spreads fully across the world, and also introduced the notion of a disease beginning with a single agent [8] instead of relying on D.

3. Behavioral Response to Awareness (e.g. Hiding, Fleeing, Ignoring). Epstein modeled behavioral responses to awareness and captured how these responses modified the spread of the disease [8]. The population of agents is divided into $H/100$ % hiders, $I/100$ % fleers, and $J/100$ % ignorers (such that $H + I + J = 100$). If ignorers become aware, they move randomly as they normally would. Hiders self-isolate for the duration of their awareness. By contrast, fleers move as quickly as possible to a new randomly-selected location a distance of K units away if they become aware.

3.2   Augmenting the CC Model with Media

4. News Media Broadcasts May Spread Awareness. This is the first novel modification of the CC model, corresponding to our observation that news media coverage correlates with flu awareness. Specifically, raw counts of flu stories from the news aggregator http://www.newslibrary.com/ have a distinct peak that coincide with measured influenza awareness, with little activity at other times [14]. We therefore simulate news media coverage by broadcasting awareness to a percentage of agents, L. Agents that listen to the news media become aware of the flu. We broadcast once the disease incidence has reached a percentage S, similar to how news media might react.

5. Social Media Connections May Spread Awareness. A second novel addition to the CC model model is a communication network through which agents share simulated social media (e.g., Tweets). Since it is not a requirement that two agents that communicate via social media be geographic neighbors, the interaction network superimposed on the world allows for connections between agents that are otherwise not physically adjacent. We introduce another dimension of agents: whether or not they use social media, with initial density M. Membership in the network is constant throughout the model runs, with $M\%$ of agents included as vertices. In our model, the degree distribution of social media users follows a power law, consistent with the findings of Saroop et al. [13]. After agents move and possibly spread awareness and infection to their local neighbors with probabilities E and B respectively, agents that use social media transmit awareness with probability N to their respective "neighbors" along this social media layer.

3.3 Automated Sensitivity Analysis

We use NetLogo's built-in BehaviorSpace tool [15] for automatic sensitivity analysis of the augmented model. We vary (from 0 to 100 by steps of 10) the behavior parameters H, I, J, the network membership parameter M, and the news media broadcast parameter S, keeping all other parameters identical to the runs reported in Epstein's CC paper [8]. Note that for simplicity we assume a constant L in these automatic runs, specifically 100%. We run each parameterization ten times, and we record the random seed used in each run along with the following metrics: (a) when the media broadcast occurred; (b,c) the maximum disease incidence and time; (d,e) the maximum awareness incidence and time; (f) the reproductive rate of the disease (R_0) [5,8]; (g) the reproductive rate of awareness [8]; (h) the time taken for the disease to cross the model's world fully.

4 Results

4.1 Replicating the CC Model

Our reimplementation of the CC ABM, using every parameter specified in the paper by Epstein et al. [8], replicates their findings. Without behavior, awareness leads infection [8]. When behavior is included, such as when 90% of agents ignore and 10% of agents hide, the rate of disease spread and total incidence both decrease (Fig. 2). The curves in these cases are qualitatively similar to each other; they do not display differences in timing and width. Other examples of how behavior influences trends are as follows: more fleers increases the rate of spread of disease, and more hiders decreases disease spread but also limits the spread of awareness [8].

4.2 Augmenting with News Media

Because we successfully recreate the CC model, we now consider our augmentations. When news media broadcasting is added, overall, it makes peak awareness earlier and more of a difference between awareness and infection. Figure 3 shows a typical result using the 90% ignore, 10% hide parameterization previously

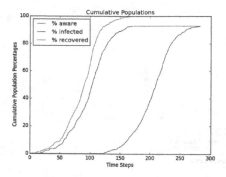
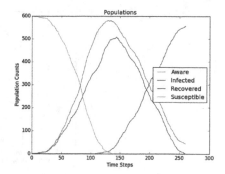

Fig. 2. Example results from step 3: awareness with behavior, with 90 % ignorers (no behavior) and 10 % hiders. Awareness and infection have distinctly different-sized peaks in upper-right because the hiders contribute to the reduced cumulative incidence (see left)

cited. With 10 % of agents listening to news media (L), peak awareness is earlier and higher compared to the current state of infection, but the infection curve seems to be unaffected.

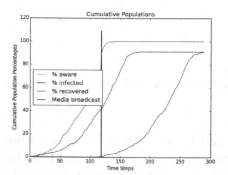
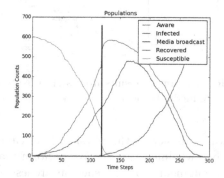

Fig. 3. Example results from step 4: example of media broadcast (vertical line on plots) affecting cumulative trends (left) and population trends (right)

In a manual sensitivity analysis these results are robust to variation in the L parameter, but depend on S. Thus, news media broadcasts appear to influence awareness. The automated sensitivity analysis confirms this and the connection to disease. It reveals that (averaged across all other parameters) the later the broadcast, the faster the disease reproduces (R_0) and the higher maximum disease incidence; see Fig. 4.

4.3   Augmenting with Social Media

We next include social media networking. In a typical run, even when only 10 % of participants use social media, we see that the network acts to spread

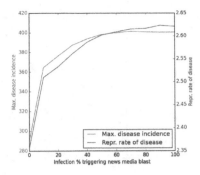

Fig. 4. Plot of the infection percentage triggering the news media broadcast vs the maximum disease incidence (red) and the reproductive rate of the disease (blue)

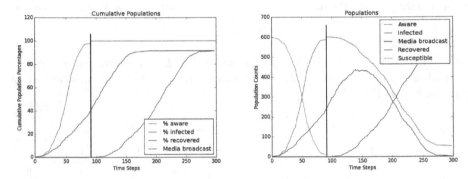

Fig. 5. Example results from step 5. Note influence of network on awareness trends, before media broadcast.

awareness across the physical landscape and the effect of media broadcasting is overwhelmed (Fig. 5). We perform manual sensitivity analysis, and find that this augmentation influences awareness more as the size of the network increases. Specifically, the width of the awareness peak widens and increasingly precedes that of the disease.

In our automated sensitivity analysis (once again averaged over all other parameters), we see that as membership in the social network increases, both the reproductive rate of disease and the maximum disease incidence decrease. However, social media membership has a positive, somewhat erratic, relationship with the epidemic length. These relationships are displayed in Fig. 6.

5   Discussion

Adding Internet-enabled news media coverage and social media connections to the CC model is a conceptual advance that may help explain awareness of disease in modern society. We augment the CC model to show how behavioral reactions to awareness influence the spread of disease, and show how news media and social

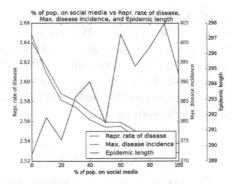

Fig. 6. Plot of the percent of population on social media vs reproductive rate of disease (red), maximum disease incidence (blue), and epidemic length (green)

media influence the spread of awareness. Both news media and social media each enable awareness to potentially spread much more quickly than infection. A key insight is the decoupling of geographic location with awareness transmission. If used to spread awareness, news media and social media provide powerful tools for officials to influence a population, and thus (depending on how the population reacts) also to potentially influence the spread of disease itself. However, as social media membership increases so does epidemic length. This highlights the uncontrollability of social media, and the need to present and guide reactions to disease; or to speak our model's language, it highlights the need to guide agents to hide or flee in appropriate situations.

5.1   Future Work

We take inspiration from the available influenza awareness, flu infection, and news media trends, but future work is still needed to more exactly model these trends. Such work may involve discovery of any additional confounding issues or variables involved that the hypothesized model may omit, which could lead to further understanding of what drives awareness. Possible suggestions to be evaluated are differing rates of awareness transmission before / after media broadcast, having agents' centrality in the social network influence their "success" transmitting awareness, and having social media transmission depend on local incidence of the contagion(s). Obtaining empirical feedback by fitting these trends and testing against them will both link our theoretical model to practice and increase its value in real-world settings.

Incorporating spatial statistics and information like population density, disease incidence density, and distance in a map of the USA may also be worth exploring if goals align with fitting observed data. It would be also worth examining the structure of Twitter to see if most users have geographically close connections, as our work disregarded this spatial component.

Another major area for improvement would be investigating the impact of diverting awareness into separate signals meant to target possible behaviors.

To again speak our model's language, public health officials may want to incite hiding or fleeing depending on the situation. Because social media was observed to be so overwhelming, this tactical messaging would correspond directly to how media might heighten or dampen that overwhelming effect. A model could incorporate the idea that an agent listens to the type of awareness he first receives, be it tactical media or uncontrolled social media; this may offer further opportunities for influencing disease spread.

6   Conclusion

Our ABM updates the CC model to include modern awareness transmission mechanisms. Specifically, we show how news media and social network information can each theoretically influence the spread of awareness, leading to drastic increases in incidence and rate of spread therein as compared to the original CC model. This update suggests that health providers and public health officials may be able to leverage these tools to spread awareness of a disease, and allows policy makers to better understand the consequences of how media technology interacts with the spread of disease.

Acknowledgements. The authors would like to acknowledge Dr. Mark Dredze for allowing us access to the HealthTweets awareness trends, and Dr. Joshua Epstein for helpful feedback.

References

1. Smith, M., Broniatowski, D., Paul, M., Dredze, M.: Tracking public awareness of influenza through Twitter. In: 3rd International Conference on Digital Disease Detection (DDD), Florence, Italy, May 2015. [rapid fire talk]
2. Broniatowski, D.A., Paul, M.J., Dredze, M.: National and local influenza surveillance through twitter: An analysis of the 2012-2013 influenza epidemic. PLoS ONE 8(12), e83672 (2013). http://dx.doi.org/10.137%2Fjournal.pone.0083672
3. Brownstein, J.S., Freifeld, C.C., Madoff, L.C.: Digital disease detection harnessing the web for public health surveillance. N. Engl. J. Med. 360(21), 2153–2157 (2009). pMID: 19423867. http://dx.doi.org/10.1056/NEJMp0900702
4. Corley, C.D., Cook, D.J., Mikler, A.R., Singh, K.P.: Text and structural data mining of influenza mentions in Web and social media. Int. J. Environ. Res. Public Health 7(2), 596–615 (2010)
5. Diekmann, O., Heesterbeek, J.A.P., Metz, J.A.J.: On the definition and the computation of the basic reproduction ratio r0 in models for infectious diseases in heterogeneous populations. J. Math. Biology 28(4), 365–382 (1990). http://dx.doi.org/10.1007/BF00178324
6. Dredze, M., Cheng, R., Paul, M., Broniatowski, D.: Healthtweets.org: A platform for public health surveillance using twitter. In: AAAI Workshop on the World Wide Web and Public Health Intelligence (2014)
7. Dugas, A.F., Jalalpour, M., Gel, Y., Levin, S., Torcaso, F., Lgusa, T., Rothman, R.E.: Influenza forecasting with google flu trends. PLoS ONE 8(2), e56176 (2013). http://dx.doi.org/10.1371%2Fjournal.pone.0056176

8. Epstein, J.M., Parker, J., Cummings, D., Hammond, R.A.: Coupledcontagion dynamics of fear and disease: Mathematical and computational explorations. PLoS ONE **3**(12), e3955 (2008). http://dx.doi.org/10.1371%2Fjournal.pone.0003955
9. Hatfield, E., Cacioppo, J.T., Rapson, R.L.: Emotional contagion. Cambridge University Press, New York (1994)
10. Kermack, W.O., McKendrick, A.G.: A contribution to the mathematical theory of epidemics. Proc. Royal Soc. London Math. Phys. Eng. Sci. **115**(772), 700–721 (1927). http://rspa.royalsocietypublishing.org/content/115/772/700
11. Kremer, M.: Integrating behavioral choice into epidemiological models of the aids epidemic. Technical report, National Bureau of Economic Research (1996)
12. Lee, K., Agrawal, A., Choudhary, A.: Real-time disease surveillance using twitter data: Demonstration on flu and cancer. In: Proceedings of the 19th ACM SIGKDD International Conference on Knowledge Discovery and Data Mining, pp. 1474–1477. KDD 2013, NY, USA (2013). http://doi.acm.org/10.1145/2487575.2487709
13. Saroop, A., Karnik, A.: Crawlers for social networks amp; structural analysis of twitter. In: 2011 IEEE 5th International Conference on Internet Multimedia Systems Architecture and Application (IMSAA), pp. 1–8, December 2011
14. Smith, M., Broniatowski, D.A., Paul, M.J., Dredze, M.: Towards real-time measurement of public epidemic awareness: Monitoring influenza awareness through twitter. In: AAAI Spring Symposium on Observational Studies through Social Media and Other Human-Generated Content (2016)
15. Wilensky, U.: Netlogo. http://ccl.northwestern.edu/netlogo/, Centerfor Connected Learning and Computer-Based Modeling, Northwestern University, Evanston, IL (1999). http://ccl.northwestern.edu/netlogo/
16. Yan, Q., Tang, S., Gabriele, S., Wu, J.: Media coverage and hospital notifications: Correlation analysis and optimal media impact duration to manage a pandemic. J. Theor. Biology **390**, 1–13 (2016). http://www.sciencedirect.com/science/article/pii/S0022519315005366
17. Yang, C; Wilensky, U.: Netlogo epidem basic model. http://ccl.northwestern.edu/netlogo/models/epidembasic, Center for ConnectedLearning and Computer-Based Modeling, Northwestern University, Evanston, IL (2011). http://ccl.northwestern.edu/netlogo/models/epiDEMBasic

An Agent-Based Framework for Active Multi-level Modeling of Organizations

Geoffrey P. Morgan[✉] and Kathleen M. Carley

Carnegie Mellon University, Pittsburgh, PA, USA
{gmorgan,kathleen.carley}@cs.cmu.edu

Abstract. Agent-based models of organizations have traditionally had a single level of agency, whether at the individual or organizational level, but many interesting organizational phenomena, including organizational resilience and turnover, involve agency at multiple organizational levels. We propose an extensible multi-modeling framework, realized in software, to model these phenomena and many more. Two applications will be given to demonstrate the framework's versatility.

Keywords: Agent-based models · Active multi-level modeling · Organizational modeling · Multi-modeling

1 Introduction

It is a given, in the agent-based modeling world, that agents, defined at a single granularity level, interact with each other to produce emergent system-level effects not obviously reducible to individual actions [1]. This is a powerful perspective: allowing people interested in various organizational and sociological problems to explore phenomena such as segregation [2], hierarchical structures [3], the development of organizational knowledge [4], and social change processes [5]. This modeling approach will continue to find traction for the foreseeable future.

Yet, various organizational phenomena, what we call multi-level phenomena, elude this modeling methodology, such as organizational resilience and turnover, because these phenomena arise from actions and reactions at multiple agent granularities. Such phenomena are common; one such phenomena is change resistance: company introduces change C to process P to achieve a forecasted result R, employees resist change C to process P, the forecasted result R is not achieved, change C is discarded. See Maurer [6] for multiple organizational change resistance examples.

Why do employees resist Change C? The reasons are varied and numerous, but often boil down to no clarity in how Change C improves either their personal work environment and outcomes [7], or no clarity in why Change C is consonant with the company's identity [8]. The organization pursues Change C due to its nominal results for other 'similar' organizations, but fails to take into account how its own employees, and the organization culture the organization has itself worked to instill in those employees, may make the proposed change impractical, difficult, or at least require careful handling. In a later section, we provide more detail for two other multi-level phenomena.

© Springer International Publishing Switzerland 2016
K.S. Xu et al. (Eds.): SBP-BRiMS 2016, LNCS 9708, pp. 272–281, 2016.
DOI: 10.1007/978-3-319-39931-7_26

In this article, we lay out a framework for an active multi-level modeling framework, realized in software, intended to allow the modeling and study of such multi-level phenomenon. This work, we feel, is an extension in multi-modeling [9–11]. The software framework is designed to allow for element replacement as needed or desired by the modeler.

We understand an aversion to the parsimony loss a multi-level framework requires, but argue it is necessary for examining and understanding multi-level phenomenon. Prior approaches to these problems have often relied on static group behavior assumptions, or by assigning group-level characteristics to individuals. In this work, we summarize advances in multi-modeling, how we leverage those advances, and detail phenomena requiring this approach. We continue with two scenarios, and detail changes in framework elements to support those modeling scenarios.

2 Related Work

As the introduction described, we will review recent thought in multi-modeling and meta-modeling and how this work contributes to the active multi-level modeling framework. We will also review two prominent organizational phenomena to establish their multi-level nature.

2.1 Multi-modeling

Multi-modeling is the use and leverage of multiple models. Models can be arranged in multiple configurations. The earliest agent-based multi-modeling example published was called simulation alignment, and anecdotally "docking" [12]. Burton [13, 14] introduced "informal docking" and its prevalence in the organizational literature, where prior models are compared to new models in concepts, ideas, and outputs. Simulation alignment is, by comparison, much less common.

But this alignment of models, where both inputs and outputs are made similar and outputs compared at multiple fidelity levels, is only one way to leverage multiple simulation models. Other multi-model configurations include: (a) collaboration, (b) inter-operability, and (c) federation [11]. See Table 1 for details.

Table 1. Multi-model configurations

Multi-modeling technique	Inputs	Outputs
Docking (A‖B)	Aligned	Aligned
Collaboration (A + B)	Aligned	Not Aligned
Inter-operability (AB)	Model output	Not Aligned
Federation ([A→B])	Integrated	Integrated

Multi-Modeling when both model inputs and outputs are aligned is called docking; the usual goal is to do a validation and model comparison exercise and inherit the validity, where appropriate, of the earlier model. Collaboration is when two or more models use the same input to produce different kinds of outcomes. This is useful when you're trying to leverage the same input source for multiple purposes. Inter-operability is when a model uses as at least one input a different model's output. Similarly, model federations, such as the C2 Wind Tunnel [15], take an input set and produce an output set with a system envelope responsible for input and output handling and digestion for individual models within the envelope. These models may also have significant interactions within the federation envelope, such as the PSTK (Power Structure Tool-Kit) and its internal lines of influence connecting multiple models [16].

Meta-Modeling is the idea that multiple models exist for similar purposes, with meta-modeling analysis used to provide the "best-fitting" model given data and goals [17]. In the future, one can imagine frameworks performing sophisticated input data analysis and then selecting the best available model given a modeler's purpose (brainstorming, evaluation, prediction, etc.). This would require a well-established meta-ontological layer and universal standards, and so instead propose a fixed and explicit set of model elements/purposes, each of which themselves could be described as their own model, that work together within the Active Multi-Level Modeling framework. By giving explicit functions to each element, we limit the models that can be repurposed to work within the framework, but make it easier to build new models and expect them to interoperate with each other. This work can learn from various agent-frameworks [18, 19], but is unique in its conception of the organization as both a collection of agents and as an agent in its own right.

The Active Multi-Level Framework thus inherits key multi-modeling and meta-modeling concepts in its conception and design. The next section concerns multi-level phenomena, and why these phenomena require an active multi-level modeling approach.

2.2   Example Phenomena of Interest

In this section, we review two multi-level phenomena, organizational resilience and turnover, describe important prior work and its findings, and try to establish the case that modeling at a single agent granularity would not be sufficient.

Organizational Resilience. Resilience, as a single-level construct, has been studied in both individuals [20] and organizations [21]. However, rather than focusing on these separate research threads, we instead want to focus on the scholarship investigating how organizations take action to stay resilient while also inculcating resilient behaviors in individuals. Sutcliffe and Vogus [22] argue organizations can inculcate individual resilience by making sure they have resources to do their jobs, and groups are more resilient if they are more heterogeneous. How an organization reacts to bad news, including disciplinary actions on individuals, affects its overall resilience to bad news [23]. And when the organization makes prescriptive guidance while ignoring how individual work is done, it exposes the organization to significant risk [24]. Meyer [25] found that while the organization's structure constrains its ability to respond to events, the response's shape is influenced by the organization's ideology, which is itself generated through interactions of people and the organization [4].

Thus, organizations can promote individual resilience, and an individual's acts can limit an organization's resilience in crisis.

Turnover. As with resilience, there have been significant research efforts in understanding both organizational [26] and individual [27] turnover factors. We are, as before, interested in these factors' intersection and thus focus on concepts such as job-embeddedness [28] and fit [29]. Job-embeddedness can be summarized as "fit, links, and sacrifice" [28, p. 1108], where embedded people have jobs well-suited to their life-spaces, have many ties to others at their work, and cannot break links without significant social cost. High embeddedness lowers an individual's intentions to quit, which has a direct relationship to quitting [30]. Allen [31] explored whether different HR practices for socializing newcomers influenced job-embeddedness, while earlier work [32] find organizations that offer more social support experience less turnover. Meanwhile, Kristof-Brown's and colleagues' [29] definition of fit examines the individual's compatibility with their work-group on multiple dimensions, including beliefs and tasking. Here again, organizations and individuals are intertwined.

3   The Proposed Framework

In this section, we will describe and detail the Active Multi-Level Modeling framework. The Active Multi-Level Modeling framework assumes two different agent cycles, an individual agent cycle and a group agent cycle. In Fig. 1, individual agent-cycle elements (I) are ovoid, while group agent-cycle elements (G) are rectangles. Elements that cross boundaries (C) are indicated with compound lines, specifically "Aggregation" and "Event Generation".

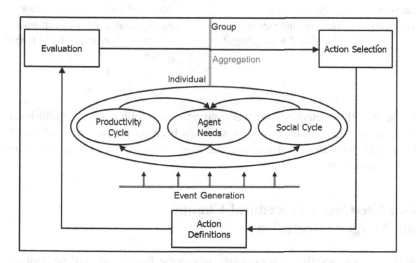

Fig. 1. The active multi-level modeling framework concept

This framework is designed to support multi-level phenomena through the dynamic interactions of individual and organizational agents. The framework has multiple elements; we describe each element in Table 2.

Table 2. Active multi-level modeling framework elements

(Area) Element	What?	Why?
(I) Agent needs	Is the agent's need for social interaction or work more pressing?	Individuals need to feel both productive and socially supported. Not all social needs are met by work, nor is all socialization work-producing.
(I) Social cycle	Social communication and interaction with other agents.	Socialization processes and modalities are often distinct from work interactions
(I) Productivity cycle	Logic required for the agent to do work	How work is done can often benefit from being explicitly modeled
(O) Evaluation	Used by the organization to evaluate current state	There are many variations on how the organization is evaluated, from throughput to agent competencies.
(O) Action selection	How the organization chooses what to do.	Some organizations address the biggest most-obvious need, other models have had advanced look-ahead procedures
(O) Action definitions	What can the organization do?	Depending on the theoretical (or applied) perspective taken, what the organization can do to its members may vary.
(C) Event generation	Agents react to events from the environment.	Many organizational phenomena have significant environmental factors.
(C) Aggregation	How are individuals affiliated with groups?	Organizational topologies influence outcomes. Agents can be in many groups.

We believe these elements are necessary for developing useful multi-level phenomena models, but because the discussion has been so abstract, we will inform the discussion with two applied use-cases, and how those use-cases will be instantiated within the framework.

4   Two Use-Cases: Procedural Change and Merger Integration

In this section, we describe two use-cases where the framework will be applied, with sub-models for the elements above used to instantiate the simulation. The two use-cases are FAA procedural change, and a multi-national's horizontal merger. We start by describing both models and then outline the extensions required for the frame-work. Note again the framework exists as an extensible set of Java libraries (compiled using Java v7), and we will extend each element to support these needs. Eventually, we plan to build a Java GUI to allow for direct model selection from all known models and use reflection to inform the GUI.

4.1 Modeling Procedural Change at the FAA

The FAA is trying to implement new policies for Air Traffic Controllers (ATCs). These changes are part of a large set of technology and procedural changes included in the NextGen system. With NextGen, the FAA is hoping to be able to take advantage of the slack capacity in US air infrastructure, and support more flights with the same number of ATCs. However, NextGen and Current ATC policies will be available for the foreseeable future, as existing airframes will not be required to have NextGen technology, and the technology is expensive to integrate onto an existing airframe. Further, although NextGen technology supports voice protocols, much of the forecasted benefits come from macro-enabled texting. ATCs, however, enjoy voice communications as a non-harmful distraction from their overwatch tasks (a cognitively stressing vigilance task).

Within the FAA model, individual agents are air traffic controllers responding to realistic event distributions over their nominal day. They respond to each event using the protocol that best suits their purposes. They maintain overwatch over flights within their airspace, but can take breaks when necessary by handing flights over to another team-mate. However, they can only respond to events when they have finished responding to a prior event, and they cannot change protocol midway through the exchange. In this model, we assume that although flights can be re-routed due to weather or other disturbances (rocket launches, etc.), no flights interact disastrously.

The organizational agents, meanwhile, review performance metrics such as "average time for control handoff" and "typical overwatch load" over nominal days and adjust people-power and the local facility's pressure to use NextGen to suit the organization's needs and their found experiences. They move people-power from hours where metrics are strong to hours where metrics are weak, attempting to have good performance relative to adjacent facilities across the day.

The model's goal is to support FAA decision making by identifying potential areas of concern not uncovered via other brain-storming processes. The FAA source-data includes activity data from over 3000 facilities, with 400 k cross-facility interactions captured, and staffing data from ∼ 1000 facilities.

4.2 Modeling a Multi-national's Horizontal Merger

The Multi-National simulation inherits ideas and processes from the Mutual Learning Simulation [4], GARCORG [3], Construct [33], and OrgAhead [34]. However, the organization of interest is one where a large-scale horizontal merger has taken place. Each individual agent is seeded, based on collected email data, with interaction partners, knowledge, and tasks. A partial organization chart is used, along with network clustering, to identify team, work-group, and division memberships. Group priorities and organizational knowledge is inferred, via a March-like process, from member contributions. Note: sub-groups within the larger organization may conflict with each other.

Individuals are expected to respond to changes as they occur and pass knowledge updates to their working peers. Each individual has a capacity for accepting updates, and updates beyond the limit will be ignored. Individuals complete tasks based on knowledge linked to the task, and may not always succeed. Each person also has need for social support and interaction, and they choose to interact with people based on shared beliefs and knowledge. Individuals can leave at will.

Organization agents, which operate on time-scales in correlation to their size (large groups have long time-scales), evaluate the group member output over time and compare it to expected outputs based on agent composition. The group can emphasize tasks, change work tempo, fire low performers, and hire replacements.

The model's goal is to be able to predict points of conflict and areas of high turnover in the merger process, and will use collected empirical longitudinal data to validate the model. The source data has three elements: a longitudinal survey administered to more than 5000 organization members, structural meta-data of 15 M emails over 2 years, and sanitized English content for 1.2 M emails from the same time-span. Figure 2 is a network visualization of a subset of the structural data.

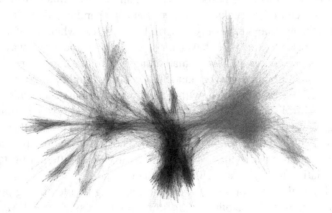

Fig. 2. Illustrative structural dynamics from a month of email data - luxuryco (dark gray nodes) acquired standardco (light gray nodes) a year before.

4.3   Uniting Models Under One Framework

By placing both models within the same framework, we believe both require an active multi-level approach. In Table 3, we go through each framework element and provide a brief description of its implementation plan.

Table 3. Framework elements for each use-case

(Area) Element	FAA procedural change	Horizontal merger
(I) Agent needs	Vigilance task demands continue until agent requires a break	Agents have innate randomized socialization and productivity goals
(I) Social cycle	Time away from vigilance tasks refreshes cognitive stocks	Similar to Construct (Carley 1991)
(I) Productivity cycle	ATC responds to event with either NextGen or Current procedures	Agent sends updates to frequent work collaborators on perceived environment changes
(O) Evaluation	Uses metrics to understand weak time-slots and relative performance	Compares expected/actual output to evaluate tasking and agent performance
(O) Action selection	Most obvious need	Forecasting a la OrgAhead
(O) Action definitions	Move employees in time, increase/reduce pressure to use NextGen	Emphasize/deemphasize tasking, hire more people, fire low performers
(C) Event generation	Activity Data from FAA informs stochastic process	Environment (inferred from data) changes stochastically
(C) Aggregation	People-Power Data from FAA	Partial org chart from Organization

5 Discussion and Conclusion

In this short paper, we have introduced the Active Multi-Level Modeling Framework for exploring and understanding multi-level phenomena such as change resistance, organizational resilience, and turnover. We introduce prior multi-modeling and meta-modeling work that influenced the development of the framework. We believe a framework like this one is necessary for exploring multi-level phenomena, and provided brief literature reviews of specific phenomena to support this perceived need. We introduced two use-cases to demonstrate the framework's versatility and wide applicability, outlining model execution within the Framework and a brief description of source-data for each use-case. We conclude with an appeal to the need for this framework.

We plan to make the framework and its extensions available via public software repositories.

References

1. Goldstein, J.: Emergence as a construct: history and issues. Emergence **1**, 49–72 (1999)
2. Schelling, T.C.: Dynamic models of segregation. J. Math. Sociol. **1**, 143–186 (1971)
3. Carley, K.M.: Measuring efficiency in a garbage can hierarchy. In: March, J.G., Weissinger-Baylon, R. (eds.) Ambiguity and Command, pp. 165–194. Pitman, Boston (1986)

4. March, J.G.: Exploration and exploitation in organizational learning. Organ. Sci. **2**, 71–87 (1991)
5. Carley, K.M., Martin, M.K., Hirshman, B.R.: The etiology of social change. Top. Cogn. Sci. **1**, 621–650 (2009)
6. Maurer, R.: Using resistance to build support for change. J. Qual. Particip. **19**, 56 (1996)
7. Sheldon, K.M., Turban, D.B., Brown, K.G., Barrick, M.R., Judge, T.A.: Applying self-determination theory to organizational research. Res. Pers. Hum. Resour. Manage. **22**, 357–394 (2003)
8. Hynes, T., Prasad, P.: Patterns of 'Mock Bureaucracy' in mining disasters: an analysis of the Westray coal mine explosion. J. Manage. Stud. **34**, 601–623 (1997)
9. Levis, A.H., Zaidi, A.K., Rafi, M.F.: Multi-modeling and meta-modeling of human organizations. In: Advances in Design for Cross-Cultural Activities, p. 148 (2012)
10. Levis, A.H., Jbara, A.A.: Multi-modeling, meta-modeling, and workflow languages. In: Theory and Application of Multi-Formalism Modeling (2013)
11. Carley, K.M., Morgan, G.P., Lanham, M., Pfeffer, J.: Multi-modeling and socio-cultural complexity: reuse and validation. Adv. Des. Cross Cult. Activities **2**, 128 (2012)
12. Axtell, R., Axelrod, R., Epstein, J.M., Cohen, M.D.: Aligning simulation models: a case study and results. Comput. Math. Organ. Theor. **1**, 123–141 (1996)
13. Burton, R.M.: The challenge of validation and docking. In: Workshop on Agent Simulation: Applications, Models, and Tools, pp. 216–221. Citeseer (1999)
14. Burton, R.M.: Computational laboratories for organization science: questions, validity and docking. Comput. Math. Organ. Theor. **9**, 91–108 (2003)
15. Roth, K.E., Barrett, S.K.: Command & control wind tunnel integration and overview. In: 2009 SISO European Simulation Interoperability Workshop, pp. 45–51. Society for Modeling & Simulation International (2009)
16. Taylor, G., Bechtel, R., Morgan, G., Waltz, E.: A framework for modeling social power structures. Ann Arbor **1001**, 48105 (2006). Soar Technology
17. Levis, A.H.: Multi-modeling and meta-modeling of adversaries and coalition partners. George Mason University (2010)
18. Ferber, J., Gutknecht, O.: A meta-model for the analysis and design of organizations in multi-agent systems. In: International Conference on Multi-Agent Systems, pp. 128–135. IEEE (1998)
19. Wooldridge, M., Jennings, N.R., Kinny, D.: The Gaia methodology for agent-oriented analysis and design. Auton. Agents Multi-Agent Syst. **3**, 285–312 (2000)
20. Tusaie, K., Dyer, J.: Resilience: a historical review of the construct. Holist. Nurs. Pract. **18**, 3–10 (2004)
21. Lengnick-Hall, C.A., Beck, T.E.: Adaptive fit versus robust transformation: How organizations respond to environmental change. J. Manage. **31**, 738–757 (2005)
22. Sutcliffe, K.M., Vogus, T.J.: Organizing for resilience. Positive Organ. Sch. Found. New Discipline **94**, 110 (2003)
23. Bies, R.J.: The delivery of bad news in organizations a framework for analysis. J. Manage. **39**, 136–162 (2013)
24. Nathanael, D., Marmaras, N.: The interplay between work practices and prescription: a key issue for organizational resilience. In: Proceedings of 2nd Resilience Engineering Symposium, pp. 229–237 (2006)
25. Meyer, A.D.: Adapting to environmental jolts. Adm. Sci. Q. **27**, 515–537 (1982)
26. Hausknecht, J.P., Trevor, C.O.: Collective turnover at the group, unit, and organizational levels: evidence, issues, and implications. J. Manage. **37**, 352–388 (2011)

27. Griffeth, R.W., Hom, P.W., Gaertner, S.: A meta-analysis of antecedents and correlates of employee turnover: update, moderator tests, and research implications for the next millennium. J. Manage. **26**, 463–488 (2000)
28. Mitchell, T.R., Holtom, B.C., Lee, T.W., Sablynski, C.J., Erez, M.: Why people stay: using job embeddedness to predict voluntary turnover. Acad. Manage. J. **44**, 1102–1121 (2001)
29. Kristof-Brown, A.L., Zimmerman, R.D., Johnson, E.C.: Consequences of individuals' fit at work: a meta-analysis of person-job, person-organization, person-group, and person-supervisor fit. Pers. Psychol. **58**, 281–342 (2005)
30. Hom, P.W., Tsui, A.S., Wu, J.B., Lee, T.W., Zhang, A.Y., Fu, P.P., Li, L.: Explaining employment relationships with social exchange and job embeddedness. J. Appl. Psychol. **94**, 277 (2009)
31. Allen, D.G.: Do organizational socialization tactics influence newcomer embeddedness and turnover? J. Manage. **32**, 237–256 (2006)
32. Allen, D.G., Shore, L.M., Griffeth, R.W.: The role of perceived organizational support and supportive human resource practices in the turnover process. J. Manage. **29**, 99–118 (2003)
33. Carley, K.M.: A Theory of Group Stability. Am. Sociol. Rev. **56**, 331–354 (1991)
34. Lee, J.S., Carley, K.M.: Orgahead: a computational model of organizational learning and decision making. CASOS Technical report. Carnegie Mellon University (2004)

Identifying Political "hot" Spots Through Massive Media Data Analysis

Peng Fang[1(✉)], Jianbo Gao[1,2(✉)], Fangli Fan[1], and Luhai Yang[1]

[1] Institute of Complexity Science and Big Data Technology, Guangxi University,
100 Daxue Road, Nanning, Guangxi 530005, People's Republic of China
xinluo.fang@qq.com
[2] PMB Intelligence LLC, Sunnyvale, CA, USA
jbgao.pmb@gmail.com
http://www.gao.ece.ufl.edu

Abstract. Political processes generated by daily events around the world are highly complicated. Using an information theoretic approach, we examine the dissimilarity of political activities of a country to different countries, and whether it is possible to identify politically "hot" countries. Using a massive political science data, the Global Database of Events, Location, and Tone (GDELT), which covers all the political event data (over 300 million) since 1979 created for studying world-wide political conflict and instability, we show that Shannon entropy for most countries does not differ substantially. Therefore, most countries have similar complexity in terms of political activities. More interestingly, we find that the relative entropies between politically very unstable countries and regular stable countries are large, and thus relative entropy can be used to effectively identify political "hot" spots.

Keywords: GDELT · Political hot spot · Shannon entropy · Relative entropy

1 Introduction

Big data have been offering unprecedented opportunities as well as challenges to both academia and industry. One of the most fascinating sources of big data is the Global Database of Events, Location, and Tone (GDELT) [1], which is a massive political science data created for studying world-wide political conflict and instability. GDELT is attracting increasing interest in the machine learning community [2] and has been extensively utilized in several applied settings covering country activity [3], domestic protests [4], finance [5] and global disasters [6]. GDELT contains more than 300-million geolocated events with global coverage from 1979 to the present. There are two major difficulties in analyzing GDELT [7]. One is the massiveness of the data. Another is the nonstationarity of the data. While being a major cause of unpredictability of some political processes, nonstationarity is also closely related to the nature of news reporting — old stories, even though important from a historical perspective, will rapidly lose value

© Springer International Publishing Switzerland 2016
K.S. Xu et al. (Eds.): SBP-BRiMS 2016, LNCS 9708, pp. 282–290, 2016.
DOI: 10.1007/978-3-319-39931-7_27

as a major news and thus will be less covered in the media as time goes by; on the other hand, irrespective of their true value, coverage on appearantly new events will always be exuberant. To overcome these difficulties, it is important to think what kind of questions can be asked about GDELT, and what kind of new answers can be obtained from GDELT. Here, we specifically ask how different the political activities engaged by different countries are. To gain insight into this question, we will use Shannon entropy and relative entropy.

The remainder of the paper is organized as follows. In Sect. 2, we briefly describe GDELT. In Sect. 3, we compute Shannon and relative entropies from GDELT, and show that in terms of Shannon entropy, political activities engaged by different countries are not very different. However, relative entropies between politically very unstable countries and regular stable countries are large, and thus relative entropies can be used to effectively identify political "hot" spots. Concluding discussions are in Sect. 4.

2 Brief Description of GDELT

Political event data are nominal codes that record who did what to whom and when (i.e., codings based on units of analysis consisting of subject-verb-object (S-V-O) phrases), coded from news media [8]. Each event is recorded with a set of attributes, including the interval-level Goldstein conflict-cooperation scale value [9]. GDELT events are drawn from a wide variety of news media, both in English and non-English, from across the world, ranging from local to international sources in nearly every country. These data are produced by the TABARI automated coding software (http://eventdata.psu.edu/software.dir/tabari.html) and a series of ancillary programs used to establish geolocation and improve the named entity recognition of TABARI. TABARI works with CAMEO [10], a very large set of verb-phrase (>15,000 phrases) and noun-phrase (>40,000 phrases) dictionaries in combination with shallow parsing of English-language sentences to identify grammatical structures such as subject-verb-object phrases, compound subjects and objects, and compound sentences. CAMEO is an update of earlier (1960 s) event coding taxonomies, with changes introduced by automated coding and new behaviors, such as suicide bombings. CAMEO provides a detailed and systematic taxonomy for coding contemporary political actors, including international, supranational, transnational, and internal actors. An earlier version of this system recently was successfully employed in the DARPA ICEWS project [11] to code 25 gigabytes of Asian news reports involving more than 6.7 million stories, which provided the key input for forecasting models with accuracy, sensitivity, and specificity all exceeding DARPA's pre-set criteria. Currently GDELT generates between 1.0 and 1.5×10^5 event codings per day. The data are updated daily and are open access at http://gdelt.utdallas.edu; tools for working with the data are discussed both on that web site and at http://gdeltblog.wordpress.com.

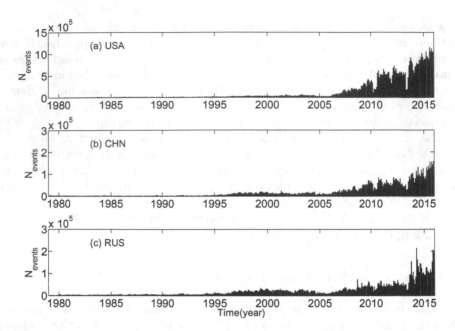

Fig. 1. Monthly total number of political events in (a) USA, (b) China, and (c) Russia.

3   Information Theoretic Analysis of GDELT

To compute Shannon and relative entropies, we first need to compute the probability that each political event occurs for a given country. To facilitate this task, in Sect. 3.1, we describe the set of political events designed by political scientists in the last few decades. Then in Sect. 3.2, we examine country-wise Shannon entropies, and in Sect. 3.3, we compute relative entropies between two arbitrary countries. Finally, in Sect. 3.4, we examine whether relative entropies can be used to effectively identify political "hot" spots.

3.1   Political Events

In the celebrated CAMEO-code book, there are 20 classes of political events and there are a number events in each class of actions.

Note the set of political events recently has been slightly expanded, to include contemporary events such as suicide bombings. Also recall that each event is assigned a conflict-cooperation score, called Goldstein-scale intensity [9]. Figure 1 shows the total monthly number of political events in USA, China, and Russia from 1979 to 2015. We observe that this number recently has been increasing rapidly, as a consequence of rapid advancement of technology.

3.2   Shannon Entropy of Country-Wise Political Activities

For a given country, assume there are E_i, $i = 1, \cdots, m$, political events, each event E_i occurring n_i times. The probability p_i for E_i to occur is

$$p_i = n_i / N_{total}, \quad N_{total} = \sum_{i=1}^{m} n_i \tag{1}$$

Then Shannon or information entropy is given by

$$H = - \sum_{i=1}^{m} p_i \log p_i \tag{2}$$

By convention, $p_j \log p_j = 0$ if $p_j = 0$. Note that if there is only one of p_i's is 1 while all others are 0, then $H = 0$. In this case, we have a deterministic scheme, and Shannon entropy is zero. At the other extreme, when the events occur with equal probability of $1/m$, H attains the maximum value of $\log m$. We thus see that the value of Shannon entropy depends on both the number of events and the distribution of the events. For more in depth discussion of Shannon entropy, especially its role in the complexity theory, we refer to our recent survey article [12]. For later convenience, here, we use natural logarithm. In this case, the unit of entropy is baud. When base 2 is used in the logarithm, the unit of entropy is bit, which is $\log 2 = 0.6931$ baud.

We have computed Shannon entropy for the political activities of 181 countries base on the data of November 2015. In Fig. 2(a-d), we have shown the probability density function (pdf) of Shannon entropy with respect to USA, China, France, and Syria, respectively. However, entropies for the political activities of other countries are not much smaller. Indeed, the distribution for the entropy of these countries, shown in Fig. 3, is quite concentrated. Therefore, we may conclude that Shannon entropy for different countries does not differ much.

3.3   Relative Entropy of Political Activities Between Countries

As we have seen, Shannon entropy depends on both the number of events and the distribution of the events. One of the most popular means of quantifying the discrepancy between two distributions is to use relative entropy defined as [13]

$$D(p\|q) = \sum_i p_i \ln \left(\frac{p_i}{q_i} \right), \tag{3}$$

Relative entropy is also called Kullback-Leibler distance. It is always nonnegative and is zero if and only if the two distributions are the same, $p = q$. Moreover, it is asymmetric, i.e., $D(p\|q) \neq D(q\|p)$.

In order to compute relative entropy between the distributions of political events of two countries, one issue has to be taken care of. That is, one has to restrict oneself to the common subset of political activities that are engaged by both countries.

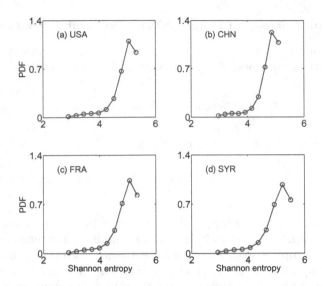

Fig. 2. Distribution of Shannon entropy for (a) USA, (b) China, (C) France, and (d) Syria base on the data of November 2015.

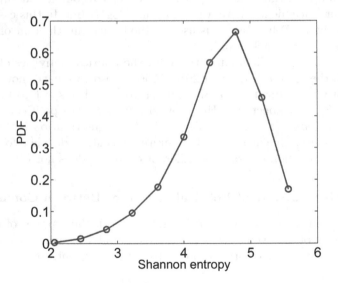

Fig. 3. Distribution of Shannon entropy of all the countries based on the data of November 2015.

We have computed pair-wise relative entropies for all the countries in the world base on the data of November 2015. Due to asymmetry, this yields about 40000 relative entropy values. In Fig. 4(a-d), we have shown the probability density function (pdf) of relative entropy of all the countries with respect to USA, China, France, and Syria, respectively. The distribution for all the pair-wise relative entropies of all the countries are shown in Fig. 5. We observe that relative entropies are generally small. Syria is clearly an exception. This has to be attributed to the constant wars there.

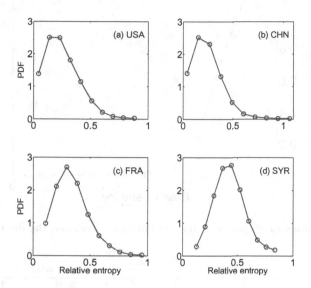

Fig. 4. Distribution of relative entropy with respect to (a) USA, (b) China, (C) France, and (d) Syria base on the data of November 2015.

3.4   Relative Entropy and Political "hot" Countries

Albeit civilization has been advancing, the world is still full of instabilities and conflicts. Our earlier observation that the distribution for the relative entropy of Syria is very different from those of other countries motivates us to develop a readily implementable scheme to identify political "hot" countries in the world. It is easier to describe our scheme by using the complementary cumulative distribution function (CCDF) instead of the pdf shown in Fig. 4(a-d). Note integration of pdf yields cumulative distribution function (CDF). CCDF is simply 1-CDF. CCDF may also be called tail probability, i.e.,

$$CCDF = Probability\{relative\ entropy \geq TH\}$$

For a given threshold value TH, a large $CCDF$ for certain country would indicate that politically that country is very different from other countries. Therefore, this CCDF may be taken as a political hotness measure of that country. Figure 6 shows examples for four different threshold values. An example of the political hotness map of the world for November 2015 is shown in Fig. 7, with the threshold value to be 0.37. The map is similar when different threshold values are used. It is interesting to note that the politically very distinct countries or entities include the familiar ones such as Iraq, Syria, Afghanistan, etc. Also note

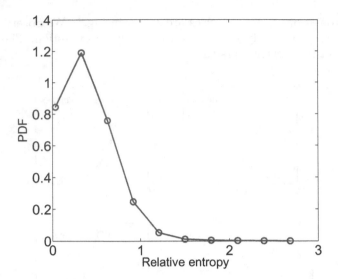

Fig. 5. Distribution of all pair-wise relative entropies based on the data of November 2015.

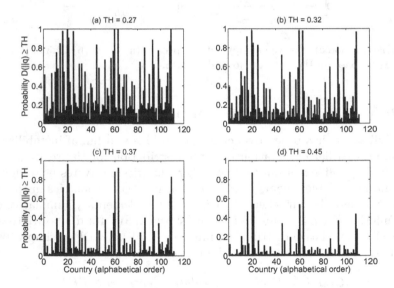

Fig. 6. Relative entropy-based hotness scores base on the data of November 2015 for alphabetically ordered all the countries in the world. Four subplots correspond to 4 different threshold values.

that the presence of France in the map as one of the hottest country is mainly due to the Paris terror attacks on Nov.13, 2015. Some of them are just because of the small size, such as Burundi — when a country is too small, its routine activities may be very different from that of a normal country.

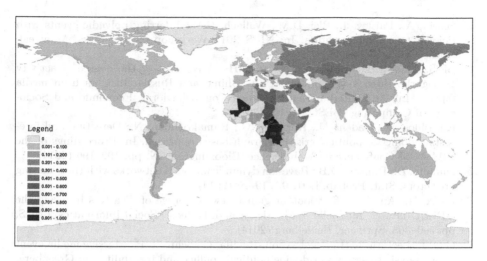

Fig. 7. A political hotness map of the world for November 2015.

4   Future Perspectives

The massive political science data, GDELT, has offered an unprecedented opportunity for researchers on big data research to develop innovative analysis methods to maximally extract useful information, especially to quantify political processes not solely from machine-learning based approaches, but also from fundamental mathematics and physics viewpoints. Here, we have examined how information theoretic approaches may shed new light on the characterization of political processes. Based on a set of about 300 unique political events carefully defined by political scientists, we have shown that Shannon entropies for the political activities engaged by different countries are close. Therefore, most countries have similar complexity in terms of political activities. More interestingly, we have found that the relative entropies between politically very unstable countries and regular stable countries are large, and thus relative entropy can be used to effectively identify political "hot" spots.

In earlier years, especially before 1990, due to scarce of events, the resolution of the map of political "hot" spots can be at best 1 year. With the rapid increase of political events in GDELT, the resolution of the map for recent years can be 1 month. It is perceivable that in future the resolution can be further improved, such as to 1 week or even 1 day.

References

1. Leetaru, K., Schrodt, P.: GDELT: Global Data on Events, Language, and Tone, 1979–2012. In: International Studies Association Annual Conference, San Diego, CA (2013), April 2013

2. Schein, A., Paisley, J., Blei, D.M., Wallach, H.M.: Inferring polyadic events with Poisson tensor factorization. In: NIPS 2014 Workshop on Networks, Montreal, Quebec, Canada (2014)

3. Bi, S., Gao, J.B., Wang, Y.D., Cao, Y.H.: A contrast of the degree of activity among the three major powers, USA, China, and Russia: Insights from media reports. In: 2015 International Conference on Behavioral, Economic and Socio-cultural Computing (BESC), IEEE (2015)

4. Keneshloo, Y., Cadena, J., Korkmaz, G., Ramakrishnan, N.: Detecting and forecasting domestic political crises: A graph-based approach. In: Proceedings of the 2014 ACM Conference on Web Science, Bloomington, IN, pp. 192–196 (2014)

5. Durante, D., Dunson, D.B.: Bayesian dynamic financial networks with time-varying predictors. Stat. Probab. Lett. **93**, 19–26 (2014)

6. Kwak, H., An, J.: A first look at global news coverage of disasters by using the GDELT dataset. In: Aiello, L.M., McFarland, D. (eds.) Social Informatics. LNCS, pp. 300–308. Springer, Heidelberg (2014)

7. Gao, J., Leetaru, K.H., Hu, J., Cioffi-Revilla, C., Schrodt, P.: Massive media event data analysis to assess world-wide political conflict and instability. In: Greenberg, A.M., Kennedy, W.G., Bos, N.D. (eds.) SBP 2013. LNCS, vol. 7812, pp. 284–292. Springer, Heidelberg (2013)

8. Schrodt, P.A.: Precedents, progress and prospects in political event data. Int. Interact. **38**, 546–569 (2012)

9. Goldstein, J.S.: A conflict-cooperation scale for WEIS events data. J. Confl. Resolut. **36**, 369–385 (1992)

10. Schrodt, P.A., Gerner, D.J., Ömür, G.: Conflict and mediation event observations (CAMEO): an event data framework for a post cold war world. In: International Conflict Mediation: New Approaches and Findings. (eds. Bercovitch, J. & Gartner, S.) (Routledge, New York) (2009)

11. O'Brien, S.P.: Crisis early warning and decision support: contemporary approaches and thoughts on future research. Int. Stud. Rev. **12**, 87–104 (2010)

12. Gao, J.B., Liu, F.Y., Zhang, J.F., Hu, J., Cao, Y.H.: Information entropy as a basic building block of complexity theory. Entropy **15**(9), 3396–3418 (2013)

13. Cover, T.M., Thomas, J.A.: Elements of Information Theory. Wiley, New York, NY (1991)

Contextual Sentiment Analysis

Will Frankenstein$^{(\boxtimes)}$, Kenneth Joseph, and Kathleen M. Carley

Carnegie Mellon University, Pittsburgh, PA, USA
{wfranken,kjoseph,kathleen.carley}@cs.cmu.edu

Abstract. This study examines the role of context in evaluating responses to social media posts online. Current sentiment analysis tools evaluate the content of posts without considering the broader context that the post comes from. Utilizing data from an in-person study, we examine differences between perceived sentiment evaluation when social media response posts are viewed in isolation and perceived sentiment evaluation when social media responses are viewed in the context of the original post. We find that evaluations of responses viewed in context change over 50 % of the time. We validate this finding by utilizing simulated data to show the result is not simply a result of data manipulation or noisy data; furthermore, we explore results of this finding with current sentiment analysis tools, examining this result with subsets of our data with high and low kappa values.

Keywords: Twitter · Social media · Sentiment analysis · Affect control theory

1 Introduction

Traditional approaches to sentiment analysis have three problems: the approaches were originally developed to analyze larger bodies of text, they ignore the social context of social media, and they are primarily focused on only one dimension of sentiment. As social media text can be extremely short, and due to the expense associated with obtaining labeled data necessary to train machine learning algorithms, most approaches to sentiment analysis today rely on extensive lexicons with the goal of having some text match words that we know map to generally positive or negative sentiment [1–3].

Most approaches to sentiment analysis in social media focus exclusively on the content of the message, ignoring the metadata and subsequent social context that the message comes out of [4–7]. For example, a user posting she is ill will receive positive, supportive posts on social media. Analyzing the social network associated with the flow of those messages would result in an incorrectly classified positive association with that sickness. While some analyses of social network sentiment incorporate analysis of a user's social media ties, these studies rely on aggregated posts and do not consider individual responses to news, topics, or events [8, 9].

Finally, sentiment is typically analyzed along a single dimension: positive and negative, with a minority of research considering objectivity [4, 10]. However, there are other dimensions to emotions, informed by cultures, which affect how individuals respond to events. Affect control theory (ACT) formalizes the way that individuals respond to events by classifying evaluation, potency, and action, allowing for cross-cultural comparisons of events [11–13]. Evaluation is the most similar dimension

© Springer International Publishing Switzerland 2016
K.S. Xu et al. (Eds.): SBP-BRiMS 2016, LNCS 9708, pp. 291–300, 2016.
DOI: 10.1007/978-3-319-39931-7_28

to most sentiment tools today: it is a spectrum from unpleasant and negative to pleasant and positive. Power reflects the social and external relations individuals have, going from weak and powerless to strong and powerful. Activity, in contrast to power, reflects internal relationships to emotion, going from unexciting and inactive to exciting and active. In this study, we utilize a recent dictionary consisting of over 2,000 terms to populate lexicons to identify messages along potency and activity [14].

The paper seeks to explore three key areas: how affect control theory can inform sentiment analysis, how individuals perceive messages seen without context differently from messages with context, and finally, the implications of context for existing tools. We examine the impact of context along all three dimensions of affect control theory, compare evaluations of messages with and without context, and compare individual ratings with automated scores given by sentiment analysis tools.

2 Data

We utilize a subset of a study where 96 individuals collectively rated 5,780 Twitter posts [15]. In the broader study, individuals were given a brief 5-minute training on the three dimensions of ACT, which can be viewed in the technical report [15]. Individuals then each rated 120 Twitter posts three times, once for each dimension of ACT. The 120 Twitter posts evaluated fall into four categories: (A) individual Twitter posts, (B) responses to Twitter posts, (C) the original post that response posts were made to, and (D) the same responses seen in category (B) – presented this time with the context of the original post. This paper focuses on the changes in response that individuals had from rating category (B) tweets to category (D) tweets.

Each set of 120 Twitter posts were evaluated twice. We only considered Twitter conversations where the original post was not a response itself. To ensure a broad diversity of topics, we chose Twitter posts from four broad areas, as outlined in the Table 1 below.

Table 1. Topic categories for data used.

	Nuclear	Arab Spring	General	Haiyan
Dates	Sep 2014 – Oct 2014	Oct 2009 – Nov 2013	Sep 2013 – Aug 2014	Nov 2013 – Dec 2013
Sample Keywords	Nuclear proliferation, uranium	Tahrir Square, Arab Spring	n/a	Haiyan, Typhoon Yolanda
Number of Posts	720	720	720	720

For "General" posts we randomly selected English-language posts from the "Gardenhose", or 10 % of the total Twitter firehose, so we did not utilize keywords to select the topics.

3 Comparing Responses with and Without Context

We first explore the data by displaying the distribution of ratings across message categories. We then perform a deeper dive into the different topics making up the dataset and show that we see the same behavior in changed evaluations across all topics. This allows us to make generalizations about the data as a whole and not limited to a subset of our data.

In the histograms below we plot the overall ratings that individuals recorded. Ratings are on a five point Likert scale from negative to positive for Evaluation, weak to strong for Power, and active to passive for Activity. We see that within Power and Activity, the overall profile of responses is consistent whether the post is the original post, the response, or the response viewed with context. The most variation appears to be within Evaluation, which sees slightly more negative posts in responses (Fig. 1).

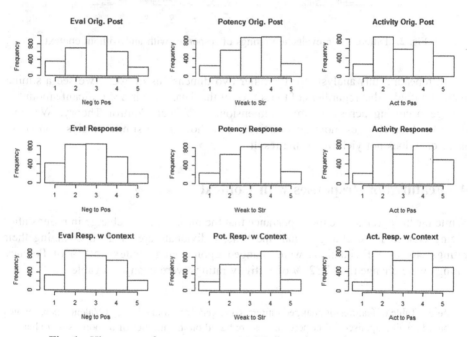

Fig. 1. Histogram of responses across ACT dimensions and post category.

There is some minor variation across topic categories, but there is significant robustness when comparing differences in the evaluation of responses with and without context (Fig. 2).

We see that in all four categories, we see substantially similar distributions of differences in evaluation across the four categories. The largest bin of changes across all four topics is no change. There is a slightly larger number of individuals changing their evaluations to more negative in Arab Spring tweets.

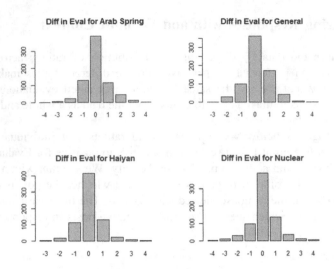

Fig. 2. Difference in evaluation ratings of responses with and without context

In repeating this analysis for the other two dimensions of ACT, we see a similar pattern unfold – that regardless of the source of the data, there is a significant amount of change occurring across all three dimensions of Affect Control Theory. We now describe these changes more quantitatively and show that a similar analysis on simulated data does not yield the same result.

4   Features of Responses with Context

While the histograms give the appearance that the most common change in ratings after seeing context is no change - half the time, individuals are, in fact, changing their ratings. 46 % of Evaluations were changed upon seeing context, 50 % of Potency ratings were changed, and 52 % of Activity ratings were changed (Table 2).

Table 2. Table of features of changed ratings. Changed Total and Changed Valence percentages are based on all responses; other percentages are based on the number of responses that changed valence.

	Evaluation	Potency	Activity
Changed Total	1,329 (46 %)	1,439 (50 %)	1497 (52 %)
Changed Valence	905 (31 %)	1140 (40 %)	1138 (40 %)
Changed to Neutral	316 (35 %)	391 (34 %)	360 (32 %)
Changed to Pos./Str./Act.	341 (38 %)	430 (38 %)	375 (33 %)
Changed to Neg./Weak/Pas.	267 (30 %)	329 (29 %)	419 (37 %)

In fact, at least 30 % of post ratings changed valence after seeing context – 40 % for Potency and Activity ratings. Since all ratings were made on a five point Likert scale, we considered all ratings to be one of 3 valences: Negative, Neutral, or Positive for Evaluation; Weak, Neutral or Strong for Potency; and Passive, Neutral, or Active for Activity.

We find that of the posts which changed valence, changes were made relatively uniformly – to either positive/strong/active, neutral, or negative/weak/passive – in overall similar numbers, with about one third of the posts that changed valence going to each category.

We investigated whether viewing context made it more likely to make a post be perceived as being more extreme or whether it largely attenuated ratings. Of posts that changed ratings, 22 %, 18 %, and 23 % of ratings respectively for Evaluation, Potency, and Activity changed to extreme positions. It appears that it is more likely to attenuate an overall rating – while there are larger numbers of neutral ratings in general, a larger proportion of those posts that changed valence across all dimensions of ACT changed to neutral as opposed to changing to a more "extreme" position on the Likert scale.

5   Validation

To validate these findings, we created two simulated datasets with similar summary properties as our data to highlight how the results we obtain are not simply due to data manipulation. Two simulated datasets were used because of uncertainty in the underlying distribution of responses. Each simulated dataset replicates one third of the responses for a given topic area, so there are 12 paired sets of 90 draws.

The first simulated dataset is drawn from a binomial distribution with four draws and a probability of success of 50 %. The second simulated dataset is drawn from a multinomial distribution with five bins with probabilities matching the distribution of categories in the Evaluative dataset. As in the original experiment, where we had two individuals evaluate the same data, we ensured our data had a similar Cohen's kappa of 0.60 by duplicating this data and randomly replacing half of the simulated data (Figs. 3 and 4, Table 3).

Table 3. Table of summary statistics comparing binomial and multinomial simulated data

	Evaluation	Potency	Activity	Binom.	Multi.
1st Quartile	2	2	2	2	2
Median	3	3	3	3	3
Mean	2.8	3.1	3.2	3.0	2.9
3rd Quartile	4	4	4	4	4
Std. Dev.	1.1	1.1	1.2	0.98	1.1

We find that when comparing our simulated data with difference ratings seen with and without context, the simulated data has a considerably larger variance. In addition to this larger variance, significantly more respondents choose not to change their rating when compared with our randomly generated data (Fig. 5).

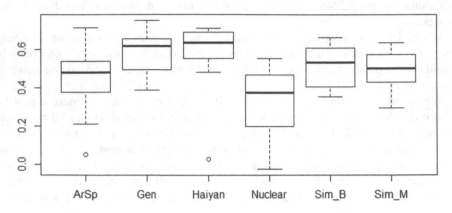

Fig. 3. Distribution of kappas across topic areas and for simulated data; 'Sim_B' indicates data drawn from the binomial distribution, 'Sim_M' indicates data drawn from the multinomial distribution.

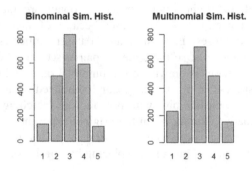

Fig. 4. Histogram of binomial and multinomial simulated data sets.

These results show that a key finding of our study – that about 50 % of all ratings change after re-evaluating the message with context – is not simply an artifact of data manipulation (Table 4).

Table 4. Table of difference statistics, compared with binomial and multinomial simulated data.

	Eval.	Pot.	Act.	Binom.	Multi.
Mean	0.10	0.04	0.02	0.03	-0.13
Variance	0.94	1.3	1.5	1.7	2.4
Mean number of 'no change' ratings across topic areas	388	360	346	217	196

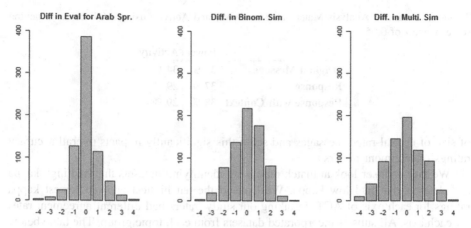

Fig. 5. Histogram of difference in evaluation ratings for Arab Spring contrasted with difference in ratings taken from simulated data.

6 Implications for Current Tools

We evaluated the implications of these findings for current sentiment analysis tools in use. We used VADER [16], as well as the most recent ACT lexicon [14] and the CASOS Universal Thesaurus to create a simple sentiment analysis tool that matched n-gram expressions within the Twitter messages – all dictionary methods that are the current standard approach for sentiment analysis tools due to the problem of sparse training data given the short length of Twitter messages [3].

We found through sensitivity analysis that changing the window of what was considered a "neutral" message to being a score from (−0.1,0.1), to (−0.05, 0.05), to (−0.01, 0.01) did not significantly change overall accuracy rates of the sentiment analysis tools used. We set 0.05 as the window for neutral messages for both of the following Tables 5 and 6.

Table 5. Sentiment Analysis Tool Matching Rates for Evaluation with neutral score window of 0.05

	VADER	Universal Thesaurus	ACT
Original Message	51 %	35 %	39 %
Response	52 %	33 %	34 %
Response with Context	50 %	35 %	35 %

We see that overall sentiment analysis tool ratings appear to match response ratings – as well as original message ratings – at relatively low rates. While our data shows that individuals do change their perceptions of social media messages once they view the message in context, it is harder to draw a connection between automated evaluations of sentiment and what these perceptions are. Future work should further examine the role

Table 6. Sentiment Analysis Match rates for Power and Activity using ACT Lexicon, neutral score window of 0.05

	Power	Activity
Original Message	39 %	34 %
Response	37 %	29 %
Response with Context	38 %	29 %

of size of neutral-rated messages and see if this significantly impacts overall accuracy ratings of sentiment miners.

We take a closer look at match ratings by identifying datasets that had high kappa and datasets that had low kappa. We isolated the ten highest and ten lowest kappa ratings for each axis of ACT; in taking our study, raters had different agreement rates for each axis. All subsets incorporated datasets from each topic group. The table below shows the ranges of the kappas for the data analyzed (Table 7).

Table 7. Ranges of 10 highest and 10 lowest weighted kappas for each ACT axis.

Evaluation		Potency		Activity	
Low	*High*	*Low*	*High*	*Low*	*High*
−0.023–0.37	0.66–0.75	−0.33–0.007	0.33–0.49	−0.13–0.042	0.27–0.34

While we would expect a higher match rate for the subset with higher kappas, we find that overall match rates are identical to the overall population. These rates are not significantly improved by looking at the average rating provided by both raters; additionally, they do not change significantly looking at other dimensions of ACT (Table 8).

Table 8. Match rates for Evaluation tools, contrasting 10 highest and 10 lowest kappa datasets

	Highest Kappas			Lowest Kappas		
	VADER	*UT*	*ACT*	*VADER*	*UT*	*ACT*
Original Message	47 %	36 %	35 %	46 %	38 %	38 %
Response	42 %	41 %	42 %	47 %	34 %	32 %
Response w/Context	40 %	44 %	40 %	47 %	33 %	34 %

7 Discussion

Social media is a dynamic communication medium – useful for a variety of policy applications, from tracking extremist groups to guiding soft power efforts internationally to raising social awareness. Social media messages are inherently social – they are messages that are meant to be shared and disseminated across platforms. In this study, we have limited our analysis to short conversation snippets on Twitter, and we have only examined the text messages contained in those social media posts. However, many platforms also allow embedding more dynamic media – from GIFs to memes to YouTube videos.

Understanding social contagion and the dynamics of social movements requires understanding the context that these movements come out of. Messages are always viewed in context: for example, a popular online hashtag, #NetflixAndChill, while sounding innocuous, refers to a casual sexual encounter – and quickly served as a shibboleth for 'hip' internet users. Understanding the context surrounding the hashtag requires readers to be aware of considerably more than the current 140 characters Twitter allows in messages. If we are going to quantitatively assess these movements and understand how this change is proliferating across social media, we need to develop better tools that can capture and reflect the ratings of individuals reading and responding to these messages.

The implications of this finding on measuring soft power sentiment: additional structural considerations need to be taken when measuring and observing online discussion of topics. While it is useful to aggregate and distinguish social media posts by their immediate sentiment, additional consideration must be taken to couch posts in the structure of online conversation. If there are several unique posts about a topic, it is going to be more informative to do an analysis of the original posts instead of simply analyzing and aggregating responses to the posts, many of which may be a simple endorsement of the original message. While different social media platforms are able to provide different levels of access to their underlying social network structure, future researchers utilizing social media should try and utilize and incorporate that structure into their sentiment analysis and overall assessment of the platform.

References

1. Pang, B., Lee, L.: Opinion Mining and Sentiment Analysis. FNT Inf. Retrieval **2**(1), 1–135 (2008)
2. Grimmer, J., Stewart, B.M.: Text as data: the promise and pitfalls of automatic content analysis methods for political texts. Polit. Anal. **21**(3), 267–297 (2013)
3. Thelwall, M., Buckley, K., Paltoglou, G., Cai, D., Kappas, A.: Sentiment strength detection in short informal text. J. Am. Soc. Inf. Sci. Technol. **61**(12), 2544–2558 (2010)
4. Esuli, A., Sebastiani, F.: Sentiwordnet: a publicly available lexical resource for opinion mining. In: Presented at the Proceedings of LREC (2006)
5. Pennebaker, J.W., Chung, C.K., Ireland, M.: The development and psychometric properties of LIWC2007. In: LIWC.net, Austin, TX, USA, LIWC2007 (2007)
6. Stone, P.J.: User's Manual for The General Inquirer. MIT Press, Cambridge (1968)
7. Agarwal, A., Xie, B., Vovsha, I., Rambow, O., Passonneau, R.: Sentiment analysis of twitter data. In: Proc. ACL 2011 Workshop on Languages in Social Media, pp. 30–38 (2011)
8. Davidov, D., Tsur, O., Rappoport, A.: Enhanced sentiment learning using twitter hashtags and smileys. In: ACL Proceedings of the 23rd International Conference on Computational Linguistics Posters, pp. 241–249 (2010)
9. Bermingham, A., Conway, M., McInerney, L., O'Hare, N., Smeaton, A.F.: Combining social network analysis and sentiment analysis to explore the potential for online radicalisation. In: IEEE 2009, pp. 231–236 (2009)

10. Thelwall, M., Buckley, K., Paltoglou, G.: Sentiment strength detection for the social web. J. Am. Soc. Inf. Sci. Technol. **63**(1), 163–173 (2012)
11. Lovin, L.S.: Affect control theory: an assessment. J. Math. Soc. **13**(1), 171–192 (1987)
12. Heise, D.R.: Affect control theory: concepts and model. J. Math. Soc. **13**(1), 1–33 (1987)
13. Osgood, C.E., May, W.H., Miron, M.S.: Cross-cultural Universals of Affective Meaning. University of Illinois Press, Urbana (1975)
14. Smith-Lovin, L., Robinson, D.T.: Interpreting and Responding to Events in Arabic Culture. Office of Naval Research, Grant N00014-09-1-0556, August 2015
15. Frankenstein, W., Joseph, K., Carley, K.M.: Social Media ACTion: SOLO Data Description. Technical report, CMU-ISR-16-103 (2016)
16. Hutto, C.J., Gilbert, E.: VADER: a parsimonious rule-based model for sentiment analysis of social media text. In: Eighth International AAAI Conference on Weblogs and Social Media (2014)

Validating the Voice
of the Crowd During Disasters

John Noel C. Victorino[1](✉), Maria Regina Justina E. Estuar[1],
and Alfredo Mahar Francisco A. Lagmay[2]

[1] Ateneo Social Computing Science Laboratory,
Department of Information Systems and Computer Science,
Ateneo de Manila University, 1108 Quezon City, Philippines
{jvictorino,restuar}@ateneo.edu
[2] DOST Project NOAH, National Institute of Geological Sciences,
University of the Philippines, 1101 Quezon City, Philippines
mlagmay@noah.dost.gov.ph

Abstract. Since the late 1990 s, the intensity of tropical cyclones have increased over time, causing massive flooding and landslides in the Philippines. *Nationwide Operational Assessment of Hazards* or *Project NOAH* was put in place as a responsive program for disaster prevention and mitigation. Part of the solution was to set up *nababaha.com*(www. nababaha.com) and *FloodPatrol* which provided the public with a web and mobile phone based application for reporting flood height. This paper addresses the problem of providing an interactive and visual method of validating crowdsourced flood reports for the purpose of helping frontline responders and decision makers in disaster management. The approach involves finding the neighborhood of the crowdsourced flood report and weather station data based on their geospatial proximity and time record. A report is classified as correct if it falls within the obtained confidence interval of the crowdsourced flood report neighborhood. The neighborhood of crowdsourced flood reports are correlated with weather station data, which serves as the ground truth in the validation process. Use cases are presented to provide examples of automatic validation. The results of this study is beneficial to disaster management coordinators, first-line responders, government unit officials and citizens. The system provides an interactive approach in validating reports from the crowd, aside from providing an avenue to report flood events in an area. Overall, this contributes to the study of how crowdsourced reports are verified and validated.

Keywords: Crowdsourcing · Validation · Verification · Disaster informatics

1   Introduction

The Philippines has been consistent on the list of most disaster-prone countries [16]. In 2012, Super Typhoon Haiyan killed an estimated 6,000 to 10,000 people

© Springer International Publishing Switzerland 2016
K.S. Xu et al. (Eds.): SBP-BRiMS 2016, LNCS 9708, pp. 301–310, 2016.
DOI: 10.1007/978-3-319-39931-7_29

and left the city of Tacloban, Leyte, severely damaged [3, 16]. Historically, tropical cyclones are getting stronger since 1990s which has caused massive flooding and landslides in the Philippines, most especially in the metropolis, Metro Manila and nearby provinces [14, 15]. For instance, Tropical Cyclone "Ketsana" was only classified as a typhoon [15]. However, reports show up to 455 mm of rain fall in 24 h on September 26, 2009 [6]. This report shows one-and-a-half times the historical average of 364 mm for 1993 - 2008 for the entire month of September [15]. In as much as flooding is highly correlated with tropical cyclones, prolonged monsoons have also been causes of massive flooding in the metro.

This experience provided strong motivation for the government to put in place *Nationwide Operational Assessment of Hazards* or *Project NOAH* as a responsive program for disaster prevention and mitigation [6]. Part of this, *nababaha.com*(www.nababaha.com) was created as a crowdsourcing facility to receive, process and visualize spatiotemporal flood data from concerned citizens [10]. *Ateneo Java Wireless Competency Center* developed *FloodPatrol*, an Android mobile phone application, to extend this facility to smartphone users.

The design is based on the concept of Volunteer geographic information (VGI) which is being used as a data source, especially in disaster management and crisis management [5]. Essentially, the crowdsourced flood reports are considered VGIs. Information provides input to flood risk management such that it aids in the process of managing a situation of an existing flood risk to meet the goal of controlling and preparing for the flood and minimizing its effects [5]. Providing an overview of the present situation where flood occurs can improve the situational awareness in those areas [5]. Crowdsourcing provides an opportunity to raise awareness because of the number of volunteers who can contribute [5]. In a similar manner, QLD Flood Crisis Map, developed through the Australian Broadcasting Corporation, is a crowdmapping of the Queensland floods based on the Ushahidi platform [11]. This allows individuals to send flood-related information via email, text message, Twitter or via the website and make them available to anyone online [11]. In general, social media platforms have become an avenue to disseminate information as well as crowdsourced information [7] in events of national concern. To cite some examples, Twitter users traded information, requested help, looked for people feared to be caught at one of the attack sites, and even documented the situation in terrorist attacks in Mumbai, India in 2008 [8]. The emergence of Ushahidi platform, a Google map-based platform designed for geo-located responses to crisis [7, 8] was also used in the 2012 Haiti Earthquake. Hundreds of recruited volunteers translated messages, posted needs on the crisis map, and picked request according to their ability [7, 8]. Ushahidi focuses on public involvement which various agency maps did not offer [7, 8].

The common concern across the usage of crowdsourcing platforms in disaster management is the accuracy of volunteered information given by ordinary citizens [5, 7, 8, 11, 17]. Agencies, especially those of the police and frontline responders, need accurate and validated data to prevent actions from being taken based on inaccurate, unsourced and questionable information [8]. Users sought for tools that are authorative and provide authorative data as well because failure to do so can cost lives [8].

Method of validation and verification process continues to be an area of study in crowdsourced systems. A good percentage provides opportunity to validate the content [9]. Some platforms have provided a voting button functionality [8]. For example, Ushahidi lets users verify a report by clicking the verification button [7]. QLD Flood Crisis Map allows users to vote up and down reports [11]. This functionality depends on a large number of people validating the reports, which are only a few percentage of the crowd [7,9]. Another approach is a group of administrators manually approve reports before they are made available [5]. Developing trust management systems that can recognize and report questionable postings, then flag these reports for more verification is needed [7]. Moving forward, there is a need for an automated method of validating reports [5]. This paper addresses specifically that concern of developing an automated and interactive process of validating reports.

2   Definition and Method

Crowdsourced flood reports contain the following attributes as shown in Table 1. Using *nababaha.com* and *FloodPatrol*, the person reports the perceived flood height from the list mentioned in Table 1. The system also allows the user to geotag directly on the map. Providing a photo and a short note is optional but accepted as it provides more details to a report. The system automatically stores the date and time the report is sent as well as the latitude, longitude and the estimated flood height in meters. As reports are collected, our approach in validating them takes into consideration what other crowdsourced flood reports are saying. This follows the idea of "wisdom of the crowd" [13].

Our first task is to minimize the spatio-temporal distance between the report that needs to be validated and other reports [12]. We have to get the neighbors that are spatially and temporally close to each other for the confirmation of a report to increase [12]. To illustrate, a report from one city is closely related to another report from the same city. We cannot say the same thing for a report in another city that is too far away. This is the same in terms of time difference. Although this approach gives no guarantee of truth but this applies a heuristic principle [12]. To do this, the basic Nearest Neighbor Search algorithm and Fixed Radius Neighbor Searching variation is used. The former answers the need to find among a group of known reports, the reports closest to the report we want to validate [1]. The latter follows the same idea with the nearest neighbor algorithm, only that the search space has a fixed boundary [2,4]. Finding crowdsourced flood report neighbors is defined as follows.

$$find\_crowdsourced\_flood\_report\_neighbors(R, h, d) = \{neighbors_R\}. \quad (1)$$

find_crowdsourced_flood_report_neighbors is the function that finds the crowdsourced flood report neighbors, $\{neighbors_R\}$. This function accepts three parameters. The first parameter is the report that we want to find its neighbors, R. The second parameter is the range of time, in hours, that we want to consider, h. The last parameter is the range of distance, in meters, that we want to look into, d.

Table 1. Crowdsourced Flood Report Data Definition

Data Definition	Description
Flood_height	A nominal scale describing the flood waters' depth. This can be one of the following:
	• No Flood
	• Ankle High
	• Knee High
	• Waist High
	• Neck High
	• Top of Head High
	• 1-Storey High
	• 2-Storeys or Higher
Flood_datetime	Time and date of the flood report
Image_url	URL of the flood image.
Details	A short note about the flood.
Location	Point geometry data type of (longitude, latitude) in EPSG:4326
Estimated_flood_height	Estimated flood height in meters, derived from flood_height [10]
	• No Flood: 0m
	• Ankle High: 0.25m
	• Knee High: 0.5m
	• Waist High: 1m
	• Neck High: 1.5m
	• Top of Head High: 1.69m
	• 1-Storey High: 3m
	• 1.5-Storeys High: 4.5m
	• 2-Storeys or Higher: 6m

After finding the neighborhood of the report that needs validation, confidence interval of the neighborhood is computed to check if the report is valid or not. We say that the report is valid if it falls within the confidence interval of the neighborhood. We find the sample mean $\overline{neighbors}_R$, sample standard deviation $s_{neighbors_R}$, sample size $n_{neighbors_R}$, and t-critical to find the confidence interval. We used a t Distribution Table to look for the t-critical based on the degrees of freedom df, and the confidence level, which we set at 95 %.

$$confidence\_interval : \overline{neighbors}_R \pm \text{t-critical} * \frac{s_{neighbors_R}}{\sqrt{n_{neighbors_R}}} . \qquad (2)$$

Weather station data is incorporated in the validation process by computing for the correlation between the neighborhoods of weather station reports and crowdsourced flood reports. First, the system searches for nearby weather stations from our crowdsourced flood report neighborhood. Similar to (1), the system searches for weather station neighbors from the members of the crowdsourced report neighborhood. This is illustrated as follows.

$$find\_ws\_neighbors(neighbors_R, h, d) = \{ws\_neighbors_{neighbors_R}\} \ . \quad (3)$$

find_ws_neighbors is the function that finds the weather station neighborhood. This function also accepts the same parameters in (1) except for the crowdsourced flood report neighborhood, $neighbors_R$. If weather station neighborhood is present, the mean of the records for each weather station is computed as well as the correlation with the crowdsourced flood report neighborhood.

3   Results and Discussion

The system uses Postgres SQL 9.4.4 and PostGIS 2.1.0 for the database, Ruby and Ruby on Rails web framework for the system application and Leaflet for front-end web client mapping. This system pulls data from *nababaha.com*, *FloodPatrol*, and various weather stations across the Philippines provided by *Project NOAH*.

Crowdsourced flood reports as early as 2001 until 2015 were used for this study. The initial version of *nababaha.com* and *FloodPatrol* allowed users to set the date and time which allowed for reports before the two applications were launched. On the other hand, weather station records are checked and pulled depending on the crowdsourced flood report to be validated. Figure 1 shows the interface of the automated validation system as well as how reports from 2001 to 2015 are visualized.

To demonstrate the approach in validating crowdsourced flood report, a report that needs validation is selected. In this case, it is Flood Report 1721. The details of this report is shown in Fig. 2.

Afterwards, the system searches for its neighbors within a 1-km radius buffer and 1-hour time buffer, as show in Table 2 and Fig. 3

Table 2. Neighborhood of Flood Report 1721 within 1-km radius, 1-hour time buffer

Flood Report Flood Height	Flood Report Estimated Flood Height (in m)	Number of Reports
Neck High	1.5	23
No Flood	0.0	1
Total Neighbors		24

In Fig. 3, connections inside "Crowdsourced Flood Report Connection" represent the connection between the report that needs validation and the crowdsourced report neighbor. On the other hand, connections inside "Weather Station

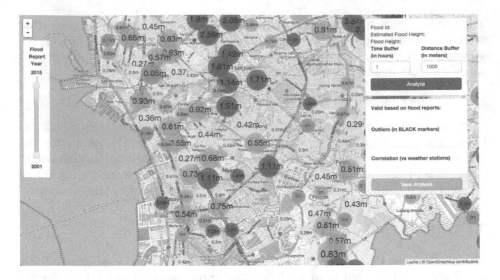

Fig. 1. Crowdsourced flood report from 2001 to 2015

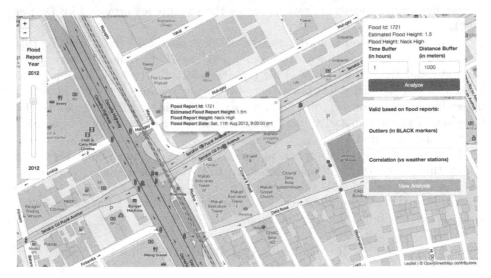

Fig. 2. Flood report 1721

Connection" represent the connection between the member of the crowdsourced flood report neighborhood and the weather station. Moreover, the flood reports were clustered according to how close they are to each other depending on the zoom level of the map. The measurement shown inside the cluster was the mean of the estimated flood height in meters. The size and color varies depending on the computed mean of the cluster. In addition, outliers, which were statistically identified, were marked with color black as shown in Fig. 3.

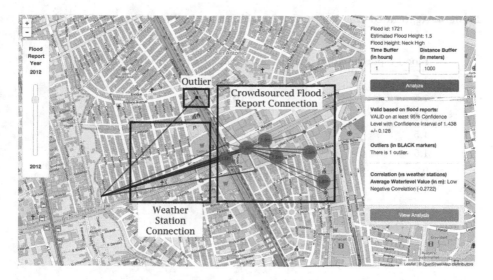

Fig. 3. Neighborhood of Flood Report 1721 within 1-km radius, 1-hour time buffer

The automated validation system considers the flood report to be valid if it falls within the computed confidence interval of the neighborhood. At 95 % confidence level, Flood Report 1721, which has an estimated flood height of 1.5m, is said to be valid because it falls within the confidence interval of 1.31 m to 1.566 m of its neighborhood.

Table 3. Flood Report 1721s neighborhood analysis at 95 % confidence level

	95.0 % Confidence Level
$\bar{x}$	1.438
s	0.306
n	24
df	23
Standard Error	0.062
t **-Critical**	2.069
Margin of Error	±0.128
Confidence Interval	1.31 - 1.566
Valid	true

Depending on the availability of weather station, the last part of the results shows the correlation of the crowdsourced flood report neighborhood and the weather station. Tthe mean of the measurement from the weather station is computed for each weather station as shown in Table 4.

The system computes for the correlation between crowdsourced flood report neighborhood and the available weather station neighborhood. In our example,

Table 4. Aggregated Weather Station Neighborhood for Flood Report 1721 Neighborhood

Flood report flood height (in m)	Average waterlevel value (in m)
1.5	10.992
1.5	10.992
1.5	10.992
1.5	10.992
1.5	11.041
0	11.041
1.5	11.041
1.5	11.041
1.5	11.041
1.5	11.041

$R = -0.272$. This means that the neighborhood of Flood Report 1721 has a low negative correlation with the waterlevel sensor available.

The validation of the voice of the crowd comes in two levels. The first level is from reports within and among the crowd. Our approach showcased the main idea of "wisdom of the crowd" [13]. The validity of a report is based on what the other nearby reports from the crowd. If members of the crowd, who are independent of each other, report similar instances of an event, most probably there is meaning in that event. This implies that a large amount of reports should be present. As part of a project that is being prepared for full adoption at the national level, there will be continuous collection of crowdsourced reports which will be used to improve the current validation technique.

Our approach also considers the difference in time and in location, which are important aspects in gathering and verifying crowdsourced reports. By developing an interactive system, user has the option to adjust the parameters then receives the validation score through confidence level. The results of the validation are presented within the system. It allows the user to provide the time buffer parameter and the distance buffer parameter. The report might not be valid within a larger time and distance buffer since other reports might be irrelevant. On the other hand, there might be a correlation between the report and the weather station within that specific range.

The second level is a comparison of weather station records as a ground truth. This is done by computing for the correlation of the crowdsourced flood reports with weather station records, which come from well-calibrated sensors. Although our approach was not able to establish whether the report is accurate in terms of ground truth, it is helpful for responders and decision makers. In this regard, further studies have to be made in maximizing these sensor data to verify crowdsourced reports.

4   Conclusion

This paper contributes to the challenge in validating crowdsourced reports. Our approach uses other reports and weather station data. The system searches for neighbors of the report by considering the difference in time and location parameters. The report is said to be valid if it falls within the confidence interval of its neighborhood. Moreover, the system presents an approach in the automated validation system that shows the network of neighbors and the possible outliers in the neighborhood.

Weather station data is used as ground truth and is compared with the crowdsourced neighborhood using correlations. Further studies have to be done in maximizing the weather station data to validate the accuracy of the crowdsourced report.

Aside from the existing capabilities of the automated validation system, these features have to be incorporated to improve the validation and verification of reports. First, a reporter should be identified in the system. Reporters, who give more trustworthy reports in the crowdsourcing community, get the higher weights. Second, each report should be allowed for calibrating of responses. This means that if a person reports a flood height of "Ankle High," the system can consider it to be a report from "No Flood" to "Knee High." This gives more allowance to the difference in people's perspective in terms of flood height. In addition, reports which are closer by distance, in theory, are more related to the report in question. Lastly, crowdsourced verification should be added. This allows the crowd to "agree" or "disagree" to a report.

Overall, our approach in validating crowdsourced flood reports uses other crowdsourced reports and weather station data as ground truth. This approach is simple yet highlights the idea of "wisdom of the crowd" [13] in an interactive system. Aside from providing an avenue to report flood events in an area, our attempt can increase analysis and trust in crowdsourcing applications because users have the capability to validate and verify reports. Users will have the capability to see which reports tell the same thing and which reports tell otherwise. Reports are further validated and verified as more people contribute to the system. Thus, a nationwide full adoption of the system is being prepared.

Acknowledgments. Acknowledgment is given to Philippine Council for Industry, Energy and Emerging Technologies Research and Development (PCIEERD), Department of Science and Technology (DOST), for funding this research; Project NOAH, Ateneo Java Wireless Competency Center and the Ateneo Social Computing Science Laboratory for providing the needed assistance and support for this paper.

References

1. Andrews, L.: A Template for the Nearest Neighbor Problem (2001). http://www.drdobbs.com/cpp/a-template-for-the-nearest-neighbor-prob/184401449, Accessed 29 March 2016

2. Bentley, J.L.: Survey of techniques for Fixed radius near neighbor searching. Technicalreport, Stanford Linear Accelerator Center, California, USA (1975)
3. Brown, S.: The Philippines Is the Most Storm-Exposed Country on Earth (2013). http://world.time.com/2013/11/11/the-philippines-is-the-most-storm-exposed-country-on-earth/, Accessed 29 March 2016
4. Chazelle, B.: An improved algorithm for the fixed-radius neighbor problem. Inf. Process. Lett. **16**(4), 193–198 (1983)
5. Degrossi, L.C., de Albuquerque, J.P., Fava, M.C., Mendiondo, E.M.: Flood Citizen Observatory: a crowdsourcing-based approach for flood risk management in Brazil. In: Reformat, M. (ed.) The 26th International Conference on Software Engineering and Knowledge Engineering, Hyatt Regency, Vancouver, BC, Canada, July 1–3, 2013, pp. 570–575. Knowledge Systems Institute Graduate School (2014)
6. Duncan, A., Hogarth, P., Paringit, E., Lagmay, A.: Sharing UK LIDAR and flood mapping experience with the Philippines. In: International Conference on Flood Resilience: Experiences in Asia and Europe, pp. 73–75
7. Gao, H., Barbier, G., Goolsby, R., Zeng, D.: Harnessing the crowdsourcing power of social media for disaster relief. Technical report, DTIC Document (2011)
8. Goolsby, R.: Social media as crisis platform: The future of community maps/crisis maps. ACM Trans. Intell. Syst. Technol. **1**(1), 7:1–7:11 (2010). http://doi.acm.org/10.1145/1858948.1858955
9. Howe, J.: Crowdsourcing: Why the Power of the Crowd Is Driving the Future of Business, 1st edn. Crown Publishing Group, New York, NY, USA (2008)
10. Lagmay, A.: Disseminating near-real time hazards information and flood maps in the philippines through web-gis. Proj. NOAH Open File Rep. **1**, 21–36 (2012)
11. McDougall, K.: Using volunteered information to map the queensland floods. In: Proceedings of the 2011 Surveying and Spatial Sciences Conference: Innovation in Action: Working Smarter (SSSC 2011), pp. 13–23. Surveying and Spatial Sciences Institute (2011)
12. Schlieder, C., Yanenko, O.: Spatio-temporal Proximity and Social Distance: A Confirmation Framework for Social Reporting. In: Proceedings of the 2Nd ACM SIGSPATIAL International Workshop on Location Based Social Networks, pp. 60–67. LBSN 2010, NY, USA (2010). http://doi.acm.org/10.1145/1867699.1867711
13. Surowiecki, J.: The wisdom of crowds: Why the many are smarter than the few and how collective wisdom shapes business. Economies, Soc. Nations **xxi**, 296 (2004)
14. Virola, R.: Statistically Speaking.. Climate Change - Will the Poor Suffer More? (2009). http://www.nscb.gov.ph/headlines/StatsSpeak/2009/030909_rav_climatechange.asp#1, Accessed 29 March 2016
15. Virola, R.: Statistically Speaking.. The Devastation of Ondoy and Pepeng! (2009). http://www.nscb.gov.ph/headlines/StatsSpeak/2009/110909_rav_mrsr_typhoons.asp#table1, Accessed 29 March 2016
16. Whiteman, H.: Philippines gets more than its share of disasters (2014). http://edition.cnn.com/2013/11/08/world/asia/philippines-typhoon-destruction/, Accessed 29 March 2016
17. Zook, M., Graham, M., Shelton, T., Gorman, S.: Volunteered geographic information and crowdsourcing disaster relief: a case study of the Haitian earthquake. World Med. Health Policy **2**(2), 7–33 (2010)

Toward a Bayesian Network Model of Events in International Relations

Ali Jalal-Kamali[✉] and David V. Pynadath

Institute for Creative Technologies, University of Southern California,
Los Angeles, USA
{jalalkam,pynadath}@usc.edu

Abstract. Formal models of international relations have a long history of exploiting representations and algorithms from artificial intelligence. As more news sources move online, there is an increasing wealth of data that can inform the creation of such models. The Global Database of Events, Language, and Tone (GDELT) extracts events from news articles from around the world, where the events represent actions taken by geopolitical actors, reflecting the actors' relationships. We can apply existing machine-learning algorithms to automatically construct a Bayesian network that represents the distribution over the actions between actors. Such a network model allows us to analyze the interdependencies among events and generate the relative likelihoods of different events. By examining the accuracy of the learned network over different years and different actor pairs, we are able to identify aspects of international relations from a data-driven approach. We are also able to identify weaknesses in the model that suggest needs for additional domain knowledge.

Keywords: Intelligent agents · Bayesian networks · Modeling and simulation · International relations

1 Introduction

Formal models of international relations have a long history of exploiting representations and algorithms from artificial intelligence [1,3,9,11]. For example, game-theoretic models have supported prescriptive analyses of foreign-policy decisions (e.g., [10]). Political scientists have also used rule-based systems to build descriptive models of the behaviors of geopolitical actors [3]. These manually created models demonstrate the value that AI methodologies can provide in the study of international relations.

As more and more news sources move online, there is increasing data that can inform the creation of such models. More importantly, these data are now in computer-readable formats that can potentially support the *automatic* creation of models. For example, the Global Database of Events, Language, and Tone (GDELT) and the iData repository of Integrated Crisis Early Warning System (ICEWS) both represent hundreds of millions of actions in over 300 categories

© Springer International Publishing Switzerland 2016
K.S. Xu et al. (Eds.): SBP-BRiMS 2016, LNCS 9708, pp. 311–322, 2016.
DOI: 10.1007/978-3-319-39931-7_30

(e.g., negotiation, accusations, military deployment) taken by geopolitical actors (countries, international organizations, etc.), often directed at other such actors[1].

A computational model of the likelihoods of different action types would be invaluable in describing the behavior of actors in international relations. For example, one might expect two allied nations to be more likely to engage in trade agreements and other cooperative actions, as opposed to nations with a more hostile relationship. The dependency may operate in the opposite direction, too, where the relationship between two nations is likely to suffer if one nation makes an accusatory statement about the other. Representing this complex interdependence among the types of actions and the actors' relationship is critical to building an accurate model.

In this work, we apply existing algorithms to learn a Bayesian network [4–6] that represents the distribution over the actions between geopolitical actors. Such a network allows us to analyze the interdependencies among events and generate relative likelihoods of different event types between two actors. We focus on our use of GDELT as a source of events that we translate into a categorical distribution of the actions two actors take toward each other[2]. Besides the events extracted by GDELT, our network model also represents the Ideal Point distances, a measure of country affinity that researchers in international relations derive from UN voting records.

By examining the accuracy of the learned network over different years and different pairs of actors, we are able to identify aspects of international relations that are quantifiable from a purely data-driven approach. By leveraging the declarative nature of our Bayesian network model, we are able to inspect its dependency structure and draw conclusions that provide insight into the types of events that are most strongly tied to geopolitical relationships. Furthermore, we can exploit existing Bayesian network inference algorithms to compute any conditional probability of interest, allowing us to analyze the joint distribution of events to gain insight into various event dependencies. We are also able to identify weaknesses in the model that suggest needs for additional domain knowledge. This work thus represents an important first step toward making use of advances in AI methods in the modeling of international relations.

2   International Relations Data

The "Global Database on Events, Location, and Tone" (GDELT) contains international events automatically extracted on a daily basis from different news sources around the world, dating back to 1979. We restrict our investigation to pairwise international relations, so we use only events that list two countries as the actors. Each event's type is in the form of a numeric code, categorizing the action according to the Conflict and Mediation Event Observations (CAMEO) framework[3]. The CAMEO event codes constitute a hierarchy, ranging

---

[1] gdeltproject.org, lockheedmartin.com/us/products/W-ICEWS/iData.html.
[2] We also used ICEWS, but omit those results for space considerations.
[3] http://eventdata.parusanalytics.com/data.dir/cameo.html.

from top-level categories like "Appeal" (02) and "Fight" (19), to intermediate categories like "Appeal for material cooperation" (021) and "Occupy territory" (192). The CAMEO framework contains 310 such categories.

For each year of GDELT events, we aggregate the events for each pair of actors (ignoring their order in this investigation). For each such pair of actors, we compute a histogram of the number of occurrences of each event type and then normalize these event counts to be a percentage. This normalization loses the potential information contained in the volume of events between actors (e.g., a qualitative difference in the relationship between the USA and the UK vs. between the USA and Turkmenistan based on only the number of actions). On the other hand, the normalization allows us to potentially generalize across relationships that are similar in character, if not in volume.

We also include *Ideal Point* distances, a measure of affinity that political scientists derive from applying a distance metric to voting in the General Assembly of the United Nations[4]. Our goal is a country-independent model of relative event likelihoods in combination with these Ideal Point distances. We collect data across pairwise relationships between countries from the years 2006–2012, which have comparable quantities of annual GDELT data, while also having available Ideal Point data. Although there are 310 different categories in the CAMEO event code list, many categories are not represented in our target data sets. To ensure consistency of representation across these years, we selected the CAMEO event codes that have appeared at least once in all of the years from 2006–2012, leaving an intersection of 133 event types. After including the Ideal Point distance, we arrive at 134 variables for each pair of countries.

To avoid distortions due to country pairs for which GDELT has very few (i.e., non-representative) events, we use a minimum threshold for actions between countries, considering both the total count of the events (to avoid small samples) and the percentage of non-zero events (to avoid highly skewed histograms) for each pair of countries. After examining the size of the data sets resulting from alternate threshold settings, we arrived at 50 for the minimum total count of events, and 5 % as the minimum percentage of the event categories present in a country relationship for every year. The resulting numbers of country

Table 1. Sample data entries from 2006.

Country Pairs		02: Appeal	021: Appeal for material cooperation	...	Ideal Point Distance
Afghanistan	China	0.017	0.027	...	0.167
Argentina	Australia	0.033	0.010	...	0.788
...	...	...	...	...	...
Yemen	Qatar	0.018	0.000	...	0.279

---

[4] https://dataverse.harvard.edu/dataverse/Voeten.

relationships for 2006–2012 are respectively: 773, 886, 1014, 1316, 1198, 1300, 1310. Table 1 shows a subset of the data for 2006.

3   Learning a Bayesian Network Model

To study the causality and dependency structure within the event categories and Ideal Point distances, we seek a model in the form of a Bayesian Network [4–6]. A Bayesian network provides a compact graphical representation of a joint probability distribution over a set of random variables. By representing such a distribution over our international relations variables, we can use standard algorithms to answer queries about the conditional probability of event categories or UN voting patterns of interest given the occurrence of other event categories. For example, we may use such a network to examine the potential likelihood of cooperation vs. conflict, contingent on the frequency of appeals, public statements, and other actions in the recent history between two countries. Unlike a classifier approach to the problem, the Bayesian network representation allows us to interchangeably treat any variable as the input or output to our queries.

In addition to performing such probabilistic inference, the Bayesian network's graphical structure can itself provide insight into the underlying process. In particular, the directed edges in the Bayesian network reflect properties of conditional independence among the variables. By studying the link structure of the network, we can get a better understanding of the causal process that generates the distribution being modeled [7]. Therefore, if we can construct an accurate Bayesian network model of the distribution over event categories and Ideal Point distances, then the resulting graph structure may reveal interesting properties underlying the behavior selection of geopolitical actors.

Another advantage of Bayesian networks is the development of algorithms that can automatically learn the best structure to represent a data set [2,5]. In this work, we use the algorithms contained in the *bnlearn* R package (www. bnlearn.com). All of our variables are continuous, so we approximate the distributions over them as Gaussians. In terms of the specific *bnlearn* algorithms, we treat all of the variables as having linear conditional Gaussian distributions, where the mean of each child is a linear function of its parents' values. While this assumption is likely to be overly strong, it provides a good first approximation. For each year's training data, after learning the network structure model, we fit the model to the data to obtain the standard deviations, intercepts, and the coefficients of each node's parents to evaluate the means. We can then measure the probability of a different year's test data given the learned network.

4   Accuracy of Bayesian Network Models

While a Bayesian network representation is capable of capturing the interdependence among the relative frequencies of different action categories, that capability will not be useful if we do not have data that supports the learning of such dependencies. To quantify the ability of the learning algorithm to capture this interdependence, we compare the performance of a Bayesian network model against a model that assumes independence among the variables. We can view the latter as learning a Bayesian network with no links among the variables. We hope that the Bayesian network with a learned link structure will provide a better explanation of our test data than one without any links.

To quantify how well a given model explains our test data, we compute a joint probability of the data set. For continuous-valued variables, we integrate the appropriate Probability Density Function (PDF) over the 1 %-wide interval in which the observed value falls. Table 2 presents the log of the mean probability of a given

Table 2. Accuracy of learned models.

log(mean(Pr(test | train)))			
Year	Independence	BN (Same)	BN (Different)
2006	−62.09	−59.26	−61.80
2007	−64.82	−62.21	−60.85
2008	−66.00	−62.87	−60.42
2009	−63.34	−59.94	−60.20
2010	−63.55	−60.73	−60.56
2011	−62.40	−59.64	−60.26
2012	−63.09	−60.68	−60.70

year's test data with respect to the models (independence in Column 2, Bayesian network in Column 3) learned from the other years' training data. The Bayesian network model consistently outperforms the independence model by multiple orders of magnitude, providing strong evidence that the dependency structure provides critical information.

5   Analysis of Bayesian Network Models

While our Bayesian networks offer predictive value, there is still room for improvement in their accuracy. However, the very significant improvement gained by the networks' dependency structure provides evidence that we can extract insight by inspecting the networks themselves. Furthermore, such examination can also inform us as to where our network models are doing well and where they are doing poorly.

5.1   Variations in Models over Time

Ideally, we would arrive at a time-invariant model, so that we can reuse the same model year after year, simply by providing it with the given year's event data. Given this goal, one question that arises is the degree to which the Bayesian network models are generalizable over time. To answer this question, we first took each of the seven network and independence models from Table 2 and measured their accuracy against the test data from *other* years as well. For each year of test data, the Bayesian networks outperform the best independence model by multiple orders of magnitude. In fact, the accuracy rank of each year's model, both for the Bayesian network and independence models, is consistent across test sets, an interesting phenomenon to study in future work.

Table 3. Common links in models over years 2006-2012.

First Event		Second Event
Host a visit	→	Make a visit
Use conventional military force	→	Express intent to meet or negotiate
Use conventional military force	→	Fight with small arms and light weapons
Abduct hijack or take hostage	↔	Return release not specified below
Accuse	↔	Make a visit
Allow international involvement	↔	Provide military protection
Allow international involvement	↔	Provide military aid
Arrest/detain/charge w/legal action	↔	Consult
Arrest/detain/charge w/legal action	↔	Engage in negotiation
Arrest/detain/charge w/legal action	↔	Express intent to meet or negotiate
Arrest/detain/charge w/legal action	↔	Make a visit
Consult	↔	Criticize or denounce
Consult	↔	Engage in negotiation
Consult	↔	Return release
Consult	↔	Use conventional military force
Consult not specified below	↔	Use conventional military force
Cooperate economically	↔	Express intent to engage in material cooperation
Express intent to cooperate	↔	Sign formal agreement
Engage in negotiation	↔	Make a visit
Make a visit	↔	Use conventional military force

We can also investigate the generalizability of the network structures by learning a Bayesian network for a given year's data (e.g., 2006), and then using the resulting structure to constrain the learning of a network for a different year (e.g., 2007). The learning of the second network can thus modify the parameters on the links, but not the link structure itself. We then evaluate this second Bayesian network on a third year's data (e.g., 2008). Table 2's fourth column presents the average probabilities over the test sets for each structure- and parameter-learning pair. Even when imposing a different year's learned structure, the resulting networks still outperform the independence model. In fact, the networks learned when using different years for the structure- and parameter-learning show much less variance than and sometimes outperform the original networks. This result show more encouraging evidence of time-invariant properties of the link structure, although further investigation is necessary.

We can also directly examine the links to see which dependencies are consistently present across the set of networks. Table 3 lists the 20 (out of a possible 8911) links that exist in each of the 7 networks, of which only the first 3 occur with the same direction. The Bayesian network structure thus provides us with potential insight into event types and interdependencies that are exhibited most frequently in the GDELT data set. It is important to note that BN links are a subset of dependencies, so non-BN methods cannot arrive at the same results.

In particular, the absence of links does not represent independence, but rather conditional independence. So the BN algorithm in a way finds the most direct influences, or causal influences.

5.2 Variations in Models over Different Countries

We seek a model that is not just time-invariant, but also actor-independent. In this section, we investigate whether there are certain actor relationships for which our models perform better than others. Table 4 shows a partial ranking of the relationships whose event histograms are given the highest and lowest probabilities by our learned networks for 2006. Table 4's event counts suggests that the highest-ranked pairs performed many more actions than the lowest-ranked pairs, even though our input data contains no information about the volume of actions between actors due to normalization. More precisely, the correlation between the number of actions between two actors and the pair's rank in our model's accuracy

Table 4. Ranking of relationships by accuracy in 2006.

Relationship	Rank	Count
USA-CAN	1	8,119
USA-UKG	2	21,176
IRN-AFG	3	2,704
USA-RUS	4	20,336
RUS-BLR	5	3,482
...	...	...
CAN-FIN	763	57
CAN-CHL	764	60
ISR-FSM	765	75
BEL-SEN	766	80
CAN-CUB	767	55

ranges over $[-0.26, -0.11]$ over the different years of data. In other words, our model more accurately predicts the action breakdown between actors for which we have more events in GDELT. This correlation is encouraging in that it suggests that a significant part of the inaccuracy of our model derives from actor relationships from which we have limited observations. In other words, we might expect our model to perform better if we were able to get a more accurate categorization of their actions by GDELT, since the algorithms used in GDELT determine the accuracy of the event categorization.

5.3 Ideal-Point-Distance Dependencies

The Bayesian network structure also allows us to look at the dependencies of specific nodes of interest. For example, we can inspect the Markov blanket of the Ideal Point distance node, i.e., its parents, children, and immediate parents of those children. Across the 2006–2012 models, there are a total of 129 events that have appeared in the Markov blanket of the Ideal Point distance variable. Table 5 contains the events that have appeared in the Markov blanket of the Ideal Point distance node at least five times. The size of the Markov blanket ranges from 10–128 nodes over all of the years, but if we ignore the 2008 network, the range narrows to 10–37, representing a much smaller subset than the 133 overall event types. There is one variable, corresponding to the event category "Make optimistic comment", that appears in the Markov blanket in all of the learned networks. There is obviously some consistency across these networks in terms of which nodes are connected to the Ideal point distance node. This consistency suggests that there is some more general dependency between actions of these identified categories and the UN voting patterns measured by the Ideal Point methodology. This dependency suggests an interesting line of investigation that can be informed by political science theories underlying that methodology.

Table 5. Markov blanket of ideal points.

Event Type	Count	Impact
Make optimistic comment	7	.0710
Meet at a third location	6	-.0237
Sign formal agreement	6	-.0592
Criticize or denounce	5	.0626
Fight with artillery and tanks	5	.0353
Provide aid	5	.0353
Make statement	5	.0327
Impose embargo, boycott, or sanctions	5	.0280
Use conventional military force	5	.0248
Demand	5	.0216
Employ aerial weapons	5	.0175
Reduce or break diplomatic relations	5	.0160

The Markov blanket also provides a sufficient subnetwork for the Ideal Point distance node, which is conditionally independent of all other nodes in the network given the variables in its Markov blanket [8]. To see how effective this subnetwork is in predicting the existing Ideal point distances from GDELT events, we learned a Bayesian network over the aggregation of the data from 2006 to 2011. We then computed the conditional probability for the Ideal Point distance for each actor pair in 2012 given the distribution of action categories and determined the probability of various intervals around the true value. We considered different size intervals (5 %, 10 %, and 20 %), and observed that our learned model computes conditional probabilities (6.9 %, 13.7 %, and 26.2 %) that exceed the baseline predictions from a uniform distribution. The prediction here is obviously very noisy, but again, it is a very encouraging sign that a purely data-driven modeling algorithm can identify an informative dependency between only a small subset of event types (e.g., those in the Markov blanket) and UN voting patterns.

6   Analysis of Probabilistic Dependencies

Even if there is no direct link between two nodes in the Bayesian network, there can still be an indirect probabilistic dependency. In this section, we analyze networks over just GDELT events, without Ideal Point distances, allowing us to use an additional two years of data for which Ideal Point data is not available. Thus, we still use the same 133 event types, but now over the data of 9 years, 2006–2014 from GDELT.

For each pair of events, A and B, we query the learned Bayesian network to compute two conditional probabilities, $Pr(B > \text{median}(B)|A > \text{median}(A))$ and $Pr(B < \text{median}(B)|A < \text{median}(A))$. The former (latter) represents the likelihood that events of type B occur with high (low) frequency when events of type A occur with high (low) frequency. We can thus roughly characterize the *impact* of A on B by the difference between these two conditional probabilities. In other words, the greater the difference in the probability, the greater impact the occurrence of A events has on the likelihood of B events. We examined the learned Bayesian networks over the 9 years and identified 626 event pairs (out of 8778 possible) that had the same direction of impact across all of them. Table 6 shows the event pairs in that set with the highest impact.

Table 6. Highest impact event pairs For GDELT.

Mean	First Event	Second Event
0.8425	Host a visit	Make a visit
0.1943	Allow international involvement	Provide military aid
0.1379	Allow international involvement	Provide military protection
0.1365	Mobilize armed forces	Provide military aid
0.1323	Mobilize armed forces	Allow international involvement
0.1291	Provide military aid	Express intent to accept mediation

7   Identifying Anomalous Events

Table 6 shows that the impact for "Make a visit $\leftrightarrow$ Host a visit" is more than quadruple the next highest value. While this result is rather intuitive (i.e., when I make a visit to you, you host a visit for me), we wished to confirm the accuracy of our intuition. To do so, we manually reviewed 58 of the news links contained in the GDELT event records to informally verify the event. As it turned out, 51 of the 58 links that were categorized as "Make a visit" were also categorized as "Host a visit". Again, this would seem as expected, but further reading of the text revealed that only 14 of the links were actually categorized correctly, while the rest were not related to either making or hosting a visit.

Similarly, the 30 strongest negative impacts all included "Mass expulsion" as one of the events, despite the relative infrequency of mass expulsions over the last decade. Examining 51 of the source news articles categorized as "Mass Expulsion", we noticed that only 13 were relevant. While this partial investigation is not necessarily conclusive, it suggests an error in the parsing of these particular event types. Fortunately, manual inspection of the news source articles showed that such systematic errors are the exception in GDELT's extraction. However, the two examples found here demonstrate an ability of our methodology to unearth such anomalies in GDELT's extraction process.

8 Dynamics of Event Interdependency

Having already examined the consistency of the learned networks and their structure, we can also examine the consistency of event interdependency within our networks by analyzing changes in the impact that event types have on each other. By treating each Bayesian network as a summarization of the data from its given year, we can extract a time series of dependency impact values. A linear regression of the impacts over time for pairs of event types reveals interesting trends in terms of how the impacts are changing over time. For example, Fig. 1a shows the change (or lack thereof) in the dependency between "Sign formal agreement" and "Express intent to cooperate". Thus, not only is the occurrence of these two event types interdependent across actor relationships, but the magnitude of that interdependency has shown to be stable over the 9 years of data. In contrast, Fig. 1b shows that the dependency between "Diminish military engagement" and "Provide military aid" has been weakening over time.

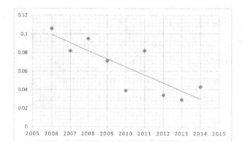 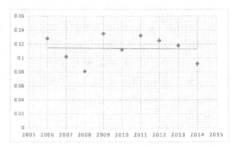

Fig. 1. (a) Impact between "Sign formal agreement" and "Express intent to cooperate". (b) Impact between "Diminish military engagement" and "Provide military aid".

9 Conclusion

In this work, we apply Bayesian network learning algorithms to available data on behavior at the geopolitical level. The link structure generated between different categories of actions provided a clear benefit in the explanation power of the models. Furthermore, the learned structures reveal qualitative properties of the relationships among action categories that can spur further investigation by political scientists.

It is important to note that, while we have limited our exploration to pairwise relationships in this paper, the Bayesian network's representation of the underlying joint distribution allows us to measure the impact of arbitrary subsets of events on other arbitrary subsets. By leveraging this representation, we can thus greatly expand the space of possible queries that can be answered. The generality of this underlying AI model and its algorithms should empower political scientists to conduct analyses that are difficult through purely statistical methods.

The analysis of our models' actor-specific performance showed that more data led to better accuracy. Exploiting additional data sources improves our models' accuracy. Use of

Table 7. Highest impact event pairs for ICEWS.

Mean	First Event	Second Event
0.2530	Conduct strike or boycott	Protest violently, riot
0.2444	Conduct strike or boycott	Coerce
0.1634	Physically assault	Employ aerial weapons
0.1625	Conduct strike or boycott	Meet at a third location

Bayesian network learning and inference algorithms makes it straightforward to incorporate such additional data. In fact, we applied our method to the iData repository of Integrated Crisis Early Warning System (ICEWS), which also uses CAMEO codes. The interesting observation was that the list of the highest impact pairs were totally different for ICEWS data, even though the methods used were the same as the ones for GDELT data, see Table 7. This observation means that we could benefit from aggregation of two data sets, in order to have a more robust prediction of the highest impact pairs. Because each data set has its own types of inaccuracies, the two data sets could potentially complement each other's shortcomings. Thus, complementary sources like ICEWS promise to increase the accuracy of our models without any change in methodology.

There are potential limits to how accurate our purely data-driven models can be. It can be impossible to distinguish some international relationships based on just event counts extracted from the news. For example, Iran's relationship with both Argentina and Israel exhibit similar percentages of "Disapprove", "Accuse", and "Reject" events, yet the two relationships would be considered very different from a political point of view. It is likely that we may need to introduce domain knowledge from the political science literature. Such domain knowledge may come in the form of hidden variables or prior structures for our Bayesian networks from which our algorithms can bootstrap.

While there remains much more work to be done, our methodology here represents an important first step toward automatically learning computational models of international relations. The ever-increasing volume of online data offers a detailed source of geopolitical behavior that can move formal modeling beyond the high-level abstractions that have been necessary in the past. With the accompanying advances in AI algorithms for constructing such models from data, there is now a valuable opportunity for a new dialog between AI researchers and political scientists.

Acknowledgments. This work was sponsored by the U.S. Army Research, Development, and Engineering Command (RDECOM). The authors also thank Torsten Woertwein, and Drs. Marco Scutari, Kalev Leetaru, Eric Voeten, Arthur Spirling, Jeffery Freiden, and Steven Brams.

References

1. Cimbala, S.J.: AI and National Security. Lexington Books, New York (1986)
2. Heckerman, D.: A tutorial on learning with Bayesian networks. Springer, Heidelberg (1998)

3. Hudson, V.M. (ed.): AI and international politics. Westview Press, Boulder (1991)
4. Jensen, F.V.: An introduction to Bayesian networks, vol. 210. UCL Press, London, UK (1996)
5. Neapolitan, R.E.: Learning Bayesian networks. Prentice Hall, New Jersey (2004)
6. Pearl, J.: Probabilistic reasoning in intelligent systems. Morgan Kaufmann, San Mateo (1988)
7. Pearl, J.: Causality; models, reasoning and inference, vol. 29. Cambridge University Press, Cambridge (2000)
8. Pearl, J.: Probabilistic reasoning in intelligent systems: networks of plausible inference. Morgan Kaufmann, San Mateo (2014)
9. Schrodt, P.A.: AI and formal models of international behavior. Am. Sociologist **19**(1), 7185 (1988)
10. Snidal, D.: The game theory of international politics. World Politics **38**(01), 2557 (1985)
11. Sylvan, D., Chan, S.: Foreign Policy Decision Making; Perception, Cognition, and AI. Praeger, New York, NY (1984)

Military and Intelligence Applications

Saint or Sinner? Language-Action Cues for Modeling Deception Using Support Vector Machines

Shuyuan Mary Ho[⊠], Xiuwen Liu, Cheryl Booth,
and Aravind Hariharan

Florida State University, Tallahassee, FL, USA
smho@fsu.edu, liux@cs.fsu.edu,
{clbl4h,ahl4r}@my.fsu.edu

Abstract. In text-based online communication, the clues available to the communicator for ascertaining the underlying intent of a message sender and discerning whether a message is deceptive are often limited to the text. Nonetheless, research has shown that it is possible to detect deception with reasonable accuracy by applying certain classification methodologies to certain observable language-action cues. This paper explores the viability of adopting support vector machines (SVMs) to develop an automated process for deception detection in computer-mediated communications (CMC). In particular, it examines the prediction accuracy of SVM models with different kernel functions on data collected from a controlled online interactive game set up on a Google + Hangout platform. The results indicate that SVM models using the radial basis function (RBF) kernel can classify the complex relationships with high accuracy between language-action cues and deception.

1 Introduction

Computer-mediated communication (CMC) technologies continue to increase the speed, geographical scope, and convenience of interpersonal communication. However, the use of CMC tools (such as chat; e-mail; social media posts) has also significantly increased users' exposure to deceptive online communications (e.g., phishing) and its attendant risks (e.g., identity theft). Despite this, CMC has become an integral part of the communications landscape, and will likely remain a key facilitator of our interpersonal communications for the foreseeable future. The challenge thus becomes how to minimize the user's exposure to these risks, so that users can continue to enjoy the benefits of CMC while their identity, online safety and security are protected. In this regard, one particularly critical question—how to evaluate the truthfulness of a statement, the authenticity of someone with whom we are communicating, and the trustworthiness of the information exchanged—must be addressed.

Communication theorists and computer scientists alike have done significant research exploring deceptive communication and deception detection both in face-to-face (F2F) and CMC environments. While it is well established that, irrespective of communication environment, certain observable language-action cues can

© Springer International Publishing Switzerland 2016
K.S. Xu et al. (Eds.): SBP-BRiMS 2016, LNCS 9708, pp. 325–334, 2016.
DOI: 10.1007/978-3-319-39931-7_31

provide insight into deceptive intent of a writer/speaker, the literature seems to have little that speaks to operationalizing these language-action cues into modeling an automated process and framework for detecting deception. This paper attempts to fill this apparent research gap by addressing the following research question: *Can we computationally classify deception in spontaneous computer-mediated communication across a pluralistic background of users*?

This study contributes uniquely to the text-based CMC literature, in that it focuses on spontaneous, synchronous communication (i.e., text/chat). The paper is organized according to the following structure. The next section discusses deception in general terms, and then specifically in the context of CMC. The third section describes our research design and methodology. The fourth section provides several illustrations of our computational analysis and results. In particular, the leave-one-out cross-validation is performed on our dataset. The final section examines the research implications and limitations, and concludes by providing some insights into future work.

2   Deception? to Tell the Truth...

There two critical truths about deception that must be appreciated in order to understand the fundamental nature of deceptive communication. First, humans are not good at detecting deception [1], and second, deception is fairly common, occurring in approximately one–quarter of all communications [2]. These two truths illustrate both our vulnerability to deception and the importance of improving our ability to detect deception.

By definition, deception is "...a message knowingly transmitted by a sender to foster a false belief or conclusion by the receiver" [2, p. 205]. This definition reveals additional features of deception. First, deception fundamentally leverages human communication—which is an interactive process, involving both a message sender (sender) and at least one message receiver (receiver). The interactive and interpersonal nature of communication, which is emphasized in Buller and Burgoon's [2] Interpersonal Deception Theory (IDT), provides the platform through which the deceiver attempts to accomplish multiple objectives—including impression management, emotion management and conversational management. In developing IDT, Buller *et al.* [3] attempted to understand how the sender of a deceptive communication) strategically shapes his/her communication by studying the perceptions and suspicions of the receiver(s). Expanding on the IDT perspective, Miller *et al.* [4] described deceptive communication as "...a general persuasive strategy that aims at influencing the beliefs, attitudes and behaviors of others by means of deliberate message distortions" (p. 99). A second feature of deception refers to the act of deception. Deceptive communication is a volitional, intentional act; it does not occur by accident or mistake [5]. While the foregoing may lead the reader to believe most, or even all, deception is planned, in fact deception can and often does occur spontaneously—with little prior thought or planning [6]. The "on the fly" deception is no less intentional—although the intention may be formed on the spot, rather than prior to the engagement. A spontaneous lie is not "unplanned;" rather, the amount of time spent planning and strategizing is minimal.

These fundamental ideas underlie the discussions in the following subsections examining how language-action cues can reveal deceptive intent in CMC.

2.1 Computer-Mediated Deception Revealed by Language-Action Cues

Our ability to detect deception, whether in a F2F communication environment or in CMC, depends on the availability of physical and verbal cues, which serve to alert the receiver to be more critical of the speaker and/or the information being imparted. The crux of the problem with deception in text-based CMC is that the availability of cues is limited, and materially reduced in quantity and quality—relative to F2F communication. The receiver in CMC has only the sender's text to assess whether the sender and/or the communication itself may be trusted. Framed alternatively in terms of IDT, the deceiver's strategy is implemented through—and can thus potentially be exposed by—his/her use of words. This underscores the importance of language-action cues in differentiating deceivers from truth-tellers [7–9].

"Language-action cues" refers to the linguistic style, phrases, or patterns in an actor's written expression, which are manifested as an indirect or subtle signal of intent to others [8–11]. Language-action cues include both words and syntax. Analysis of words as cues consists in examining the extent of use of particular types of words. Pennebaker and King [12] undertook to categorize a variety of words for their Linguistic Inquiry and Word Count (LIWC) tool [12, 13], which is widely used in linguistic analysis.

In addition to words, syntactical cues are available in CMC. For example, the amount of detail provided (less or more), consistency of detail, the use of more or fewer sensory or spatiotemporal words, and changes in the diversity and complexity of language have all been examined as syntactical indicators of deceptive intent [13]. Results of a study by Zhou and Zhang [14] suggested that deceivers tend to be wordy in their messages (compared to truth-tellers), but provide less relevant or meaningful information in them. Zhou and Zhang [14, 15] also found that the vocabulary and syntax used by deceivers tends to be more limited than that used by truth-tellers, and that the linguistic style of a deceiver tended to be more casual. Finally, Ho et al. [8] discussed latency (i.e., *timelag*), specifically measuring the length of time it took a deceiver to respond to a question concerning the subject of his/her deception and comparing it with the length of time it took a truth-teller to respond. All of these language-action cues can be used to capture and benchmark both verbal (word count and details of information disclosed) and non-verbal behaviors (latency and usage of expression words) which can then be computed and analyzed [14].

2.2 Modeling Deception

There are a number of possible approaches to analyzing language-action cues and "modeling" deception with them. Zhou et al. [16] examined, summarized and compared the effectiveness (i.e., accuracy) of four primary modeling and analysis approaches: discriminant analysis, logistic regression, neural networks and decision trees. In particular, they tested the accuracy of each of these approaches. However, although extremely informative and useful, their work does not address how well these methods (or others) may lend themselves to application as the basis for development of an automated, predictive deception detection system. In other words, the question of operationalization of language-action cues in the context of modeling deception is left unanswered. Our study attempts to address this apparent gap, building on the line of inquiry suggested by the research of Zhou *et al.* [16]. To this end, our research explores another type of classification approach—support vector machines (SVM)—to examine accuracy in detecting the potential for deception, with specific consideration given to its potential for developing an automated system.

3   A Sociotechnical Research Design

To address the overarching research question specified above, our research approach emphasizes the development of specific metrics for analyzing language-action cues and word choice as information behaviors. Our approach also focuses on identifying and analyzing certain communication patterns in order to distinguish between different communication typologies. Using an interactive online game (developed in-house[1] using Google + Hangout) that presents players with randomized interactive interpersonal scenarios in which one player must guess whether the other player is being truthful or deceptive, we collected language-action cues from spontaneous conversation via spontaneous, synchronous chat/text-messages. This game-based approach offers both an opportunity and *a motivation* for the players to deceive (or, at least, to express an intent to deceive) [8, 9].

Game play consists of two players, who are randomly paired by the game system. Each pair of players is presented with interactive question-and-answer scenarios, in which each player is randomly assigned an *outer role* as either a "*speaker*" (i.e., sender) or a "*detector*" (i.e., receiver). Each player must ask or answer questions based on that scenario in accordance with their respective roles. The speaker initiates the scenario/chat exchange by answering (truthfully) the question on which the scenario is based (the ground-truth question). For example, the ground truth question might be "Have you ever been given a parking ticket?" If the speaker has received one, s/he would answer "yes." The detector, who also sees the question but *does not see* the speaker's answer, then asks a series of questions designed to ascertain whether or not the speaker had received a ticket. However, there is a critical twist in the game: Along with the outer role of "speaker", the speaker in each scenario is randomly assigned an *inner* role—either *saint* (truthful) or *sinner* (deceptive), and his/her responses to the detector's questions are based *on that role* rather than the ground truth. Thus, if the speaker had, in fact, never received a parking ticket but was playing the particular scenario as a "sinner" (i.e. deceiver), s/he would make every attempt to convince the detector that s/he had indeed received one. Finally, at the end of each scenario, the detector tries to determine whether the speaker was being deceptive or truthful based on

---

[1] Developed in Florida State University.

these question-and-answer exchanges. This "guess" is then compared with the ground truth collected at the beginning of the scenario, to assess the truthful or deceptive nature of the speaker's responses throughout the exchange.

Each game session lasts approximately 30 min, and consists of an average of 4 distinct scenarios/exchanges (each player changes *outer roles* such that s/he is a speaker at least twice and a detector at least twice). Each scenario lasts approximately 7.5 min, after which the roles of the players are automatically switched.

4 Data Collection and Analysis

Data[2] collection occurred from Fall 2014 through Spring 2015. The data set used for analysis included a total of 80 games sessions. There were 40 participants; 22 males and 18 female, and each pair of players (20 pairs in total) played a total of 4 game sessions. The participants' ages ranged from 18 to approximately 68 years old. Most, but not all, participants were students at Florida State University, and played for free pizza and similar (non-academically-related) incentives. Players' names were replaced with pseudo-names in order to protect their privacy.

The data was cleaned and validated. First, spelling errors were corrected. Various abbreviations, acronyms and chat terms were revised to be spelled out in full (e.g., "LOL" for "laughing out loud," "U" for "you," "2" where used to stand for "to," etc.). After this, we further ensured and validated that the corresponding instigators' *inner role* assignment (saint vs. sinner) in each session was aligned correctly, to eliminate any systemic errors. Once the data had been cleaned, the linguistic cues from the data were extracted using the LIWC tool [12, 13], and the text corpus was converted into numerical representation. The final dataset consisted of 2,196 lines of chat, and 7,271 words were processed (i.e. language-action cues extracted) using the LIWC toolkit.

4.1 Support Vector Machines

To examine the efficacy (i.e., predictive accuracy) of various language-action cues in detecting deception, we employed support vector machine (SVM) analysis (using Matlab R2015a). SVMs identify the unique decision boundary that maximizes the minimum distance of all training samples to the decision boundary to improve generalization performance [17]. Our primary reason for adopting SVM to analyze our data was exploratory: it has not been widely explored or employed previously in other deception research even though SVMs have been one of the proven approaches in pattern recognition. Matlab provides enhanced data visualization capabilities to run SVM packages as compared to other analytical approaches we had explored [8, 9]. Accordingly, we utilized these functions to visualize the decision boundaries on our dataset using various combinations of two- or three-dimensional features. Unique

[2] The Florida State University's Institutional Review Board has approved human subject data collection (Protocols #2014.13490 and #2015.15885).

classifiers were constructed and graphs were generated, using our dataset as training data, to depict the resulting decision boundaries.

Different combinations of language-action cues can be used as input for SVM analysis. We initially experimented with combinations of two cues. We paired negation (as *negate*) and latency (as *timelag*), and social (as *social*) and latency, as the language-action cue pairs of interest. We analyzed each pair of cues using first a linear kernel, which results in a linear decision-boundary, and then a RBF (radial basis function) kernel, which results in a nonlinear decision-boundary. Our linear kernel model yielded an accuracy of approximately 61.67 % accuracy for *negate* and *timelag* cues, and 71.67 % accuracy for *social* and *timelag* cues. As illustrated in Figs. 1 and 2, our training dataset using the RBF model yielded an accuracy of 100 %. In either case, the results suggest the features should generalize well using novel test samples.

We performed leave-one-out cross-validation using the linear kernel models and RBF kernel models, to estimate any potential over-fitting problem. The results using

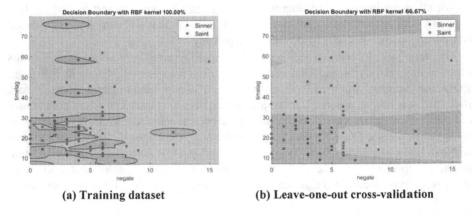

(a) Training dataset   (b) Leave-one-out cross-validation

Fig. 1. 2D SVM RBF kernels with *negate* and *timelag* cues. (a) Training dataset. (b) Leave-one-out cross-validation

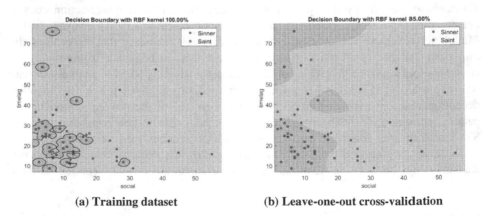

(a) Training dataset   (b) Leave-one-out cross-validation

Fig. 2. 2D SVM RBF kernels with *social* and *timelag* cues. (a) Training dataset. (b) Leave-one-out cross-validation

the linear kernel model yielded a cross validation accuracy of approximately 60 % for *negate* and *timelag* cues, and 70 % for *social* and *timelag* cues. Likewise, the results using the RBF model yielded a cross validation accuracy of 66.67 % for *negate* and *timelag* cues, and 85 % accuracy for *social* and *timelag* cues. Figures 1 and 2 illustrate the results of the 2-cue RBF kernel models with training dataset as well as the leave-one-out cross-validation.

We further examined models combining all three of these cues (*timelag*, *social* and *negate*). Our initial result generated by the three-cue linear kernel model achieved an accuracy of 70 % (Fig. 3), and the three-cue RBF kernel model yielded an accuracy of 100 % (Fig. 4). We also performed leave-one-out cross-validation on these models. This cross validation produced a total of C(60, 20) = 4.19184 × 10^{15} combinations. Figures 3 and 4 illustrate the visualizations for both three-cue linear kernel and RBF kernel models generated by leave-one-out cross-validation. The accuracy of the dataset remained at 70 % after the cross validation for the linear kernel model (Fig. 3), and an accuracy of 83.33 % after the cross validation for the RBF kernel model (Fig. 4).

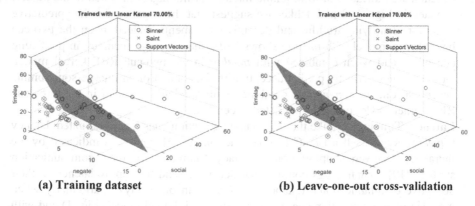

(a) Training dataset   (b) Leave-one-out cross-validation

Fig. 3. 3D SVM linear kernels with *negate*, *social* and *timelag* cues. (a) Training dataset. (b) Leave-one-out cross-validation

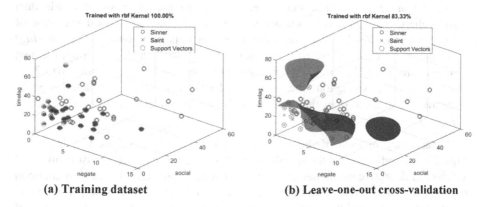

(a) Training dataset   (b) Leave-one-out cross-validation

Fig. 4. 3D SVM RBF kernels with negate, social and timelag cues. (a) Training dataset. (b) Leave-one-out cross-validation

The results show that the parameters of the classifier for the three-cue model(s) can be fine-tuned to reach an accuracy level as high as that obtained from the corresponding two-cue models. As in the cases examining pairs of cues, the decision boundary using three cues also separates the samples in the corresponding three classes with fairly high accuracy. The addition of a third cue to the model did not negatively impact its accuracy.

4.2   Language-Action Cues

The models discussed above examined two or three language-action cues. The specific language-action cues presented below were selected simply for purposes of illustration. Other combinations of LIWC cues and features could be examined using this approach, and would be expected to similarly reveal the relative importance of each of the features being examined in detecting deception.

1. **Negation**: Toma and Hancock [18] suggested that use of words associated with negation are indicative of truth-telling intent. Our findings, illustrated in the models depicted in Figs. 1, 3 and 4 likewise suggest that the negation cue has predictive value for identifying truthful and deceptive statements. Indeed, even the two-cue linear kernel model with *timelag* shows better-than-average accuracy in predicting deception, and when combined with *timelag* in the two-cue RBF kernel model, these two cues perfectly predict deception in the context of an interpersonal online deception. Likewise, *negation,* combined with *timelag* and *social* in the three-cue RBF kernel model perfectly predicted deception.

2. **Latency (Timelag)**: Unlike the other language-action cues we investigated, latency (*timelag*) is not itself a LIWC language-action cue. However, as indicated by the literature, deceivers employ nonimmediacy techniques in their communication strategy [19]. Our findings confirm that deceivers tend to employ latency in their deceptive communications. Latency (as *timelag* in our study)—particularly when modeled with the negation cue in the two-cue RBF kernel model (Fig. 1) and with *negation* and *social* in the three-cue RBF kernel model (Fig. 4) does appear to have predictive value for detecting deception in CMC.

3. **Social**: Our results suggest that deceivers tend to use more social words than truth-tellers. Thus, the language-action cue, *social,* may be a good predictor of deception in CMC. The two-cue linear kernel models using latency (i.e., *timelag*) and the social cues illustrate an accuracy of 71.67 %—which is an improvement upon the results of latency (i.e., *timelag*) and the *negate* cues at accuracy 61.67 %. After the cross validation, the two-cue RBF kernel model can achieve 85 % accuracy (Fig. 2), as does the three-cue RBF kernel model with 83.33 % accuracy (Fig. 4).

The accuracy of each of the models (≈ 70 %–100 %) indicates two important things. First, it indicates that the specific cues chosen in this case were reasonably good predictors of deceptive intent. Second, the fact that the results are an improvement upon what most human detectors could achieve (which is often less than 50 % [1]) seems to validate not only the utility of implementing an automated system, but also the use of SVM with RBF kernels.

5 Limitations

Certain limitations to our study bear at least brief discussion. Some of the limitations involve the design and implementation of the game itself. For example, a stand-alone online game application independent of Google + Hangout could be an important improvement. Other limitations involve the application of SVM analysis itself. Particularly given the complexity of the RBF Kernel models, the model needs to be refined by using data collected in subsequent game sessions, and a larger dataset.

6 Conclusions and Future Work

Our results show that the identification of key text-based language-action cues, correlated to deception, can be effectively used to develop models of behavior. These models can then be used to detect and predict communicative behavioral intent in a CMC environment. The use of SVM analysis allows for generalization of significant cues, and hence can be used in creating an automated deception detection process or system. Our study contributes to the literature of synchronous, spontaneous deceptive communication. The element of spontaneity in our study creates a true "on the fly" deception dynamics, by giving the deceiver no time to plan and prepare.

This paper presents a potential foundation for developing a machine learning system to automate the detection of deception in a spontaneous CMC environment. Future research will focus on mapping out additional deceptive language-action cues, and will employ a stand-alone platform for our interactive social media games. We are currently collecting larger datasets, and investigating other cues in different combinations and cross-validating results, with the objective of identifying additional classifiers that can separate the three classes with the greatest accuracy. Ultimately, our future research will include the design and implementation of a "live" online polygraph system that can be used to automatically detect deception in a CMC environment.

Acknowledgement. The authors wish to thank the National Science Foundation EAGER grants #1347113, 09/01/13—08/31/15, the Florida Center for Cybersecurity Collaborative Seed Grant 03/01/15—02/28/16, and the Florida State University Council for Research and Creativity Planning Grant #034138, 12/01/13—12/12/14.

References

1. Ekman, P., O'Sullivan, M.: Who can catch a liar? Am. Phychologist **46**(9), 913–920 (1991)
2. Buller, D.B., Burgoon, J.K.: Interpersonal deception theory. Commun. Theory **6**(3), 203–242 (1996)
3. Buller, D.B., Burgoon, J.K., Buslig, A., Roiger, J.: Testing interpersonal deception theory: The language of interpersonal deception. Commun. Theory **6**(3), 268–289 (1996)
4. Miller, G.R., Deturck, M.A., Kalbfleisch, P.J.: Self-monitoring, rehearsal, and deceptive communication. Hum. Commun. Res. **10**(1), 97–117 (1983)

5. Vrij, A.: Detecting Lies and Deceit: The Psychology of Lying and the Implications for Professional Practice. John Wiley & Sons Ltd., West Susses (2000). ISBN 0-471-85316-X

6. Whitty, M.T., Buchanan, T., Joinson, A.N., Meredith, A.: Not all lies are spontaneous: An examination of deception across different modes of communication. J. Am. Soc. Inf. Sci. Technol. **63**(1), 208–216 (2012)

7. Hancock, J.T., Birnholtz, J., Bazarova, N., Guillory, J., Perlin, J., Amos, B.: Butler lies: Awareness, deception and design. In: CHI 2009, pp. 517–526. ACM, Boston (2009)

8. Ho, S.M., Hancock, J.T., Booth, C., Liu, X., Liu, M., Timmarajus, S.S., Burmester, M.: Real or Spiel? A decision tree approach for automated detection of deceptive language-action cues. In: Hawaii International Conference on System Sciences (HICSS 1949), pp. 3706–3715. IEEE, Kauai, Hawaii (2016)

9. Ho, S.M., Hancock, J.T., Booth, C., Liu, X., Timmarajus, S.S., Burmester, M.: Liar, Liar, IM on Fire: Deceptive language-action cues in spontaneous online communication. In: IEEE International Conference on Intelligence and Security Informatics, pp. 157–159. IEEE, Baltimore (2015)

10. Ho, S.M., Fu, H., Timmarajus, S.S., Booth, C., Baeg, J.H., Liu, M.: Insider threat: Language-action cues in group dynamics. In: SIGMIS-CPR 2015, pp. 101–104. ACM, Newport Beach (2015)

11. Ho, S.M., J.T. Hancock, C. Booth, M. Burmester, X. Liu, Timmarajus, S.S.: Demystifying insider threat: Language-action cues in group dynamics. in Hawaii International Conference on System Sciences (HICSS 1949), pp. 2729–2738. IEEE, Kauai, Hawaii (2016)

12. Pennebaker, J.W., King, L.A.: Linguistic styles: Language use as an individual difference. J. Pers. Soc. Pyschology **77**(6), 1296–1312 (1999)

13. Newman, M.L., Pennebaker, J.W., Berry, D.S., Richard, J.M.: Lying words: Predicting deception from linguistic styles. Pers. Soc. Psychology Bull. **29**(5), 665–675 (2003)

14. Zhou, L., Zhang, D.: Can online behavior unveil a deceiver? In: HICSS. Hilton Waikoloa Village Big Island. IEEE Press, Hawaii (2004)

15. Zhou, L., Zhang, D.: Following linguistic footprints: Automatic deception detection in online communication. Commun. ACM **51**(9), 119–122 (2008)

16. Zhou, L., Burgoon, J.K., Twitchell, D.P., Qin, T., Nunamaker Jr., J.F.: A comparison of classification methods for predicting deception in computer-mediated communication. J. Manage. Inf. Sys. **20**(4), 139–166 (2004)

17. Vapnik, V.N.: The Nature of Statistical Learning Theory, p. 314. Springer-Verlag, New York (2000). ISBN 978-0-387-98780-4

18. Toma, C.L., Hancock, J.T.: What lies beneath: The linguistic traces of deception in online dating profiles. J. Commun. **62**(1), 78–97 (2012)

19. Buller, D.B., Burgoon, J.K.: Deception: Strategic and nonstrategic communication. In: Daly, J.A., Wiemann, J.M. (eds.) Strategic interpersonal communication, pp. 191–223. Lawrence Erlbaum Assoicates, Inc., Hillsdale, New Jersey (1994)

The Geography of Conflict Diamonds: The Case of Sierra Leone

Bianica Pires[1]([✉]) and Andrew Crooks[2]

[1] Biocomplexity Institute of Virginia Tech, Arlington, VA 22203, USA
bpires@vt.edu
[2] Computational Social Science Program,
George Mason University, Fairfax, VA 22030, USA
acrooks2@gmu.edu

Abstract. In the early 1990s, Sierra Leone entered into nearly 10 years of civil war. The ease of accessibility to the country's diamonds is said to have provided the funding needed to sustain the insurgency over the years. According to Le Billon, the spatial dispersion of a resource is a major defining feature of a war. Using geographic information systems to create a realistic landscape and theory to ground agent behavior, an agent-based model is developed to explore Le Billon's claim. Different scenarios are explored as the diamond mines are made secure and the mining areas are moved from rural areas to the capital. It is found that unexpected consequences can come from minimally increasing security when the mining sites are in rural regions, potentially displacing conflict rather than removing it. On the other hand, minimal security may be sufficient to prevent conflict when resources are found in the city.

Keywords: Agent-based modeling · Geographic information systems · Civil war · Conflict

1 Introduction

In the early 1990s, Sierra Leone, a small country on the western coast of Africa, entered into nearly 10 years of civil war. Sparked by an abusive government and fueled by an illicit diamond market, the decade-long war killed an estimated 70,000 and displaced another 2.6 million people [1]. It is said that the primary driver of the war was the country's most abundant and valued resource, diamonds [2]. While the resource has resulted in growth in other countries such as Botswana, Sierra Leone has experienced some of the highest levels of poverty in the world. Unlike the diamond mines of Botswana, however, the alluvial diamond mines of Sierra Leone cover widespread areas in remote parts of the country where mining areas cannot be easily fenced and security is minimal [3].

Le Billon [4] argued that the spatial dispersion of a resource is a major defining feature of a war, impacting the type of conflict that may emerge. An agent-based model (ABM) was developed of the resource-driven conflict to explore

© Springer International Publishing Switzerland 2016
K.S. Xu et al. (Eds.): SBP-BRiMS 2016, LNCS 9708, pp. 335–345, 2016.
DOI: 10.1007/978-3-319-39931-7_32

Le Billon's [4] theory. Some of the earliest ABMs of rebellion include Axelrod's [7] model of new political actors and Epstein's [8] civil violence model. More recent ABM's have explored in-group dynamics and ethnic salience (e.g., [9–11]). While the ABM presented here shares similarities with prior ABMs that have explored income, resources, and identity as drivers of conflict, it also introduces some key differences. Utilizing geographic information systems (GIS) and socioeconomic data of the country, a landscape and population that better represent the actual setting being modeled is created while the behavior of agents draws from theory. Different scenarios are run as the diamond mines are made more secure and the mining areas are moved to the capital. It is found that unexpected consequences can come from minimally increasing security over the diamond mines in rural regions. For instance, while minimal increases in government control stopped rebel activity in the south, it displaced the conflict to a district in the north, which had not seen violence in prior runs of the model.

2   Background

Theorists have pointed to opportunity, along with motivation and group identity, as indicators of war (e.g., [5,6]). Opportunity can come in the form of financing, the availability of recruits, and the ability to garner these resources with relative ease, which can be due to factors such as geography, economics, and availability. Others have focused on the financing of war through "lootable" resources (e.g., [6]). Le Billon [4], while agreeing that lootable resources are a factor in conflict, argues further that the spatial dispersion of a resource is a major defining feature of a war, impacting the type and duration of rebellion.

According to Le Billon [4], conflict characteristics are affected by two geographic factors: (1) the location of resources as they relate to the country's center (proximate versus distant) and (2) the concentration of resources (point versus diffuse). Distant resources (i.e., in remote areas) are easier for rebel forces to capture and control, while proximate resources are easier to secure and are less likely to be captured (e.g., coffee). Diffuse resources are widespread over large geographic areas, making the resource more difficult to secure (e.g., alluvial diamonds). Point resources, however, are concentrated in small geographic areas and typically require mechanized extraction (e.g., kimberlite diamonds) making them easier to secure and less likely to be exploited [4]. Assuming an environment that is ripe for conflict, the geographic features of a resource can influence the type of conflict. This relationship is illustrated in Table 1.

3   Model Development

An ABM was developed in MASON [12] to explore the role of geography in the resource-driven war. GIS data was utilized to create the modeling landscape, while socioeconomic data provided initial agent attributes. Due to the localized nature of social processes, including civil violence, ABM combined with GIS is ideal for modeling the unique environment of Sierra Leone and the long-lasting

Table 1. The relationship between the resource dispersion and conflict type [4].

Concentration/Relation to center	Diffuse Widely spread with minimal control	Point Concentrated in small areas
Distant Located in remote territories	Warlordism	Secession
Proximate Close to center of power	Rioting / mass rebellion	State control or coup

conflict it endured. For brevity, a high-level overview of the model is presented here. The detailed model description using the Overview, Design Concepts, and Details (ODD) protocol [13], the source code, and data to run the model can be downloaded from https://www.openabm.org/model/4955/. The model's initialization process is discussed in Sect. 3.1; the agents' behavior is discussed in Sect. 3.2; and the model's outputs are reviewed in Sect. 3.3.

3.1 Model Initialization

The modeling world encompasses the country of Sierra Leone, an area of approximately 71,740 km^2. Each run of the simulation begins by reading in the spatial dataset and building the environment using data from the Global Administrative Areas database [14], the Oak Ridge National Laboratory [15], the Peace Research Institute Oslo [16], and OpenStreetMap [17]. The agent population is created using data from the Republic of Sierra Leone 1985 and 2004 Population and Household Census [19] and the Oak Ridge National Laboratory [15], while socioeconomic data, which provided information on age distribution, income levels, and employment statistics, came from the Republic of Sierra Leone 2004 Population and Housing Census [22,23] and Statistics Sierra Leone's Annual Statistical Digest [18]. Due to the computational constraints of modeling the complete population of Sierra Leone (approximately 4.9 million), the population within each parcel is reclassified to equal one percent of the total population. Model runs performed at varying populations yielded similar qualitative results. Note that households are not explicitly modeled here. The idea of a household is used only to ensure that agents can be assigned an income even if unemployed. Table 2 summarizes the input parameters used in the model.

The model proceeds in one-month increments. While the decision to join the rebellion may occur in a short time period (hours or even minutes), there is a lag of weeks or even months between the time someone makes that decision (or is forced to make that decision) and the time they are actually ready for combat [24]. In addition, the war lasted years. From a modeling perspective, we are interested in capturing the dynamics of the conflict over the years, not days or hours. We also need to consider the balance between spatial and temporal computational resources. The modeling world is the entire country of Sierra Leone and simulates the dynamics of a war as it spreads a country.

Table 2. Input parameters and variables.

Parameter	Range	Default value	Reference
Agents			
Initial number of agents	1–4.9 million	49,000	[15, 19]
Percentage of population in the initial opposition group	0–1	0.005	[2]
Age	Grouped in age ranges	0–6, 7–17 18–64	[23]
Income level	1–3	1–3	[18]
Employment status	1–4	1–4	[22]
Vision	0–370	25	Authors estimation
Likelihood to mine if food poor	0–1	0.01	[5, 6]
Likelihood to mine if total poor	0–1	0.05	[5, 6]
Rebel threshold if adult and not a miner	0–1	0.1	[5, 6]
Rebel threshold if adult miner	0–1	0.01	[5, 6]
Rebel threshold if minor	0–1	0.01	[5, 6]
Parcels			
Distance to diamond mines	0–1	0–1	[14, 16]
Remoteness	0–1	0–1	[17, 25]
Government control over mines	0–1	0	[26]
Maximum parcel risk	0–1	0–1	Authors estimation

3.2   Agent Behavior

The PECS (Physical conditions, Emotional state, Cognitive capabilities, and Social status) framework is a cognitive architecture that provides a flexible framework to model human behavior [20]. Using PECS to implement agent behavior, Fig. 1 provides details on the specific motives (i.e., needs) and the set of potential actions available to the agent. The Intensity Analyzer is responsible for determining the action-guiding motive from the set of possible motives. Two sub-models discussed here – the Needs Model and the Opportunity Model – are incorporated into the Intensity Analyzer to determine agent behavior.[1]

The Needs Model. As illustrated in Fig. 1, agents can have three motives: (1) the need for basic necessities such as food and shelter, (2) the need for security of employment, housing, and financials, and (3) the need to maintain the home. These motives represent the two most fundamental levels from Maslow's [21] hierarchy of needs.[2] Agents meet these needs through a household income, whether from employment in the formal market, employment in the illicit diamond market, or employment of a "household" member. While

---

[1] A third sub-model, the Identity Model, activates the identity of the agent based on the outcome of the Needs and Opportunity Models. A detailed description of this sub-model is provided in the ODD, which can be downloaded from https://www.openabm.org/model/4955/.

[2] While the Needs Model is responsible for determining the agents' motive, and as such, could be called the "Motives Model", it was instead named after the humanistic needs theory for which it draws from to highlight the application of theory.

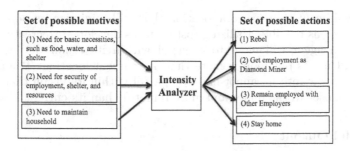

Fig. 1. Motives and actions via the Intensity Analyzer (adapted from [20]).

the Needs Model determines the action-guiding motive, the Opportunity Model helps determine the final goal and, subsequently, the final action the agent will take.

The Opportunity Model. Drawing from opportunity-based theories, which have stressed such factors as the accessibility to resources, the geographic concentration of rebels, and economic factors, agents in the model require opportunity to join the illicit mining industry or to rebel. The first factor of opportunity is the accessibility to resources. This is driven by three criteria: the presence of diamond mines, the remoteness of the area, and the level of government control (or security) surrounding the resource [4]. The second factor is economic in nature. We use a simple likelihood to mine variable to model this, where the lower income brackets are the most vulnerable to joining the conflict. The final factor is the concentration of rebels within an agent's "vision". The more geographically concentrated the rebels, the easier it is to overcome challenges of collective action and to mobilize. In the model, if the first two factors of opportunity are met, then there exists the opportunity to mine in the illicit market. If the third factor is also met, then there exists the opportunity to rebel. In the case of those agents forced to rebel, however, economic factors are not considered, as these cases were largely children abducted and violently coerced to join the conflict [1].

The Action Sequence. An agent can perform one of three activities at each time step: mine, rebel, or do nothing. If an agent stays home or works in the formal market, the agent will do nothing (agents "going to work" is not explicitly modeled). If the agent joins the illicit diamond market, that agent will leave its current employer and will join the diamond mining industry. If the agent's income level was zero, it is increased to one. An agent who becomes a rebel, on the other hand, does not work for any employer, as the agent is either being forced to rebel or is seeking to take control of a mining area for purposes beyond that of the average independent miner.

Agents that are miners or rebels will move on the modeling landscape. These agents need to be near the diamond mines, but at the same time it is assumed that they will want to move to a location that will minimize its potential level of risk as much as possible. Utilizing cost surfaces developed to create the initial landscape, an agent will move to a parcel within its vision that is closer to the diamond mines but more remote than its current location. The agent will continue to move until it cannot find any parcel within its vision that would be better (i.e., closer to the mines and more remote) than its current location.

3.3 Model Output

The model exports a set of comparative statistics, including the number of agents by a set of labor attributes and income levels. Statistics are collected at the district-level by time step so that changes in the conflict's dynamics can be assessed across time and geographic location. The spatial dynamics of the conflict as it evolves across time are observed through the interface during model runs.

4 Simulation Results

This section describes the model results. First, sensitivity testing was performed to ensure the model was working as intended and to establish qualitative agreement of model results to empirical data of the conflict. To determine initial default parameter values, the model was calibrated by adjusting parameter settings and selecting values based on observed visual results that most closely replicated the actual conflict from a qualitative perspective. Figure 2 shows average intensity levels of rebel activity. Because the model does not simulate events, intensity here is a function of that proportion of the total population that rebelled.

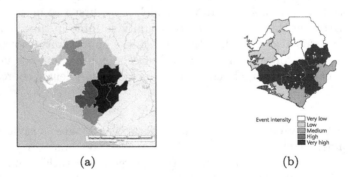

(a)   (b)

Fig. 2. A visual comparison of model results to actual events. a: Average model results using default parameter values. b: Actual event intensity [26].

Next, two experiments were performed to explore Le Billon's [4] theory. As discussed in Sect. 2, Le Billon [4] examined four types of conflicts and the environmental factors required for each to emerge. To explore this theory, two experiments were performed: (1) an experiment where resources are distant and government control (i.e., security) is varied, and (2) an experiment where resources are moved closer to the country's center and government control is varied (all other parameter values are set to the default values shown in Table 2).

The Impact of the Spatial Dispersion of a Resource on Conflict Type When Resources are Distant. To explore the potential impact on a conflict between having distant, diffuse resources, which is associated with conflicts of warlordism, and distant, point resources, which is associated with secession attempts, government control is varied in increments of 0.05 and the diamond mines, whose relation to the "center" of the country is already distant, are maintained at their current locations. Government control of zero represents the minimum securities typically found with diffuse resources while government control of one represents the increased security over point resources.

Figure 3 illustrates the spatial dynamics of rebel intensity as government control is increased. Results shown are the average rebel intensity during year 10 of the conflict. Figure 3a–b show that at lower levels of government control, the resulting violence was widespread with some regions experiencing very high levels of rebel activity. In this case, the resulting spatial dynamics of the violence was similar to the actual areas where conflict was the most intense. Because of the geographic similarities between the real-world case of warlordism in Sierra Leone and model results, the model output supports the theory that distant, diffuse resources are associated with conflicts of warlordism. As expected, Fig. 3c–d show that with increasing government control, the intensity of rebels and the geographic spread of the violence decreased. There are a few unexpected results, however. For instance, while minimal increases in government control were enough at times to stop rebel activity in the south, the conflict looked to be displaced to a district in the north. As government control was increased systematically to simulate a resource situation going from diffuse to point, rebellion occurred in smaller, more contained areas, often on the boundaries of the country. Given these spatial dynamics and the unique geographical location and size of the conflict, a situation of secession may be feasible.

The Impact of the Spatial Dispersion of a Resource on Conflict Type When Resources are Proximate. Freetown is the country's capital, most populated city, and main financial center [27]. Freetown can thus be considered the "center" of Sierra Leone. In this second experiment, the diamond mines are moved to Freetown and results are observed as government control is varied from zero to one at increments of 0.05. Figure 4 shows the spatial dynamics of rebel intensity as government control is increased. Results shown are the average rebel intensity during year 10 of the conflict.

Environments with proximate, diffuse resources are associated with conflicts of mass rebellion or riots near the center of power. When government control is low, this experiment seeks to simulate this environment, as shown in Fig. 4a. While rebel activity emerged in the model, it was largely contained to the capital and its surrounding areas. Although the resources were placed in Freetown,

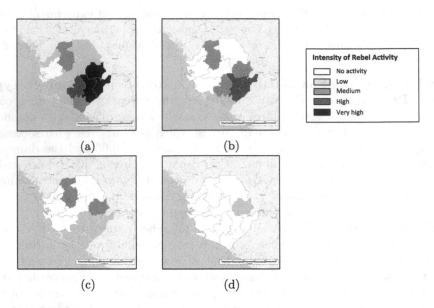

Fig. 3. Average model results in year 10 when resources are distant. a: Government control is 0.0. b: Government control is 0.2. c: Government control is 0.4. d: Government control is 0.6.

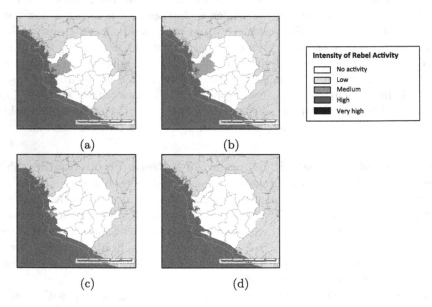

Fig. 4. Average model results in year 10 when resources are proximate. a: Government control is 0.0. b: Government control is 0.25. c: Government control is 0.35. d: Government control is 0.45.

which is located in the district of Western Area Urban, its neighboring district (Western Area Rural) actually experienced higher levels of rebel activity (see Fig. 4a–b). Given the geographic location of rebel activity and the spread of the violence in the model at low levels of government control, these results provide support to the idea that diffuse, proximate resources are associated with rebellion. From Fig. 4, we find that only minimal increases in government control are required to rapidly drop the intensity of rebel activity, supporting the idea that proximate resources are easier for the government to control. As government control was maximized, an environment with proximate, point resources is modeled, as shown in Fig. 4c–d. These types of resources are associated with conflicts of state control or coups. A coup would occur in the country's center of political power, however, at relatively low government control levels (0.25 and above), no rebel activity ensues in the capital. Thus, we cannot support or reject the notion that proximate, point resources are associated with coups.

5   Conclusion

Since diamonds were discovered in Sierra Leone, the government has been unable to control the activity and provide residents with the benefits of having the resource [27]. Through the interplay of ABM and GIS, the model presented explores Le Billon's [4] theory and the impact that the unique environmental and socioeconomic attributes of a region and its population can have on the onset of conflict. The resulting intensity and spatial characteristics of conflict in the model provided support to Le Billon's [4] theory that the spatial dispersion of a resource can lead to warlordism, secession, and mass rebellion. However, the model did not implement the necessary detail to support Le Billon's [4] claim that proximate, point resources lead to a coup. Furthermore, in future work, agent movement could be empirically calibrated to the displacement levels of the population. Nevertheless, by applying simple behavior we were able to explore theory and test "what if" scenarios. When an environment is ripe for conflict, this type of model could potentially provide insight into the locations most prone to conflict and the characteristics of a conflict. Different conflict types may require unique intervention strategies [28], an important consideration for policy.

References

1. UN Development Programme: Case Study Sierra Leone: Evaluation of UNDP Assistance to Conflict-Affected Countries. UNDP, Evaluation Office, New York, NY (2006)
2. Leoa, I.: Youth marginalisation and the burdens of war in Sierra Leone. In: Freedom From Fear, pp. 26–28 (2010)
3. Goreux, L.: Conflict Diamonds, Africa Region Working Paper Series, No. 13. The World Bank, Washington, DC (2001)
4. Le Billon, P.: The political ecology of war: natural resources and armed conflicts. Polit. Geogr. 20, 561–584 (2001)

5. Fearon, J.D., Laitin, D.D.: Ethnicity, insurgency, and civil war. Am. Polit. Sci. Rev. **97**, 75–90 (2003)
6. Lujala, P., Gleditsch, N., Gilmore, E.: A diamond curse? civil war and a lootable resource. J. Confl. Resolut. **49**, 538–562 (2005)
7. Axelrod, R.: A Model of the Emergence of New Political Actors. Working Papers 93-11-068, Santa Fe Institute, Santa Fe, NM (1993)
8. Epstein, J.M.: Modeling civil violence: an agent-based computational approach. PNAS **99**, 7243–7250 (2002)
9. Bhavnani, R., Miodownik, D.: Ethnic polarization, ethnic salience, and civil war. J. Confl. Resolut. **53**, 30–49 (2009)
10. Miodownik, D., Bhavnani, R.: Ethnic minority rule and civil war onset how identity salience, fiscal policy, and natural resource profiles moderate outcomes. Confl. Manage. Peace Sci. **28**, 438–458 (2011)
11. Pint, B., Crooks, A.T., Geller, A.: Exploring the emergence of organized crime in Rio de Janeiro: an agent-based modeling approach. In: 2010 Second Brazilian Workshop on Social Simulation (BWSS), pp. 7–14. Sao Paulo, BR (2010)
12. Luke, S., Cioffi-Revilla, C., Panait, L., Sullivan, K., Balan, G.: MASON: a multi-agent simulation environment. Simulation **81**, 517–527 (2005)
13. Müller, B., Bohn, F., Dreßler, G., Groeneveld, J., Klassert, C., Martin, R., Schluter, M., Schulze, J., Weisse, H., Schwarz, N.: Describing human decisions in agent-based models ODD+D: an extension of the ODD protocol. Environ. Model. Softw. **48**(1), 37–48 (2013)
14. GADM, Global Administrative Areas. http://www.gadm.org/countryres. Accessed Jan 2013 (2009)
15. Oak Ridge National Laboratory. Landscan Global Population Dataset. OakRidge National Laboratory, Oak Ridge, TN (2007)
16. Gilmore, E., Gleditsch, N., Lujala, P., Rod, J.K.: Conflict diamonds: a new dataset. Confl. Manage. Peace Sci. **22**, 257–272 (2005)
17. OpenStreetMap, CloudMade-Map data CCBYSA 2010 (2010). http://downloads. cloudmade.com/africa/sierra_leone#downloads_breadcrumbs. Accessed Dec 2010
18. Statistics Sierra Leone, Annual Statistical Digest 2005/2006. Statistics Sierra Leone, Freetown, Sierra Leone (2006a). http://statistics.sl/final_digest_2006.pdf. Accessed Jan 2014
19. Statistics Sierra Leone, Final Results: 2004 Population and Housing Census. Statistics Sierra Leone, Freetown, Sierra Leone (2006b). http://www.sierra-leone.org/Census/ssl_final_results.pdf. Accessed Apr 2014
20. Schmidt, B.: The Modelling of Human Behaviour. BE: Society for Computer Simulation International, Ghent (2000)
21. Maslow, A.H.: Motivation and Personality. Harper, New York (1954)
22. Braima, S.J., Amara, P.S., Kargbo, B.B., Moserey, B.: Republic of Sierra Leone 2004 Population and Housing Census: Analytical Report on Employment and Labour Force. Statistics Sierra Leone, Freetown, Sierra Leone (2006)
23. Thomas, A.C., MacCormack, V.M., Bangura, P.S.: Republic of Sierra Leone: 2004 Population and Housing Census: Analytical Report on Population Size and Distribution Age and Sex Structure. Statistics Sierra Leone, Freetown, Sierra Leone (2006)
24. BBC, Children of Conflict. BBC World Service (2014). http://bbc.in/1umgSiP. Accessed Apr 2014
25. Commonwealth Department of Health and Aged Care: Measuring Remoteness Accessibility/Remoteness Index of Australia (ARIA). Dept. of Health, Canberra, Australia (2001)

26. Le Billon, P.: Diamond wars? conflict diamonds and geographies of resource wars. Ann. Assoc. Am. Geogr. **98**, 345–372 (2008)
27. Campbell, G.: Blood diamonds: Tracing the Deadly Path of the World's Most Precious Stones. Westview Press, Cambridge (2004)
28. Le Billon, P.: Fuelling War: Natural Resources and Armed Conflict. Routledge, London (2005)

From Tweets to Intelligence:
Understanding the Islamic Jihad
Supporting Community on Twitter

Matthew Benigni[✉] and Kathleen M. Carley

Center for Computational Analysis of Social and Organizational Systems (CASOS),
Institute for Software Research, Carnegie Mellon University,
5000 Forbes Avenue, Pittsburg, PA, USA
{mbenigni,kathleen.carley}@cs.cmu.edu
http://www.casos.cs.cmu.edu/

Abstract. ISIS' ability to build and maintain a large online community that disseminates propaganda and garners support continues to give their message global reach. Although these communities contain trained media cadre, recent literature suggests that large numbers of "unaffiliated sympathizers" who simply retweet or repost propaganda explain ISIS' unprecedented online success [1, 2]. Tailored methodologies to detect and study these online threat-group-supporting communities (OTGSC) could help provide the understanding needed to craft effective counter-narratives however continued development of these methods will require collaboration between data scientists and regional experts. We illustrate the potential of this partnership using two ongoing projects at the Center for Computational Analysis of Social and Organizational Systems (CASOS) at Carnegie Mellon University. First we present the CASOS Jihadist Twitter Community (CJTC), an online community of over 15,000 Twitter users that support one or more of the Islamic extremist groups engaged in the ongoing conflicts in Northern Iraq and Syria. We briefly discuss the methods used to detect and monitor these communities and highlight forms of information that can be extracted from them. We then present an active social botnet that attempts to elevate the social influence of users supportive to Jabhat al-Nusra's agenda. In each case we highlight the ability of these methods to incorporate regional expertise for better performance and recommend future research.

Keywords: Threat network detection · Community detection · Social media intelligence · Online social networks · Social bots · ISIS · Radicalization

1 Introduction

Extremist groups' powerful use of online social networks (OSNs) to disseminate propaganda and garner support has motivated intervention strategies from industry as well as governments however early efforts to provide effective

© Springer International Publishing Switzerland 2016
K.S. Xu et al. (Eds.): SBP-BRiMS 2016, LNCS 9708, pp. 346–355, 2016.
DOI: 10.1007/978-3-319-39931-7_33

counter-narratives have not produced the results desired. Mr. Michael Lumpkin, the director of the United States Department of State's Center for Global Engagement, is charged with leading efforts to "coordinate, integrate, and synchronize government-wide communications activities directed at foreign audiences in order to counter the messaging and diminish the influence of international terrorist organizations" [3]. In a recent interview, Mr. Lumpkin expressed the need for a new approach:

So we need to, candidly, stop tweeting at terrorists. I think we need to focus on exposing the true nature of what Daesh is.

Mr. Michael Lumpkin, NPR Interview March 3, 2016

A logical follow-up question to Mr. Lumkin's statement would be "Expose to whom?" Recent literature suggests that "unaffiliated sympathizers" who simply retweet or repost propaganda represent a paradigmatic shift that partly explains the unprecedented success of ISIS [1,2] and could be the audience organizations like the Global Engagement Center need to focus on. Gaining understanding of this large population of unaffiliated sympathizers and the narratives most effective in influencing them motivates methods to detect and extract information from large online threat-group-supporting communities (OTGSC). However, detecting, monitoring, and data-mining targeted OTGSs requires novel methods, and development must include both data science and regional expertise. We define data science as a set of fundamental principles that support and guide the principled extraction of information and knowledge from data, and in this paper we present the CASOS Jihadist Twitter Community (CJTC), a online community of over 15,000 Twitter users who support one or more of the radical groups engaged in the ongoing conflicts in Northern Iraq and Syria. We describe how large OTGSCs can offer unique insights into the unaffiliated supporters who appear critical to ISIS' success. We then provide an example of one method used to excite and grow these OGTSCs in the form of an active social botnet. The botnet attempts to elevate the social influence of users supportive to Jabhat al-Nusra's agenda, while encouraging following ties amongst botnet followers. Our goal is to present two novel examples of social computing applied to counterterrorism, and motivate the continued interdisciplinary collaboration required to gain understanding of large online communities and effectively counter extremist propaganda.

2   The CASOS Jihadist Twitter Community (CJTC)

On November 13, 2015 much of the world watched as terrorist launched a series of coordinated attacks in Paris killing 130 people. In near real-time social media erupted with support for the victims of these attacks, but some online communities viewed the attacks as cause for celebration. In fact, passive supporting but unaffiliated social media users have become an essential element of groups like ISIS and Jabhat al-Nusra's recruiting strategy, possibly aid the motivation and resourcing for attacks like those seen in Paris [1]. Large online social networks like Twitter offer a means to generate large online communities, and many of the members appear to be "unaffiliated supporters." In fact, Twitter has suspended over 125,000 ISIS-supporting accounts from August to December of 2015 [4]. As ISIS recruiters identify community members who show increasing levels of radicalization, small teams of social media cadre have been observed lavishing attention on these recruitment targets and subsequently move the conversation to more secure online platforms [2]. Less secure but large open platforms like Twitter enable extremist groups propaganda to gain broad reach. Denying this *key terrain* requires novel methods designed specifically to identify and analyze threat-group-supporting communities embedded in OSNs. Information like key users, powerful narratives, and advanced dissemination methods can all be extracted from OTGSCs to inform messaging and intervention strategies. Benigni et al. present Iterative Vertex Clustering and Classification [5], a novel method to detect large, ideologically organized online communities, using both agent level attributes and network structure. We briefly present the methodology, introduce the CJTC, provide illustrative analysis of the network, and share ongoing research goals in this section.

2.1 Background: From Community Detection to Threat Network Detection

The application of network science to counter-terrorism has a long history [6,7]; however, the rise of social media and online social networks (OSNs) has motivated methods to apply network science theory to networks at much larger scale. Community detection attempts to identify groups of vertices more densely connected to one another than to other vertices in a network, but networks extracted from OSNs present unique challenges due to their size and high clustering coefficients. Furthermore, an individual's social network also often reflects his or her membership to many different social groups. Thus in many instances algorithms that use only network structure do not provide the precision needed to identify OTGSCs [5]. A sub-class of community detection methods has emerged that attempts to leverage node attributes and network structure called community detection in annotated networks. These methods have been shown to perform well with OSNs because of their ability to account for a great variety of vertex features like user account attributes while still capitalizing on the information provided by the structure of the graph; they also perform well at scale [8,9]. However, we find that effective OTGSC detection requires information from users' following, mention, and hashtag behaviors as well. Benigni et al. present IVCC, an community detection method designed to extract OTGSCs by modeling users within a heterogeneous graph structure with annotated nodes [5] (Fig. 1).

2.2 Iterative Vertex Clustering and Classification

Iterative Vertex Clustering and Classification (IVCC) is conducted two phases, and often iteratively. In Phase I, unsupervised clustering methods like Newman and Louvain grouping are used to both identify positive cases labels and

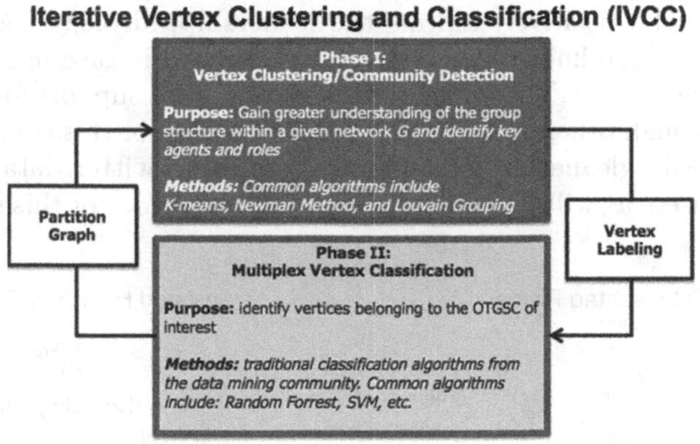

Iterative Vertex Clustering and Classification (IVCC)

Fig. 1. IVCC is an online threat-group-Supporting community (OTGSC) detection methodology conducted in two phases. In Phase I either community optimization or vertex clustering algorithms are used to identify positive and negative case examples to facilitate supervised detection in Phase II.

remove noise. This pre-clustering facilitates supervised classification of OTGSC members in Phase II. At the core of the methodology is the use of both user level features and rich multiplex network structures offered by OSNs. First the authors construct $U_{u \times a}$ consisting of a numeric user attributes where u is the total number of users or nodes in the network. Examples of such attributes are follower count, number of posts, or creation date. Node attributes could also be developed from other sources of intelligence. Spectral methods are used to dimensionally reduce network data like following, mention, or hashtag behaviors. By constructing symmetric graphs of users' following F and mention M relationships, and a weighted bipartite graph H of hash tags in a user's timeline, lead eigenvectors can then be extracted from each graph and concatenated with U to form a feature space for classification. Although IVCC is presented using Twitter data [5], similar methods could be used more generally with large heterogeneous networks.

Benigni et al. collected a two-hop snowball sample of five popular ISIS propagandists presented in [10], resulting in approximately 120,000 Twitter users. With two iterations of IVCC, they removed accounts with high following counts (i.e. politicians, news media members, celebrities, etc.), and extracted a network of nearly 23,000 *ISIS supporters*. The results of this initial work form the seed accounts for the CJTC.

2.3 Threat Network Analysis: The CJTC

CASOS is currently extending IVCC to dynamically monitor threat-group-supporting online communities. By using historical results and active learning, we update the CJTC based on the recent community activity. Currently the community contains just over 15,000 supporters, where we define a supporter as a Twitter user who positively affirms the leadership, ideology, fighters, or call to Jihad of any of the known Jihadist groups engaged in ongoing operations in Northern Iraq and Syria. The majority of tweeters voice support for ISIS or Jabhat al-Nusra though other groups are present. The size of this community offers insights not easily gleaned from randomly sampled Twitter data or manually developed datasets as will be highlighted in the remainder of this section.

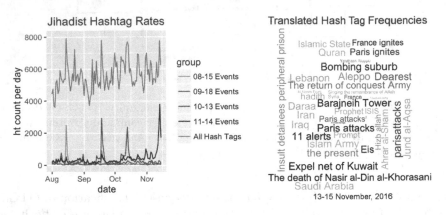

Fig. 2. The left panel depicts the volume of hashtags used within the CJTC from AUG-NOV 2015. The right panel highlights the hashtags most explanatory of the increased activiy on November 14, 2015.

Though many demographical analyses could be useful, for conciseness we will use temporal network activity patterns to illustrate information extraction from OTGSCs. The Twitter REST API limits collection to a tweeter's last 3,200 posts which forces us to normalize daily volume. Some tweeters have more than 3,200 posts in the past 6 months, and quite a few of our tweeters have not posted in over 90 days. Identification of dormant users could provide insight into the radicalization process, but will not be analyzed or discussed in this work. We estimate the CJTC's daily volume by normalizing based on the number of tweeters in our dataset who have a collected tweet before and after any given day which often highlights current events that stimulate this community. A simple news search of events on days of increased activity often reveals operational events in Syria, Northern Iraq, or large scale terror attacks. Similar analysis of hash tag trends often provides richer insight. Figure 2 highlights temporal analysis of CJTC hash tag use. The left panel of depicts hash tag frequencies over time, while the right panel depicts trending hashtags on 13–14 November, 2015. Size in the word cloud connotes frequency, and color denotes how anomalous a particular tag's frequency was when compared to a 6 months average. The community's reaction to the 13 November, 2015 Paris attacks is illustrated with both increased volume and trending hashtags. Increased hash tag volume depicted in the left panel of Fig 2, coupled with the corresponding hash tag trends in the right panel give startling insight into the unique nature of this online community. Ongoing operations in Syria provide another example. The hash tag ڊﺋﻮﯾﺒﻪﮐﺑﻜﺒﺟﯿﻪﺑﺋﻢ, translated Zabadani, increases tenfold in terms of daily frequency on 15 August and 18 September, 2015. Both dates refer the breakdown of ceasefire agreements in the region [11]. With proper subject matter and language expertise, similar analysis can identify changes in popularity of leaders, organizations, or narratives over time.

2.4 Moving Forward

As a supervised learning methodology, IVCC lends itself to leveraging regional expertise by learning patterns based on examples. Active-learning refers to supervised algorithms that iteratively select examples to be labelled by experts, and have been found substantially increase performance with far fewer labelled instances. Such methods could enable regional expertise to be incorporated into the classifier at minimal cost. Furthermore, a user-oriented, server-based interface could enable the regional expert to contribute to the set of annotated instances while conducting his or her own exploratory data analysis. As the set of annotated examples or "training set" grows new, more nuanced classifiers could be trained. Due to the size and diversity of these online communities, exploration and interpretation of results is likely a research area unto itself. One could identify the news sources or propagandists most influential within these communities, and develop more-informed counter-narratives and strategic communications strategies. The challenge in developing tools and methods to facilitate OTGSC analysis lies in the novelty of the analytical task. Regional experts cannot yet articulate exactly what they want methods to provide, and researchers are challenged to understand what information extractions are most useful to senior leader information requirements. Establishing online tools that provide illustrative analyses and capture feedback while end users to explore large communities would likely accelerate research efforts aimed at countering groups like ISIS.

3 The FiribiNome Social Botnet: Sophisticated Promotion of Propaganda to Excite a Community

While analyzing the CJTC, as well as a similar dataset focused on online dialogue focused on the Russian occupation of Crimea, we observe accounts that tweet with high daily volume, but each tweet or retweet simply contains a string of @mentions. In this section we analyze a network of social bots used to promote specific online activists or propagandists.

Social bots, software automated social media accounts, have become increasingly common in OSNs. Though some provide useful services, like news aggregating bots, others can be used to shape online discourse [12]. ISIS' use of bots has been well documented [13], and their competitors are following suit. Social botnets are teams of software controlled online social network accounts designed to mimic human users and manipulate discussion by increasing the likelihood of a supported account's content going viral. The use of bots to influence political opinion has been observed in both domestically [14] and abroad [15], the use of social bots has been documented in the MENA region [12], and ISIS use of them motivated a DARPA challenge to develop detection methods [16]. In isolation, these accounts appear to be producing spam and relatively harmless, however they are examples of a sophisticated strategy to promote specific accounts while remaining undetected by Twitter.

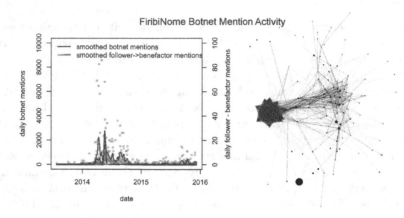

Fig. 3. Depicts mention behaviors and their effects within the FiribiNome Social Botnet. The left panel depicts two scaled time series. The red circles and smoothed trend line depict the number of daily mentions by botnet members. The blue circles and corresponding trend line depict botnet followers' mentions of benefactor accounts. The association between the two series implies the botnet was able to generate discussion about benefactor accounts among its followers. The right panel depicts the mention network of the FiribiNome social botnet. The vertices are user accounts. The plot depicts how *botnet members*, red vertices, are used to increase the social influence of *benefactors*, black vertices, by promoting them to *botnet followers*, blue vertices. Vertices are scaled by follower count (Color figure online).

3.1   CJTC Botnet Analysis

Figure 3 depicts the mention activity associated with the a Jabhat al-Nusra supporting social botnet designed to increase the social influence of a specific set of accounts and encourage following connections between Jabhat al-Nusra supporting tweeters. The botnet consists of two types of accounts. *Botnet members*, are depicted by red vertices in the right panel of Fig. 3, and consist of 74 accounts exhibiting near identical behavior. Each account follows between 116 and 134 accounts, most of which are *botnet members*. Their following counts vary from 142 to 322 accounts of which many appear to be real tweeters. They come online for 38–58 days, tweet between 71 to 170 times, then go dormant. This behavior can clearly be seen by the red trend line in Fig. 3. Their tweets consist of original posts or retweets containing strings of @mentions of other *botnet members*, but occasionally mention or retweet content from what we call *benefactor accounts* (depicted by black vertices in the right panel of Fig. 3). The *botnet* account FiribiNome20 illustrates this behavior. In isolation, these accounts appear to be producing spam and relatively harmless, however our analysis indicates the network of *botnet members* increases the social influence of *benefactor accounts*. The blue series in the left panel of Fig. 3 and corresponding blue vertices in the right panel depict the mention activity of the 843 active botnet followers as of February 2016. The left panel depicts *follower* accounts' mentions of *benefactor* accounts and the temporal relationship betwen the activity associated with each account type implying the botnet effectively promotes discussion of *benefactor* accounts. How much discussion is generated remains an open question. Due to the large number of extremist accounts suspended by Twitter, the number of *botnet followers* active in the summer of 2014 was likely much larger. This mention behavior exhibited by *botnet members* could also trigger Twitter's recommendation system to recommend following ties between *botnet followers*,or encourage *botnet followers* to follow *benefactors*.

Examples of *benefactor accounts* are depicted in Table 1; each representing a slightly different style and type of messaging commonly observed in the CJTC. Dr. Hani al-Sibai is a London-based radical Islamic Scholar cited by Ansar al-Sharia as one of five influential motivators of Tunisian terrorists [17]. *@ba8yaa* or "Daesh are the Enemy" attempts to discredit ISIS through satire and counter-propaganda and could prove informative in development of counter-narratives. There are also many accounts that present the appearance of reporting near-real-time news like *@Ghshmarjhy*, while other accounts promote third party applications like @Almokhtsar and @FiribiNome12. We have found some of these applications request permission to tweet or follow users on the tweeter's behalf. These highly followed and highly mentioned accounts each could offer insight into the sophisticated methods used to leverage social media.

3.2   Moving Forward

It is possible that botnet structures with similar mention behavior could be developed in a more sophisticated manner. Larger networks with more human-like behavior would be much more challenging to detect. The FiribiNome botnet

Table 1. Depicts four account promoted by the FiribiNome social botnet. Each account represents a slightly different style and type of messaging.

Account	Follower Count	Messaging Type
@Hanisibu	104K	Islamic Scholar
@ba8yaa	1,272	anti-ISIS satire/propaganda
@Ghshmarjhy	6,644	Syrian revolution updates
@Almokhtsar	164K	app: MENA news feed

could simply represent a proof of concept explaining its lack of activity since 2014. Although simple heuristics like average mentions per tweet enabled us to detect the botnet, more advanced detection strategies are needed to determine if more sophisticated botnets are influencing the CJTC. Methods of operationalizing this type of intelligence are worth exploring as well. It is possible that similar mention behaviors could be used to target specific online communities with counter-narratives. Again, the need for an interdisciplinary collaboration between the data scientist, regional expert, and decision maker is needed to identify opportunities for useful intelligence extraction.

4   Conclusion

We have highlighted the potential of extracting intelligence from large online threat-group-supporting communities (OTGSCs) and presented illustrative examples with a goal of motivating continued interdisciplinary collaboration. We have also presented the CJTC dataset as an example of an OTGSC to emphasize how detecting and monitoring OTGSCs can be an important tool in understanding the passive support structure essential to the distribution of extremist propaganda. Furthermore, these methods could facilitate identification of sophisticated dissemination techniques used in these communities and inform our own information operations. Our goal is to refine these methods and grow a consortium of data scientists, regional experts, and strategic decision makers by hosting, curating, and reporting on datasets like the CJTC.

Acknowledgements. This work was supported in part by the Office of Naval Research (ONR) through a MINERVA N000141310835 on State Stability. Additional support for this project was provided by the center for Computational Analysis of Social and Organizational Systems (CASOS) at CMU. The views ond conclusions contained in this document are those of the authors and should not be interpreted as representing the official policies, either expressed or implied, of the Office of Naval Research or the U.S. Government.

References

1. Yannick Veilleux-Lepage. Paradigmatic Shifts in Jihadism in Cyberspace: The Emerging Role of Unaffiliated Sympathizers in the Islamic State's Social Media Strategy (2015)
2. Berger, JM.: Tailored Online Interventions: The Islamic States RecruitmentStrategy. Combating Terrorism Center Sentinel
3. Dozier, K.: Anti-ISIS-Propaganda Czars Ninja War Plan: We Were Never Here, March 2016
4. Isaac, M.: Twitter Steps Up Efforts to Thwart Terrorists' Tweets. The New York Times (2016). http://www.nytimes.com/2016/02/06/technology/twitter-account-suspensions-terrorism.html
5. Benigni, M., Joseph, K., Carley, K.: Threat Group Detection in Social Media: Uncovering the ISIS Supporting Network on Twitter. Submitted to Plos One

6. Krebs, V.: Uncloaking terrorist networks. First Monday **7**(4), 215–235 (2002)
7. Carley, K.M.: A Dynamic Network Approach to the Assessment of Terrorist Groups and the Impact of Alternative Courses of Action. Technical report, October 2006
8. Tang, L., Liu, H.: Leveraging social media networks for classification. Data Min. Knowl. Discov. **23**(3), 447–478 (2011)
9. Binkiewicz, N., Vogelstein, J.T., Rohe, K.: Clustering, Covariate Assisted Spectral (2014). arXiv preprint arXiv: 1411.2158
10. Carter, J.A., Maher, S., Neumann, P.R.: #Greenbirds Measuring Importance and Influence in Syrian Foreign Fighter Networks. International Centre for the Study of Radicalization Report, April 2014
11. Syria ceasefire ends, fighting resumes. Reuters, August 2015
12. Abokhodair, N., Yoo, D., McDonald, D.W.: Dissecting a Social Botnet: Growth, Content and Influence in Twitter, pp. 839–851. ACM Press (2015)
13. Berger, J.M.: How ISIS Games Twitter. The Atlantic, June 2014
14. Ferrara, E., Varol, O., Davis, C., Menczer, F., Flammini, A.: The rise of social bots (2014). arXiv preprint arXiv: 1407.5225
15. Forelle, M., Howard, P., Monroy-Hernndez, A., Savage, S.: Political Bots the Manipulation of Public Opinion in Venezuela. arXiv: 1507.07109 [physics]. arxiv: 1507.07109, July 2015
16. Subrahmanian, V.S., Azaria, A., Durst, S., Kagan, V., Galstyan, A., Lerman, K., Zhu, L., Ferrara, E., Flammini, A., Menczer, F., Waltzman, R., Stevens, A., Dekhtyar, A., Gao, S., Hogg, T., Kooti, F., Liu, Y., Varol, O., Shiralkar, P., Vydiswaran, V., Mei, Q., Huang, T.: The DARPATwitter Bot Challenge. arXiv: 1601.05140 [physics]. arxiv: 1601.05140, January 2016
17. Ansar al-Sharia Tunisias and Long Game. Dawa, hisba, and jihad (2013)

Be Alert and Stay the Course: An Agent-Based Model Exploring Maritime Piracy Countermeasures

Ciara Sibley[✉]

Warfighter Human System Integration Laboratory,
Naval Research Laboratory, Washington, DC, USA
Ciara.Sibley@nrl.navy.mil

Abstract. Since its origin nearly forty centuries ago maritime piracy continues to threaten global trade, impacting thousands of merchant vessels each year. Increased security on board merchant ships as well as more costly military countermeasures have been deployed with great success around the Horn of Africa, yet the number of incidents occurring within the Gulf of Guinea have recently drastically increased. This paper will present an agent-based model (ABM) simulating merchant vessels sailing within an environment in which a number of pirates and Naval warships are present. Results demonstrate how ABMs can be used to explore the impact of various piracy countermeasures and discover potentially counter-intuitive consequences of different macro-level policies and micro-level decisions. Ultimately this work serves as a proof of concept for using ABMs to assess the efficacy of strategies for combatting maritime piracy and lays the foundation for future models to inform policy and tactics.

Keywords: Piracy countermeasures · Agent-based model · Maritime security

1 Introduction

The last two decades have seen a large upsurge in maritime piracy, threatening the global shipping economy and impacting thousands of vessels each year [3, 9, 10]. One estimate contends that piracy costs the international economy between $7 and $12 billion per annum when accounting for the cost of ransoms, insurance premiums, re-routing, security equipment, Naval forces, prosecutions, policy organizations and regional economy impacts [3]. Although costly to implement, various countermeasures, such as increased security on board merchant ships and military involvement, have been deployed with great success around East Africa, decreasing the number of Somalia-based piracy and armed robbery incidents from 78 in 2007 to 20 in 2013, with no reports of merchant ships being hijacked in the Somalia-based High Risk Area [9]. Unfortunately, however, this decrease in piracy on the East coast of Africa has been paralleled with an increase in piracy related incidents on the West coast, in particular within the Gulf of Guinea [7]. The question remains as to which countermeasures were the most effective in reducing piracy around East Africa?

© Springer International Publishing Switzerland 2016
K.S. Xu et al. (Eds.): SBP-BRiMS 2016, LNCS 9708, pp. 356–365, 2016.
DOI: 10.1007/978-3-319-39931-7_34

An abundance of work has been conducted building models to optimize ship routing and develop piracy risk maps [1, 4, 5, 12]. This research has generally taken the approach that avoiding certain high threat locations is the best way for vessels to stay safe from pirates. Only a handful of researchers have applied bottom-up, agent-based modeling (ABM) methodologies, however, which allow for other countermeasures and what-if analyses to be explored. Experimentation with micro-level behaviors informs understanding of the processes which impact outcomes and has the possibility of uncovering emergent phenomena [2]. For example, Tsilis [15] developed an ABM to assess which factors were most important for ensuring merchants can safely sail through the Gulf of Aden when they have access to an escort warship. By simulating 300,000 missions he was able to identify the critical values for vessel speeds, positions relative to warships and pirate identification distances which countered the threat of attack by pirates.

Vaněk et al. [16] developed an ABM to explore whether employing a transit corridor system within the Indian Ocean, as was done in the Gulf of Aden, would effectively counter piracy attempts. After calibrating their model with real world data, they concluded that while effective within the Gulf of Aden, the same strategy would not be helpful within the vast expanse of the Indian Ocean. The authors also considered alertness level of merchant vessels and calibrated this parameter to fit the real world data. While the authors acknowledged that increased security and alertness levels is one possible countermeasure, they did not present any findings as to how increasing or decreasing alertness might impact piracy rates. Success of an attack in their model was a function of a merchant's speed and alertness state (which was binary) and influenced by alertness level, which was operationalized as the frequency a merchant checked whether a pirate was within its vicinity.

This paper presents an ABM which was developed to explore the implications of different countermeasures for combatting piracy, assessed by the number or merchants hijacked, pirates arrested and additional fuel consumed. In the remainder of the paper, the ABM and experimental design will be introduced (Sect. 2), after which the results will be presented (Sect. 3), findings discussed (Sect. 4) and finally concluded (Sect. 5).

2 Method

2.1 Model Design and Agent Behaviors

Netlogo 5.2.1 [17] was used to develop an agent-based model to simulate merchant vessels sailing from a stylized west coast to various port destinations on the east coast amidst the threat of piracy. See Fig. 1 for a visual depiction of the model's Graphical User Interface. This simplified, abstracted representation was utilized in order to aid in understanding the relationships and interdependencies among the various agents [6]. Specifically, the model includes three different agent types: Merchant vessels, Pirate vessels and Naval warships. Merchant vessels have the goal of sailing to their destination port via the most direct path while also avoiding pirates. Pirate vessels have the goal of attacking and hijacking merchant ships while also evading Navy ships. Lastly, Naval warships have the goal of finding and arresting pirates.

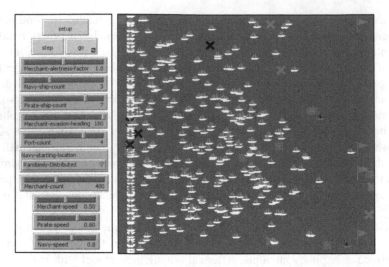

Fig. 1. Model parameters and visualization depicting Merchant Vessels, Pirates and Navy ships. World size is 32 X 30 cells. Brown and white ships signify merchant ships; black and red ships signify pirates; and red shields signify Naval warships. Green flags represent destination ports. Red X's denote pirates who were arrested. Black X's denote merchants who were hijacked. (Color figure online)

Merchant vessels are randomly assigned destination ports and departure times for when to begin their journey and will sail the shortest path to their destination unless they observe a pirate nearby. When or if a merchant first observes a pirate is dependent on their alertness setting, which assigns the maximum radius distance at which merchants detect pirates and subsequently modify their route to evade the pirate. This alertness value varies by each merchant vessel and represents security measures employed onboard each ship. This is based upon data suggesting that the target of piracy attacks have increased in range of vessel type, from cargo ships, tankers, sailing yachts to tugboats, each of which employ variable levels of security measures onboard [13].

Pirate vessels are randomly distributed throughout the environment and identify potential targets based upon proximity, such that the closest merchant ship is targeted. Once targeted, pirates begin pursuit of merchants. If a pirate vessel gets close enough to a merchant vessel, pirates will board and hijack the merchant vessel after 4 times steps, unless the Navy shows up to prevent the hijacking. If pirates observe a Naval warship within 5 cells of them during their hijack attempt, the pirates will immediately re-board their ship and begin evading the Navy. After a successful hijack though, the merchant ship disappears and a subset of the pirates re-board their pirate vessel and are available to attack more merchants. Merchant vessels that are boarded or hijacked by pirates are incapacitated for the rest of that particular simulation run.

Naval warships are initialized at various locations within the environment at the beginning of each simulation, depending on the user-defined selection. Once the simulation begins Naval warships wait until they either see a pirate within their alertness radius of 4 cells, or until a merchant vessel calls them for help. If a Naval warship gets within one cell of a pirate the pirate is subdued and the warship is immediately available to target more pirates or assist merchant vessels, as it is assumed the warship deploys a different team to oversee the arrest. Lastly, Naval warships sail on average faster than pirate vessels, which sail on average faster than merchant vessels. The fast speeds of pirates is intended to mimic the fact that most pirate attack groups deploy smaller fast attack crafts, equipped with sophisticated weaponry, when in direct pursuit of targeted merchant ships [13]. Cruising and evasion speeds settings for this experiment are summarized in Table 1, in addition to alertness levels for each agent type.

Table 1. Merchant, Pirate and Naval Warship speeds and alertness levels

	Crusing Speed	Evasion Speed	Alertness Level
Merchant Vessels	Normally distributed among agents with a mean of 0.5 and standard deviation of 0.1	Cruising Speed * 1.3	Randomly distributed among agents between 1 - 5 patches
Pirate Vessels	Normally distributed among agents with a mean of 0.6 and standard deviation of 0.1	Cruising Speed * 1.3	All agents can see Naval Warships up to 5 patches away
Naval Warships	Normally distributed among agents with a mean of 0.8 and standard deviation of 0.1	Same as Crusing Speed	All agents can see pirates within 4 patches or are called by merchant vessels for help

Verification of the model was conducted to ensure that all agents were behaving as intended. This was performed through a combination of print lines and systematic manipulation and inspection of individual agent behaviors and trait values at each time step. Additionally, use of extreme parameter value settings were employed, such as vessel speeds and counts in order to verify outcomes were as expected, given the intention of the code [14].

2.2 Model Parameters

In order to investigate the impact of varying levels of merchant alertness, a parameter called the Merchant Alertness Factor was created which multiplies the alertness level radius by the designated factor, which varies from 0–2 at increments of 0.5. For example, a Merchant Alertness Factor of 2 would double the distance at which all merchant vessels first detect pirates. A setting of 0 would eliminate any ability to detect pirates in nearby waters.

The number of pirate vessels and Naval warships deployed were also parametrized in order to investigate the impact of varying ratios of Pirate: Navy presence. Furthermore, initial starting location of Naval warships was manipulated to assess whether particular starting locations resulted in higher numbers of pirate arrests and lower numbers of merchant hijackings. Initial pirate locations were either randomly distributed; initialized near east and west coast ports; initialized randomly along the horizontal world midline; or initialized randomly along the vertical world midline.

Additionally, to investigate whether it is safer for merchants to travel more closely together the number of destination ports was parameterized. Fewer destination ports generally forced more clustered, group movement, while higher numbers of ports forced more distributed movement patterns.

Lastly, the evasive tactics, which merchants engage in once a pirate is detected, are driven by a parameter which sets the path heading. This parameter allows exploration of the impact of different evasive maneuvers. For this experiment, headings were set to 180 and 90° away from pirates, but random paths or alternative tactics could also be implemented and explored in future experiments, as outlined in the Discussion section. Pirate vessels always take the shortest, most direct path available sailing towards the merchants and directly away from Naval warships.

2.3   Design of Experiment

In this specific experiment, six parameter settings were varied such that a 5 X 6 X 5 X 2 X 2 X 4 full factorial design was conducted, as presented in Table 2, with each combination repeated 20 times. This resulted in a total of 48,000 simulation runs. The number of merchant vessels was held constant at 400, in addition to the mean speed settings for each vessel type (Merchants, Pirates and Navy warships). In order to assess the efficacy of various countermeasures, two primary outcome measures were gathered at the end of each simulation run: Percentage of Merchants Hijacked and Percentage of Pirates Arrested. Additionally, data was collected on the distance that merchant ships sailed, given different parameter settings.

Table 2. Parameters manipulated and their associated settings. Each combination was run 20 times, for a total of 48,000 runs.

Parameters Manipulated	Setting Values
Merchant Alertness Factor	0, 0.5, 1, 1.5, 2
Number Navy Ships	0, 1, 3, 5, 7, 9
Number of Pirates	1, 3, 5, 7, 9
Merchant Evasion Heading	90, 180
Number of Ports	1, 5
Navy Starting Location	Near Ports, Middle-Vertical, Middle-Horizontal, Randomly Distributed

3   Results

Analysis revealed that the two primary outcome variables of interest, Percentage of Merchants Hijacked and Percentage of Pirates Arrested were strongly correlated, r (47998) = −.67, $p < .001$. As such, a six-way multivariate analysis of variance

(MANOVA) was conducted to assess the effect of each parameter on these outcome measures. Furthermore, in accordance with Kleijnen [11], who asserts that higher order effects are hard to interpret and often negligible in magnitude, only the second-order interactions with significant effect sizes are reported.

Results revealed a statistically significant and large main effect for both the Number of Naval Warships deployed, Pillai's Trace = 1.02, F(10, 95706) = 9967.8, p < .001, η2 = 0.40, and the Number of Pirates present, Pillai's Trace = 0.64, F(8, 95706) = 5588.5, p < .001, η2 = 0.23, on the outcome measures. A statistically significant and small main effect was found for both Evasion Heading, Pillai's Trace = 0.11, F(2, 47852) = 2941.5, p < .001, η2 = 0.01, and for Merchant Alertness, Pillai's Trace = 0.34, F(8, 95706) = 2470.3, p < .001, η2 = 0.01, on the outcome measures. Lastly, a statistically significant but negligible effect size was found for the Number of Ports, Pillai's Trace = 0.01, F(2, 47852) = 272.1, p < .001, η2 = 0.001, in addition to Navy Starting Location, Pillai's Trace = 0.005, F(6, 95706) = 42.3, p < .001, η2 = 0.0006, on the outcome measures.

In addition, results showed a statistically significant and medium interaction effect size for the Number of Pirates and Naval Warships present, Pillai's Trace = 0.61, F(40, 95706) = 1061.6, p < .001, η2 = 0.11, on the outcome measures. Figure 2 demonstrates part of this effect, shown by the percentage of Merchants Hijacked precipitously dropping but then experiencing diminishing returns as the Naval presence increases.

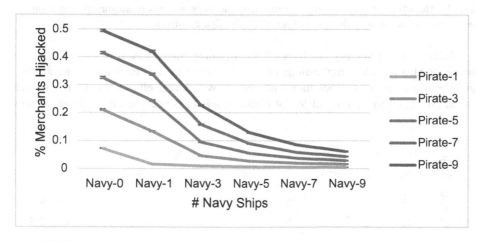

Fig. 2. The effect of varying numbers of pirate ships and naval warships on the percentage of merchants hijacked. (Color figure online)

A statistically significant and medium interaction effect was also found between the Merchant Alertness Factor and the Number of Naval Warships present, Pillai's Trace = 0.53, F(40, 95706) = 870.1, p < .001, η2 = 0.07, on the outcome measures. Figure 3 demonstrates part of this finding via the reversing effect of Merchant Alertness on the Number of Merchants Hijacked once three or more Naval Warships are present.

A final statistically significant and small interaction effect was found between the Evasion Heading and the Number of Naval Warships present, Pillai's Trace = 0.11, F (10, 95706) = 580.2, p < .001, $\eta2$ = 0.02, on the outcome measures. This effect is primarily driven by significantly more merchants being hijacked at low levels of Naval Warship presence when the Heading was set to 90°, but differences essentially disappear when more than 5 Naval Warships were present.

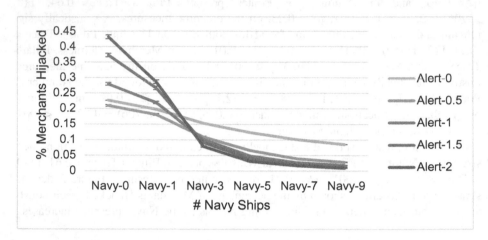

Fig. 3. The effect of varying levels of merchant alertness and naval warship presence on the percentage of merchants who were hijacked. (Color figure online)

Additional exploratory analysis of the distance merchants traveled in different scenarios also revealed implications of varied evasion tactics, in that evading at 180° greatly increased the distance merchants sailed; with larger effects (i.e. more distance sailed compared to evading at 90) at higher levels of pirate presence. However, this

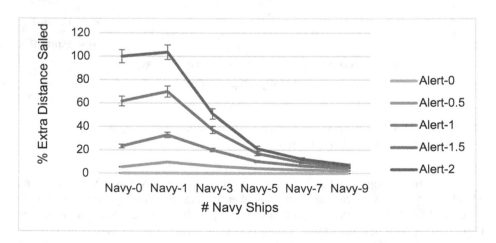

Fig. 4. The effect of varying levels of merchant alertness and naval warship presence on the additional distance sailed by merchants evading pirates. (Color figure online)

effect was also accompanied by upwards of a 7 % reduction in hijackings at high levels of pirate presence. In addition, Fig. 4 demonstrates how high Merchant Alertness levels had a large impact on the additional distance sailed by Merchant vessels when the Naval Warship presence was low.

4 Discussion

4.1 Review of Results

As observed in Fig. 3, output from this ABM revealed a counter-intuitive effect whereby higher levels of merchant alertness actually lead to more hijackings when there are only one, or zero, Naval Warships present in the environment; only when three or more Naval warships are deployed do higher level of alertness actually lower the percentage of merchants hijacked. Observation of the merchant's behavior suggests this effect is partially attributable to high alertness levels causing merchants to begin evading too soon and results in numerous merchants clustering in confined spaces near the environment's perimeters. This enables pirates to converge from different directions and pick merchants off one by one, appearing as though the pirates had coordinated their behavior despite having no communication. Additionally, as pirate vessels are faster on average than merchant vessels, it typically is futile for a merchant to attempt to outrun a pirate unless they haven't yet been seen by the pirate (which is unknown to the merchant), or they can avoid being attacked until a Naval warship can intervene. For this reason, merchant alertness and early evasive maneuvers only become beneficial once Naval presence increases to at least 1:3 Naval warships to Pirate vessels. Of course, these findings only apply to the assumptions of this model; in order to inform actual decisions the model would need to be extended to include more complex evasion tactics and sophisticated merchant behaviors, such as armed personnel who can choose to combat the pirates rather than evade.

This ABM's results demonstrate some of the factors that a merchant or policy maker might be interested in weighing when deciding how to respond to the threat of piracy. For example, Fig. 4 demonstrates how higher levels of alertness can result in large fuel expenditures, as some highly alert merchants sail up to twice their originally planned distance while attempting to evade pirates. This additional distance reduces with increased Naval presence, however even a 10 % increase which occurs with high levels of Naval presence can have large consequences in terms of fuel costs. Additionally, Fig. 2 illustrates how even high ratios of Naval warships to Pirate vessels (i.e., 1:1) still amount to approximately 5 % of merchant vessels being hijacked. As such, if maritime intelligence suggests a greater than 1:1 ratio of Navy:Pirate ships and the extra cost of fuel is less than the cost associated with being hijacked, it could be advantageous for merchant ships to increase security alertness measures and invest the extra fuel required to evade pirates until the Navy can arrive. These results also suggest that merchant vessels should stay to their original course and not attempt to evade if pirates are detected and there are fewer than 3 Naval warships present in the environment. This can be confirmed by comparing alertness levels of 0 to higher alertness levels in Figs. 3 and 4 when there are fewer than 3 Naval warships.

Results of this model were consistent with findings of Vaněk et al. [16], suggesting that the difference between one and five ports did not have a significant effect on the number of merchants hijacked nor on the number of pirates arrested. Fewer destination ports generally forced more clustered movement, reminiscent of a coordinated transit corridor, while higher numbers of ports forced more distributed movement patterns, however a more systematic approach exploring different environmental layouts and actual shipping lanes should be explored in future model versions. Finally, the initial starting location of Naval warships did not have a significant impact on the number of merchants hijacked or pirates arrested, however future models should incorporate factors such as intelligence reports (information sharing among merchants and Navy agents) and learning to see if these have an impact on hijackings.

4.2   Implications and Future Work

The results of this analysis serve as a proof of concept for demonstrating how an ABM might be used to inform questions regarding effective piracy countermeasures. These findings are not intended to represent reality, but rather establish the utility in building a more realistic model. In order to inform real decisions and increase validity of the model's output, realistic environmental models (employing GIS and oceanographic data) and calibrated agent behaviors (e.g., vessel speeds, tactics) are required. For example, the Evasion Heading parameter enabled investigation of tactics for preventing hijackings as well as what the time/fuel costs associated with various combinations of tactics and alertness levels, however an accurate investigation of the impact of other more realistic tactics, such as zig-zagging to cause wash waves and prevent pirates from easily boarding should be implemented in higher fidelity environment models.

Additionally, providing agents access to intelligence reports and calibrating sensor ranges for identifying threats should be included and assessed within different maritime environments, i.e., open expanses of ocean, littorals. Furthermore, embedding armed security personnel and more sophisticated defensive behavior would allow additional countermeasure scenarios to be assessed (e.g. evading vs. fighting back). Furthermore, a calibrated model would provide the ability to assess costs associated with implementing different countermeasures, such as the deployment of Naval warships, on-board security personnel, or various evasive tactics and their associated fuel costs. Nonetheless, it is important to heed Einstein's advice that "everything should be made as simple as possible, but not simpler" and preserve the simplicity of the model in order to aid in understanding and interpreting results.

5   Summary

As the threat of piracy has greatly reduced along the East coast of Africa, but is on the rise on the West coast, it would be beneficial to understand which factors contributed most to the rapid decrease in piracy within the Gulf of Aden. Decision makers should consider using ABMs to assess the efficacy of various piracy countermeasures applied in specific regions before investing large amounts of time and resources in potentially costly or marginally effective strategies. The ABM presented here serves as a proof of concept and could be used as the foundation for more complex and calibrated models.

References

1. An, W., Ayala, D.F.M., Sidoti, D., Mishra, M., Han, X., Pattipati, K.R., Hansen, J.: Dynamic asset allocation approaches for counter-piracy operations. In: 2012 15th International Conference on Information Fusion (FUSION), pp. 1284–1291. IEEE (2012)
2. Bonabeau, E.: Agent-based modeling: Methods and techniques for simulating human systems. Proc. Nat. Acad. Sci. **99**(suppl 3), 7280–7287 (2002)
3. Bowden, A., Hurlburt, K., Aloyo, E., Marts, C., Lee, A.: The economic costs of maritime piracy. One Earth Future Foundation (2010)
4. Caplan, J.M., Moreto, W.D., Kennedy, L.W.: Forecasting global maritime piracy utilizing the risk terrain modeling (RTM) approach to spatial risk assessment. In: Crime and Terrorism Risk: Studies in Criminology and Criminal Justice. Routledge (2011)
5. Christiansen, M., Fagerholt, K., Nygreen, B., Ronen, D.: Ship routing & scheduling in the new millennium. Eur. J. Oper. Res. **228**(3), 467–483 (2013)
6. Epstein, J.M.: Why model? J. Artif. Soc. Soc. Simul. **11**(4), 12 (2008)
7. Felbab-Brown, V.: The Not-so-jolly roger: dealing with piracy off the coast of somalia and in the gulf of guinea. In: The Brookings Institution, Africa Growth Initiative, pp. 5–8 (2014)
8. Fu, X., Ng, A.K., Lau, Y.Y.: The impacts of maritime piracy on global economic development: the case of Somalia. Marit. Policy Manage. **37**(7), 677–697 (2010)
9. International Maritime Organization: Reports on Act of Piracy and Armed Robbery Against Ships, Annual Report – 2013 (2013)
10. Jesus, H.J.: Protection of foreign ships against piracy and terrorism at sea: legal aspects. Int. J. Mar. Coast. Law **18**(3), 363–400 (2003)
11. Kleijnen, J.P.: An overview of the design and analysis of simulation experiments for sensitivity analysis. Eur. J. Oper. Res. **164**(2), 287–300 (2005)
12. Marchione, E., Johnson, S.D., Wilson, A.: Modelling maritime piracy: a spatial approach. J. Artif. Soc. Soci. Simul. **17**(2), 9 (2014)
13. Onuoha, F.: Sea piracy and maritime security in the Horn of Africa: the somali coast and Gulf of Aden in perspective. Afr. Secur. Stud. **18**(3), 31–44 (2009)
14. Rykiel, E.J.: Testing ecological models: the meaning of validation. Ecol. Model. **90**(3), 229–244 (1996)
15. Tsilis, T.: Counter-piracy escort operations in the Gulf of Aden (2011). (Doctoral dissertation, Monterey, California. Naval Postgraduate School)
16. Vaněk, O., Jakob, M., Hrstka, O., Pěchouček, M.: Agent-based model of maritime traffic in piracy-affected waters. Transp. Res. Part C Rmerg. Technol. **36**, 157–176 (2013)
17. Wilensky, U.: NetLogo. http://ccl.northwestern.edu/netlogo/. Center for Connected Learning and Computer-Based Modeling, Northwestern University, Evanston, IL (1999)

How People Talk About Armed Conflicts

Jeremy R. Cole$^{(\boxtimes)}$, Ying Xu, and David Reitter

College of Information Science and Technology, The Pennsylvania State University,
University Park, PA, USA
{jrcole,ying.xu,reitter}@psu.edu

Abstract. Armed conflicts around the world produce displacement, injury, and death. This study examines how anonymous and pseudonymous Internet commenters discuss such conflicts. Specifically, we ask how permissible it is to express positive or negative sentiments about these conflicts as a function of variables including region, conflict nature, and severity. Data from the Armed Conflicts Database is aggregated to identify a number of potential factors that may influence views on acceptable sentiments. We used sentiment analysis to code a large-scale sample of the Reddit corpus. We judged permissibility using the Reddit voting features. This revealed that positive sentiments are found not permissible for higher numbers of fatalities, and that negative sentiments are found to be more permissible for certain regions and older conflicts, but less permissible for territorial conflicts. Thus, this study provides evidence that many features help construct public perception of a conflict.

Keywords: Behavioral and social sciences · Corpus linguistics · GLMM · Armed conflicts · Public opinion

1 Introduction

According to the Armed Conflict Database [1], there are 42 active conflicts around the world which annually cause 180,000 fatalities and have resulted in more than 12 million refugees. For people who live in relatively peaceful areas, such as in North America, their perception, opinion and action towards these international crises are important. Collectively, these perceptions are influential in shaping their countries' foreign policy [6].

Traditional media and social media interact to help form these perceptions, while consumers of these media implicitly or explicitly rate the quality of the content produced. Contributors and evaluators naturally have biases that can build on each other, if contributions that follow these biases are rated more positively than those that do not. The question is what are those biases to begin with?

In particular, we seek to answer this question in the context of armed conflicts. For instance, how do people talk, evaluate, and contribute to the discussion of various armed conflicts? Specifically, we want to understand when and why people accept negative discussions and positive discussions. We examine

© Springer International Publishing Switzerland 2016
K.S. Xu et al. (Eds.): SBP-BRiMS 2016, LNCS 9708, pp. 366–376, 2016.
DOI: 10.1007/978-3-319-39931-7_35

the effect of features such as the number of fatalities, refugees and Internally Displaced People (IDP). Further, we investigate a variety of characteristics that may influence objectivity, such as the location and age of the conflict. We hope to discover the relationship between these features and the perception of specific armed conflicts.

To accomplish this, we turn to an Internet forum, Reddit. Reddit is unique in that a wide range of subjects are discussed there, but its users possess a relative homogeneity in both demographics and opinion that give it certain advantages over other social media. As will be explored later, much of Reddit consists primarily of young Western people. Further, we will also demonstrate how their opinions on these armed conflicts are fairly homogeneous. This gives us an ideal dataset to reflect on a specific culture's judgments of acceptability.

2 Background

2.1 Public Perception of Armed Conflicts

Public opinion on armed conflicts from World War II to Vietnam, the Gulf, Afghanistan and Iraq war have also been investigated to find patterns of public responses to international conflicts [2]. However, with prevalent access to the Internet and social media, public opinions are becoming more influential in political decision-making; they are even changing the political cycle [14]. However, systematic studies of public opinion from social media on armed conflicts is limited. Studies focusing on social media in armed conflicts often regard social media tools as agents or platforms to express views and coordinate actions: for example, arranging protests and organizing uprisings [8].

Public perception of armed conflicts through both traditional and social media suffers from several biases. Advanced communication technologies make it possible to disseminate information about various conflicts around the world [12]. However, government and news organizations cover and frame conflicts in a certain way that adds their own bias. For example, when covering foreign events, reporters largely focus on war, terrorism and political violence [11]. Even still, most of the world's conflicts are largely unreported by the media [7]. Even if it is unintentional, this causes a significant limitation for the public to be able to obtain the broad-spectrum of information they need to evaluate conflicts objectively [9]. News agencies in general focus on news in their target audience's country or nearby countries, shrinking the amount of time focused on international coverage [13].

Despite these biases, a growing number of studies are covering the relationship between the characteristics of armed conflicts and the public perception of them. Berinsky has found that public reactions to conflicts have been shaped less by their defining characteristics, such as fatalities and resource costs, than by one's political affiliations [2]. For instance, if someone's political party supports a war, then she most likely does as well. Furthermore, Gartner et al. found that marginal fatalities, which are the number of fatalities that occurred that year, are more important in explaining opinion than the cumulative number of fatalities [4]. They also studied the relationship between race and opinion towards the Vietnam war, finding that people were likely to view the conflict more negatively based on the number of people who died that were in the same locale, regardless of race [5]. Surprisingly, they found that Asian Americans had greater support for the Vietnam War; this could be because some of them were fleeing communism themselves [5].

2.2 Social Computing

The usage of large-scale datasets from social media has produced a flood of research and discourse in areas like sociology, political science, and psychology. Researchers have extensively studied the role of social media as a platform in coordinating activities before, during, and after conflicts. For example, time-series analysis of social media datasets has offered an ever-evolving account of public opinion and attention on a variety of issues, such as economic and social welfare, foreign affairs, and environmental issues [10]. While studies have looked at public perception of individual armed conflicts, we are not aware of any studies that have leveraged the mass of social media data to examine what causes perceptual differences among them.

2.3 Reddit Corpus

Reddit is an ideal corpus for such a systematic investigation for a few reasons. First, Reddit is partially driven by news articles. Most Reddit submissions start with a user submitting a hyperlink, giving discussants a clear shared context.

After the submission, people can reply with their thoughts in a *comment*. Users can also comment on these comments or provide feedback with Reddit's curation system. Reddit users can *upvote* a comment or submission to express approval or *downvote* a comment or submission to express disapproval.

Both submissions and comments are sorted by *score*[1], the number of upvotes minus the number of downvotes. If a comment is sufficiently downvoted, it will be hidden. Hidden comments can still be read, replied to, upvoted, and downvoted, but a user has to manually display hidden comments. When a comment receives both upvotes and downvotes, it is considered *controversial*. In the case of armed conflicts, this is vanishingly rare: less than 1 % of comments are considered controversial.

Reddit is organized into a highly expandable number of smaller forums, normally referred to as subreddits. Each of these subreddits can be about any given topic, general or specific. For instance, it is possible to have subreddits about science, biology, and genetics. These three subreddits can be completely independent, as they are not organized hierarchically.

---

[1] While comments are hierarchical, at each level of the hierarchy, they are sorted by score.

Besides the organizational ways that Reddit is different than Twitter, it is also more culturally homogeneous. The vast majority of the discussions are in English. Over half of the users are in the United States, and the majority are fairly young[2]. Among those not in the United States, the next most contributing countries are primarily East Asian, Australasian, Western European, and Northern European[3].

2.4 Armed Conflict Database

The Armed Conflicts Database is developed by the International Institute for Strategic Study, containing various indexes of armed conflicts around the world [1]. Conflicts are sorted into several regions: Caribbean and the Americas, East Asia and Australasia, Europe, Middle East and North Africa, Russia and Eurasia, South Asia, and lastly, Sub-Saharan Africa. The Armed Conflict Database contains data on fatalities, Internally-Displaced People (IDP), and refugees from a conflict: both by year and in total. It additionally lists the year the conflict started and a variety of factors that relate to the conflict's origin, such as ethnic violence or terrorism. Lastly, they rate the current Intensity of the conflict, which can be Archived, Low, Medium, or High. The Database keeps track of approximately ninety conflicts, the plurality of which are now Archived. There are presently 42 active conflicts monitored by the Database.

3 Research Questions

Our research questions aim to discover how people perceive and discuss armed conflicts. We suspect some of the same biases that affect traditional media will also affect social media. This may be especially true for Reddit due to its news-driven process. Nonetheless, it is unlikely that such different sources would produce exactly the same phenomena.

Our data set consists of approximately 426 GB of Reddit data, ranging from the year 2012 to the year 2014. We cross-referenced this with data from the Armed Conflicts Database, collecting a list of the 48 conflicts that are considered by the Database to have been active in at least one of those years. The active status is determined by experts working for the Database who are monitoring trends of armed conflicts worldwide. We used active conflicts to ensure none were seen by commenters as purely historical. See Fig. 1 for a summary of where the conflicts occurred.

We gathered Reddit comments that are relevant to each of the 48 conflicts by searching for comments in every single subreddit. We compiled sets of keywords for every conflict then collected comments which matched them. For instance, if a comment contained the phrase "Syrian civil war", we would mark that comment

---

[2] Source–http://www.pewinternet.org/2013/07/03/6-of-online-adults-are-reddit-users/.

[3] Source–http://www.redditblog.com/2013/12/top-posts-of-2013-stats-and-snoo-years.html.

as relevant to the conflict in Syria. Most of our keywords were fairly specific; nonetheless, we sought to counter false positives by having a sufficiently large data set. The biggest source of comments was the "worldnews" subreddit, with many of the others coming from similar subreddits.

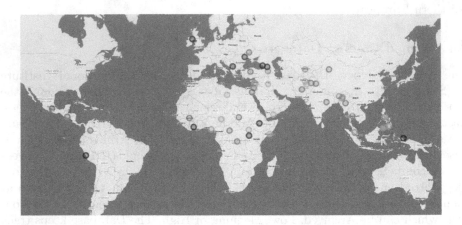

Fig. 1. A map of the 48 conflicts. Black, yellow, orange and red circles indicate the current level of intensity as rated by the Armed Conflict Database: archived, low, medium and high, respectively (Color figure online).

We then analyze comments based on two main features: *sentiment* and *acceptability*. We rated over 25,000 comments using the StanfordNLP Sentiment Analyzer [15]. The acceptability refers to how the community perceives the comment and its sentiment. In this case, an acceptable comment is one that the community would give upvotes, and an unacceptable comment is one the community would give downvotes.

We wanted to examine several characterizing features of conflicts that affect how they are discussed by traditional news agencies.

1. **Severity.** This includes the total number of people who were killed, made refugees, or internally displaced as a result of the conflict. While common sense would suggest that severity should play a large role in perception, prior work suggests it plays a very modest role in public perception of conflicts.
2. **Region.** There are six major regions that the Armed Conflict Database groups conflicts into. We hypothesize that regions that share more cultural similarity would be viewed differently than regions which share less. More specifically, regions where Reddit users are common would be perceived differently than regions where they are not.
3. **Marginal Severity.** This includes the number of people who were killed, made refugees, or internally displaced the same year as the comment was made. From previous research, we suspected that marginal severity would play a larger role than total severity.

4. **Age.** The number of years that have passed since the conflict started. We hypothesized age could be significant due to waning interest. After decades of conflict, it is perhaps difficult for some users to still empathize with ongoing tragedy.

5. **Nature.** This includes a set of attributes: Separatism, Terrorism, Foreign Antagonism, Territorial Disputes, Criminal Violence, and Ethnic Violence. We hypothesized that attributes that are easier for Westerners to empathize with due to history, such as Separatism or Foreign Antagonism, might be treated differently. Further, attributes that seem very foreign or 'uncivilized', such as Ethnic Violence or Territorial Disputes, may also be seen differently. Importantly, a conflict can have more than one of these attributes.

6. **Expert Perception.** The Armed Conflict Database rates conflicts for their current level of intensity. While this is obviously correlated with deaths, refugees, and IDP ($p < 0.0001$), the average person may be more likely to get their ideas of severity from experts, rather than numbers. This could also influence the selection process and tone of traditional media articles.

4 Methods

4.1 Data Preparation

Interestingly, the sentiment analysis provided nearly twenty times more negative comments than positive comments. It is possible that this is due to the subject matter; in general, pity and sadness might be more common responses to tragedy than optimism and hope. It is also possible that some positive comments actually reflected misanthropic views about violence. Due to the difficulty of interpreting neutral sentiment comments, they were excluded.

As Reddit manipulates how likely it is for a user to see a comment, we did not consider the actual number of upvotes or downvotes a comment received to be useful information. We thus coded anything upvoted at all as acceptable, and anything downvoted at all as unacceptable. We excluded comments that received no votes.

Our two filters resulted in 781 positive comments and 14289 negative comments. If the Positive Sentiment Model is substantially different than the Negative Sentiment Model, that suggests sentiment, perhaps through interaction with other variables, does influence acceptability. It is possible negative comments can be interpreted as the norm; in that case, that model would reflect when posters are more likely to submit something that is unacceptable in general.

4.2 Model

We use a Logistic Linear Mixed Effects Model (GLMM) to attempt to explain which sentiments were acceptable as determined by the conflict. Due to missing

Table 1. The **positive** sentiment models. Region was fitted as a factor. Intensity was fitted as an ordinal variable where only the linear effect is reported. Significant variables are marked with an asterisk.

	Estimate	Std. Error	z value	Pr($< |z|$)
Intensity	0.3561	0.6668	0.5340	0.59300
Fatalities	0.1091	0.001	118.0000	0.00001*
IDP	0.3545	0.2568	1.3800	0.16700
Refugees	0.1361	0.3371	0.4040	0.68600
Separatism	−0.5505	< 0.001	−632.0000	< 0.00001*
Criminal Violence	0.5866	0.5738	1.0220	0.30700
Ethnic Violence	−2.4295	3.2196	−0.7550	0.45000
Terrorism	−0.8033	< 0.001	−861.0000	< 0.00001*
Territorial Dispute	−1.1330	1.2790	−0.8850	0.37600
Foreign Antagonism	1.0048	0.5995	1.6760	0.09380
Marginal Fatalities	0.3844	0.2611	1.4720	0.14100
Marginal Refugees	0.2577	0.3425	0.7520	0.45200
Marginal IDP	0.3223	0.2868	1.1240	0.261000
Age	−0.2801	0.3285	−0.8530	0.39400
Region-EastAsia/Australasia	−2.4860	2.9980	−0.8290	0.40689
Region-Europe	−0.4437	0.8270	−0.5360	0.59162
Region-MiddleEast/NorthAfrica	−0.2767	0.8544	−0.3240	0.74601
Region-Russia/Eurasia	−1.1910	1.6430	−0.7250	0.46847
Region-SouthAsia	−2173.0000	> 100.0000	0.0000	0.99996
Region-Sub-SaharanAfrica	2.0970	2.3000	0.9120	0.36199

data for certain features (such as the marginal Refugees for any given year) and a high number of predictors, we decided to fit individual models for each predictor. This allows us to also see precisely how well each variable explains the variance. However, as we are making more than one comparison, we have to adjust our test of significance to avoid false positives. We use the Bonferoni correction [3], resulting in a significance threshold of $p = 0.0025$.

As every comment is coded as either unacceptable (1) or acceptable (0), positive β relates to downvotes and negative β relates to upvotes, which correspond with disapproval and approval effectively.

Random intercepts grouped by Author and by Subreddit were fitted to account for partial interdependence. As Reddit is pseudononymous, some authors may have a reputation for making consistently good or bad posts, and different communities may have different standards for acceptability.

5   Results

5.1   Positive Comments

For positive comments, there are reliable effects due to Fatalities, Separatism, and Terrorism. A conflict having more Fatalities increases the likelihood that a comment is considered unacceptable, while the presence of Terrorism and Separatism decrease that likelihood. See Table 1.

5.2   Negative Comments

For negative comments, there are reliable effects for Territorial Disputes, Age, and for the regions of Europe and East Asia / Australasia. Conflicts in both Regions make the comments less likely to be disapproved of, as does an older age. The conflict being a Territorial Dispute, on the other hand makes the comment more likely to be disapproved of. See Table 2.

Table 2. The **negative** sentiment models. Region was fitted as a factor. Intensity was fitted as an ordinal variable where only the linear effect is reported. Significant variables are marked with an asterisk

	Estimate	Std. Error	z value	$Pr(<|z|)$
Intensity	0.1525	0.0697	2.1870	0.028700
Fatalities	0.0105	0.0247	0.4240	0.67200
IDP	0.0190	0.0332	0.5740	0.56600
Refugees	−0.0679	0.0397	−1.7100	0.08730
Separatism	−0.1036	0.0566	−1.8310	0.06710
Criminal Violence	0.0988	0.0559	1.7680	0.07700
Ethnic Violence	−0.2560	0.1605	−1.5950	0.11100
Terrorism	0.1249	0.0563	2.2160	0.02670
Territorial Dispute	0.4590	0.0798	5.7530	$< 0.00001*$
Foreign Antagonism	0.0958	0.0606	1.5820	0.11400
Marginal Fatalities	0.0097	0.0309	0.3140	0.75400
Marginal Refugees	−0.0615	0.0413	−1.4880	0.13700
Marginal IDP	−0.0501	0.0385	−1.3030	0.19200
Age	−0.1146	0.0277	−4.1400	$< 0.00010*$
Region-EastAsia/Australasia	−0.5248	0.1515	−3.2500	0.00116*
Region-Europe	−0.4437	0.8270	−0.5360	$< 0.00001*$
Region-MiddleEast/NorthAfrica	0.1383	0.0750	1.8420	0.06542
Region-Russia/Eurasia	−0.1895	0.0891	−2.1250	0.03355
Region-SouthAsia	−0.6498	0.2974	−2.1850	0.02891
Region-SubsaharanAfrica	0.1725	0.1578	1.0930	0.27448

6   Discussion

There are several ways to interpret the rate of unacceptable posts for any given conflict. As the majority of Reddit users are fairly homogeneous in opinion, it is possible that people vary how likely they are to make posts about certain topics when they know that many people will disapprove. This could stem from *passion* about a topic: someone feels it is more important to speak their mind than to have a popular post. Alternatively, it could reflect *dissent*; while there might not be enough dissenters to affect the gatekeeping system, the willingness to express something others disapprove of at all suggests there are some who disagree with the majority opinion.

6.1   Negative Sentiment

Due to the vast majority of comments being of negative sentiment, we assume it is the default way to respond to a conflict. Thus, we will consider the Negative Sentiment model the same as the Default model.

The Regions of East Asia/Australasia and Europe both have reliable negative predictors. This implies the rate of disapproval when discussing these topics is very low. This is somewhat unsurprising given the general demographics of Reddit. Users from those regions, due to comparable socioeconomic status or military alliances, may see themselves as similar, prompting a homophily effect. The lack of significant effects for the other regions could be because commenters conflate areas with which they are less likely to empathize, such as the Middle East/North Africa and Sub-Saharan Africa.

The negative effect of Age makes sense from the perspective of passion. While many people may have an opinion on older conflicts, these feelings may be less immediate due to the numbness or weariness of prolonged violence.

Territorial disputes on the other hand, are logically controversial. To those whose country plays a role or are immediately affected by them, they may seem existential. However, to those farther away, they could seem like petty bickering.

Interestingly, the objective variables, such as Fatalities and IDP, played almost no role in the model. Absence of evidence is not evidence of absence; however, this integrates well with previous work [2], which likewise found social connections to be more predictive than fatalities or cost of a conflict.

6.2   Positive Sentiment

We can assume positive sentiment corresponds to hope, optimism, or perhaps even sarcastically phrased misanthropic sentiments. For instance, in the case of comments about conflicts with higher Fatalities being more likely to be disapproved of, it might be the latter.

On the other hand, positive sentiment surrounding conflicts containing notes of Terrorism or Separatism may correspond to hope. For instance, they could be expressions of hope for those attempting to separate from a regime where they do not feel represented, or wishes for those who are suffering terrorism to remain steadfast. These should both be uncontroversial ideas, so it is unsurprising they are more likely to be considered acceptable.

6.3 Future Work

Some of the interpretation is ultimately speculative. A more fine-grained model of sentiment could potentially help us determine with more authority what is the true cause of these effects. We would further like to tie these ideas into more general cognitive and perceptual biases. For instance, one possibility of the low effect of negative externalities could be due to poor estimates of those values. Lastly, it would be interesting to see how these results generalize to communities besides Reddit.

7 Conclusion

This paper sought to add to a growing body of work about media perception of armed conflicts by systematically investigating a large sample of Reddit data and cross-referencing the Armed Conflict Database. The vast majority of discussions are negative in tone, which is logical given the somber nature of violence. Among these, comments were less likely to be disapproved of if they were from the same demographic as the majority; they were more likely to be disapproved of if they concerned a Territorial Dispute or were older. We consider this study as preliminary evidence of the effect of perceptual biases in viewing conflicts. These biases are important due to the public's role in shaping foreign policy.

Acknowledgements. This research was supported by the National Science Foundation under grants titled "Updating the Militarized Dispute Data Through Crowdsourcing" (SBE-SES-1528624) and "Alignment in webforum discourse" (CISE-IIS-1459300).

References

1. Armed Conflict Database: Monitoring Conflicts Worldwide (2016). https://acd.iiss.org/en
2. Berinsky, A.J.: In Time of War: Understanding American Public Opinion from World War II to Iraq. University of Chicago Press, Chicago (2009)
3. Dunn, O.J.: Estimation of the medians for dependent variables. In: The Annals of Mathematical Statistics, pp. 192–197 (1959)
4. Gartner, S.S., Segura, G.M.: War, casualties, and public opinion. J. Conflict Resolut. **42**(3), 278–300 (1998)
5. Gartner, S.S., Segura, G.M.: Race, casualties, and opinion in the vietnam war. J. Politics **62**(1), 115–146 (2000)
6. Gelpi, C., Feaver, P.D., Reifler, J.: Paying the Human Costs of War: American Public Opinion and Casualties in Military Conflicts. Princeton University Press, Princeton (2009)

7. Hawkins, V.: Media selectivity and the other side of the CNN effect: the consequences of not paying attention to conflict. Media War Confl. **4**(1), 55–68 (2011)
8. Lim, M.: Clicks, cabs, and coffee houses: social media and oppositional movements in Egypt, 2004–2011. J. Commun. **62**(2), 231–248 (2012)
9. Nelson, T.E., Clawson, R.A., Oxley, Z.M.: Media framing of a civil liberties conflict and its effect on tolerance. Am. Polit. Sci. Rev. **91**(3), 567–583 (1997)
10. Neuman, W.R., Guggenheim, L., Jang, S.M., Bae, S.Y.: The dynamics of public attention: agenda-setting theory meets big data. J. Commun. **64**(2), 193–214 (2014)
11. Nossek, H.: Our news and their news: the role of national identity in the coverage of foreign news. Journalism **5**(3), 343–368 (2004)
12. Sacco, V., Bossio, D.: Using social media in the news reportage of war & conflict: opportunities and challenges. J. Media Innovations **2**(1), 59–76 (2015)
13. Seib, P.: Beyond the Front Lines: How the News Media Cover a World Shaped by War. Palgrave Macmilan, New York (2004)
14. Shirky, C.: The political power of social media. Foreign Aff. **90**(1), 28–41 (2011)
15. Socher, R., Perelygin, A., Wu, J.Y., Chuang, J., Manning, C.D., Ng, A.Y., Potts, C.: Recursive deep models for semantic compositionality over a sentiment treebank. In: Proceedings of the Conference on Empirical Methods in Natural Language Processing (EMNLP), vol. 1631, p. 1642

Sensing Distress Following a Terrorist Event

Xidao Wen and Yu-Ru Lin[✉]

University of Pittsburgh, Pittsburgh, USA
{xidao.wen,yurulin}@pitt.edu

Abstract. Terrorism aims to cause the psychological instability in the targeted population. Social psychological research through traditional methods, like surveys or interviews, have provided insights for understanding the psychological effects of such traumatic events. However, these studies are costly and usually are reported with a significant delay after the events. With the allure of social media, we are provided with a unique opportunity to gather timely psychological signals from publicly available Twitter data. In this study, we collected more than 4 million tweets from 16 K Paris users. We present our analysis of the immediate emotional response as well as the subsequent recovery process following the Paris attacks. Our analysis shows that, immediately after the attacks, a greater level of anxiety was associated with locations closer to the attack site. Users' emotional shift gradually returned to the pre-impact status over days to weeks, while the emotional trajectories vary with the degrees of users' social interactions. The analysis further reveals a significant impact of media exposure on the recovery process. This study provides both theoretical and practical implications for understanding users' emotional vulnerability and resilience concerning mass violence events.

Keywords: Social media · Stress response · Terrorism · Disaster resilience

1 Introduction

On November 13th, 2015, a series of terrorist attacks occurred at multiple sites in Paris. In addition to the physical damage, terrorism injected substantial adverse outcomes to the targeted population and beyond, including psychological instabilities, as reflected in the emotional expressions on social media in related to Paris attacks. The development of effective coping strategies and interventions after the terrorist attacks requires a thorough understanding of individual variations in the temporary public reactions and chronic responses.

Much physiological research (e.g., [1]) provides the understanding of the individual differences in the face of acute stress situations. There have been fewer attempts though on understanding the short-term emotional response and trajectories. We believe through the understanding of the individual differences in response and the whole recovery process, we could better reduce the psychological damage and prepare for future similar incidence.

© Springer International Publishing Switzerland 2016
K.S. Xu et al. (Eds.): SBP-BRiMS 2016, LNCS 9708, pp. 377–388, 2016.
DOI: 10.1007/978-3-319-39931-7_36

In this study, we employ Twitter data to track individual recovery process on the Paris attack. We collect about 4 million historical tweets from 16 K Paris users and analyze the impacts of different characteristics, specifically, gender, social support, media coverage, geographic proximity, and linguistic styles, on different immediate emotional reactions to the acts of terrorism. We further investigate the interplays between individual differences and their short-term recovery patterns.

The contributions of this study include:

- We provided the first large-scale study of the emotional trajectories in the aftermath of a massive violence event through social media.
- We proposed a novel framework to investigate the immediate emotional response and recovery process, based on which we re-examined the factors that associate with particular responses or response growths.
- We found the emotional trajectories vary with users' tendency of social interactions and the degree of media exposure to the attacks while a higher post-geographic proximity tends to uprise the level of anxiety as an immediate response.

2   Related Work

Social media has enabled a wide range of diverse research to harvest a large-scale social interaction data. One area of research focuses on studying individuals' information behaviors on social media, which has been served as a "backchannel communication" in emergency circumstances and allows users to obtain timely, accurate and local information [2]. Other studies attempt to understand the spread of rumor and misinformation in situations of crisis with the aim of improving the information accuracies [3]. Less attention has been paid to the emotional responses in the aftermath of the emergencies. One relevant research to our study is [4], in which authors tried to identify the sentimental content in social media during the disaster events. However, it does not explore the reasons for the volume of the emotional content. A more relevant study is [5], who explains the variations in the shared sentimental expressions after the Boston bombing by different factors, including geographical proximity and social interactions. However, the study was conducted at the community level. There hasn't been a study to probe the individual differences in recovering disaster situations.

On the other hand, physiological psychology research has provided more grounded understanding of the individual differences in explanation of the different outcome on psychological consequences and resilience, in the face of life stressors. Regarding the psychological effects, research has documented some adverse sequelae of traumatic events, including depression, demoralization, stress, sleep disruption, etc. (see [6] for a systematic review of the spectrum of consequences). For the resilience indicators, studies have shown that gender, social support, exposure to the events [1], are of great importance in predicting the individual differences (see [7] for a detailed review). While social psychological experiments or clinical research usually involves a much higher expense and significant delay to study the effect of disasters on individuals, social media provides a cost-effective way to collect a substantial amount of data to validate existing theories, or uncover new patterns.

This paper builds upon these advances in both areas. We combine the understanding of factors of psychological vulnerability and resilience from psychology research with the rich data from social media. We seek to connect our results with previously established theories based on the observations from social media.

3   Research Questions and Hypotheses

This work aims to understand the individual differences in responding to the terrorist attacks as well as in the various recovering trajectories. Prior work has shown that psychological consequences of the terrorist acts can vary widely based on demographic variables, the level of exposure, the impact of the events, experience, and perceived social support [8]. Therefore, we take into account all of these established factors.

Geographic Proximity. Psychological consequences vary across the population with a different degree of exposure to traumatic events [9]. In studying the emotional public response to Boston bombing in 2013, Lin and Margolin [5] found that the expressions of fear on Twitter are stronger in the cities that are geographically closer to Boston. With these evidence from the literature, we derive our first hypothesis:

> **H1:** *The higher geographic proximity users have to the attack sites, the higher level of distress users exhibit as an immediate response and persist in recovery.*

Media Exposure. Schlenger et al. [10] found a significant correlation between the amount of time spending for watching TV and the symptoms of stress, for adults following the 911 attacks. Therefore, we derive the following hypothesis:

> **H2:** *The higher media exposure of users to the attacks, the higher level of distress users exhibit as an immediate response and persist in recovery.*

Social Support/Interactions. Research suggests that social support "buffers" the impact of distress on individuals and therefore indirectly affects the emotional well-being [11]. Helgeson [12] further defines the structural measures of social support as the number of friends a person has, and the frequency of interactions with friends or family. Based on these two observations, we derive the following hypotheses:

> **H3a:** *The smaller ego-network size users have, the higher level of distress users exhibit as an immediate response and persist in recovery.*
> **H3b:** *The lower frequency of interactions users have with others, the higher level of distress users exhibit as an immediate response and persist in recovery.*

Linguistic Indicators. Previous research suggests that certain linguistic indicators have notable effects in the face of the stressful situations [13]. Psychological distancing measures how people separate themselves from the present situation. High psychological distancing in the process of experiencing difficulties has been linked to a reduced feeling of difficulties [13] by leaning back from the painful contemporary situations. The second indicator is cognitive complexity, measured by the amount of precise distinction words used in the speech. Individuals who speak with more cognitive complexity are reported to be more flexible and resilient under life stressors [14]. These findings lead to the following hypotheses:

H4a: *Cognitive complexity has adverse effects on the level of distress users exhibit as an immediate response and persist in recovery.*

H4b: *Psychological distancing has negative effects on the degree of distress users exhibit as an immediate response and persist in recovery.*

Gender. The gender difference has been suggested as predictive of the adverse outcomes of the traumatic events (e.g., Schlenger et al. [10]). In particular, females have been associated with worse short-term outcomes. Therefore, we derive the following hypothesis:

H5: *Female users are associated with a higher level of distress response as immediate effect/in recovering process.*

4   Our Approach

4.1   Data Collection

To examine the above hypotheses, we collected Twitter data related to the Paris attacks. Our data collection needs to track the behavioral change of individuals in Paris across multiple time points and infers the users' distress response based on their characteristics. To minimize the selection bias, we tracked a large set of geo-coded users with complete tweet history during the study period. The idea is based on the *computational focus group* method [15].

The data collection process is built as follows. We first constructed a *computational focus group* defined by users' geo-locations. In particular, with a time window for four weeks after the attacks, we obtained 16,950 users who posted geo-tagged tweets from Paris area, through Twitter API. Then, for every user in our focus group, we traced back his/her full historical tweets through Twitter REST API. In total, we acquired 15,509 unique users whose first tweet in our collection was posted before Oct. 1, 2015. In this way, we have their complete tweets since Oct. 1 2015. Our final data collection includes 3.4 million tweets, among which 219,786 have geo-coordinates (Fig. 1 shows a random sample of 2000 tweets posted around the six attack sites in two days after the event). Our dataset indicates that there was a burst in the number of users and tweets on the day of the attacks. The average volume of tweets before, during and after the attacks are 33,793 tweets/day, 59,247 tweets/day, and 47,386 tweets/day.

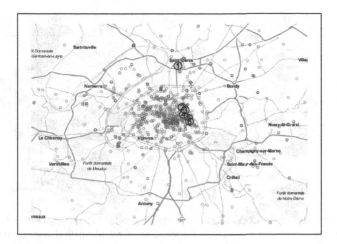

Fig. 1. A random sample of 2000 tweets posted around the 6 attack sites (yellow dots 1–6) within two days after the event. (Color figure online)

4.2 Analysis Framework

In this section, we describe the analysis framework for analyzing the distress response after the attacks. There are several ways to measure the psychological response to disasters or traumatic events in general. Arguably, one of the most accurate measurements is through interviews or questionnaires. However, the analysis through this approach has been hard to implement due to the difficulty of gathering a sufficient number of respondents, significant cost and a severe delay in the data collection [16]. On the other hand, computer-based linguistic measures, such as Linguistic Inquiry and Word Count (LIWC) [17], have been widely adopted to infer personalities, identify depressive, or other health-related symptoms [18]. One of the pioneer work in this field uses LIWC as linguistic markers of the psychological change surrounding September 11, 2001 [19].

In this study, we quantified the distress responses for Paris attacks from three affective dimensions, namely, *anxiety*, *sadness*, and *anger*. Each corresponds to one subset of the dictionary in LIWC. We selected LIWC as it is one of the few lexicons that support a variety of languages (e.g., English, French, etc.). This is particularly helpful in our analysis since more than 60 % of our data are in French. We applied LIWC-based linguistic measures as follows. For every user, we counted the number of tweets contained at least one word in the respective lexicon and divided by the total number of tweets by this user on the same day. That is, we measured individuals' daily distress response as the rate of tweets having, at least, one word in each of the three negative emotional dictionaries (normalized by the total number of tweets posted by her within the same day). In this study, we focused on French and English tweets. Figure 2(a) shows the overall decreasing trend of distress signals over a period of two weeks after the attacks.

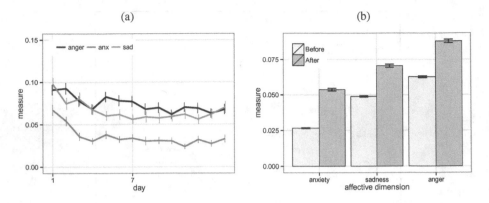

Fig. 2. (a) The trend of distress response. Day 1 on x-axis indicates Nov. 14, 2015. Each data point shows the average sentiment measure of all identified users with errorbars indicating standard errors. (b) The distress responses before and within four days after the attacks.

Immediate Response. We analyze whether there exists a distress breakdown and if so what are the potential factors associated with it. To identify the breakdown, we set a time window, based on which we examine users' emotional responses. Bunney [20] suggested that the breakdown time should be around, at least, two days after the traumatic events, and up to four weeks. We empirically defined the analysis time span of immediate response as four days as we observed a drop within four days after the attack (as shown in Fig. 2(a)). We selected users based on criteria as follows: they have, at least, five tweets (excluding retweets) within four days after the attacks; they need to have geo-tagged tweets in both periods. This process resulted in 1,168 users. Then, we computed the immediate distress response in three affective dimensions based on *aggregated observations within four days* after the attacks for every user, respectively. Figure 2(b) shows the comparison of the immediate acute responses and the ones before the attacks (computed based on all tweets since Oct. 1, 2015). We observe significantly elevated levels of distress responses in all three dimensions.

Short-term Recovery. Another observation from Fig. 2(a) is that the level of distress gradually decreases and tends to stabilize within one week after the attacks. For this short-term recovery analysis, we selected the set of users (1,121 users in total with at least ten tweets) except the second time period was defined as ten days. We then took their *daily observations in one week* as the repeated measures of the emotional responses.

Individual Covariates. To test our hypotheses, we computed the following variables:

– **Geographic distance.** To test H1, we measured the inverse of geographic proximity, geographic distance, by computing the median distance between the coordinates extracted from each user's geo-tagged tweets in the last

45 days and the closest attack sites. These coordinates of each user in the last 45 days serve as a means to estimate the geographic range of his/her life-hood activities towards the attack sites ($M = 3.99$ km, $SD = 3.26$). Similarly, we computed the geographic distance after the attacks ($M = 3.67$ km, $SD = 3.43$ within four days; $M = 3.75$ km, $SD = 3.35$ within seven days).

– **Media exposure.** In the context of Twitter, we could approximate the level of media exposure by counting the number of tweets that contain URL and have keywords "attack", or "terro*", divided by the number of tweets users posted on the same day. To test H2, We computed the content-specific URL rate for each user as a proxy to estimate the extent to which users are immersed by news/articles surrounding Paris attacks in each period (an aggregated measure in 4 days, and a set of repeated measures in 7 days). We should note that this proxy might underestimate the amount of media content users have over the event since they might just read but not share on Twitter.

– **Social network.** To test H3a, we acquired the size of her ego network on Twitter (*friends count*, and *followers count*) through Twitter API. We assumed that users' network size did not differ much over the course of our analysis. To test H3b, we counted the number of tweets one user has mentioned others at least once and normalized by the number of tweets he or she posted during the same day (*communication rate*), in each period.

– **Linguistic indicators.** To test H4, we measured two linguistic markers (*cognitive complexity* and *psychological distancing*). We followed the LIWC based measurement described in [14,21] to compute for each tweet of the users, and then take the average based on all tweets one posted in each period.

– **Gender.** Since there is no gender information on Twitter user profile. To test H5, we used *genderize* API to obtain the estimated gender-based Twitter users' name. We were able to infer the gender for 11,107 users (47.8 % of the users are female).

Analytic Methods. To address the above hypotheses, we constructed two sets of regression models—a set of multiple linear regression models for predicting the immediate response and a set of latent growth mixture models for predicting the recovery process.

In answering the research questions concerning the immediate response, we built three linear regression models, in which each dependent variable is the level of distress in one of three affective dimensions. Our independent variables are the ones previously listed as individual covariates, controlled by the number of tweets within four days in the aftermath and the distress level prior the attacks.

To address the question about the recovery process, we need to estimate the influence of the covariates on the growth of each affective dimension. Particularly, we have two types of covariates—time-invariant covariates (*TICs*) and time-varying covariates (*TVCs*). Our TICs include gender, prior and post geographic distance, friends count, and followers count. The TVCs include the media coverage, social interactions, and two linguistic markers. We need to note that some users might not tweet on certain days in our study period. Therefore, we have partially-missing observations. We selected Latent Growth Mixture Models (*LGMMs*) as they are featured in incorporating missing data, TICs, and TVCs in the model, compared to other longitudinal statistical models, such as ANOVA (see [22,23] for comprehensive comparisons and examples). In this study, TICs address the hypotheses whether characteristics of one individual are predictive of different distress levels on the first day after the attacks (*intercept*), or steeper or less steep rates (*slope*) of changes in the distress level over the week. TVCs directly predict the repeated measures of the distress level while controlling for the influence of the growth factors [22].

LGMMs analysis was done in three steps for each of the affective dimensions. First, single class growth models were determined to estimate the growth parameters. Second, successive class models were built, by adding one additional class each time (from one- to five-class), and then compared based on Bayesian (BIC) information, to determine the optimal number of latent growth trajectories. In the selection of an optimal number of classes, we also enforced the criteria that the class with a minimum percentage of users should be higher than 3 %, for the sake of meaningful interpretations. In the last step, we regressed the models again with all covariates, controlled by the pre-impact emotional level and the number of tweets at each time point.

5 Empirical Results

We convert the values of the covariates to z-scores before running each model. The first three columns of Table 1 reports the coefficients from the linear regression models. The Adjusted R^2 of the models for anxiety, sadness, anger is .07, .10, .13 respectively. The rest of Table 1 presents the estimated coefficients of *intercept* and *slope* for TICs, and the estimated coefficients of *slope* for TVCs from the latent mixture models. The optimal number of classes for the trajectories of anxiety, sadness, and anger is 2 ($BIC = -9101.32$), 2 ($BIC = -6727.45$), and 2 ($BIC = -6001.15$) based on the criteria discussed above. As this study aims to associate the factors with the growth patterns, we did not report on the different classes of trajectories.

In H1, we hypothesize that the smaller geographic proximity (or, a higher geo-distance) to the closest attack sites, the lower level of distress response. Our model suggests that prior distance is insignificant in the immediate response model. Interestingly, we find a rather mixed effect of prior distance in the recovery model. The prior distance is significant and negatively associated with the rate of change in sadness ($\beta = -.016$, $p < .05$) while positively associated with the rate of changes in anger ($\beta = .02$, $p < .01$). Besides, the model also reveals a certain trend toward significantly negative association between the post distance and the degree of anxiety ($\beta = -.004$, $p = .08$) in the immediate response model.

In H2, we expect a positive relation between the media exposure and level of distress response. As a result from our models, the coefficients of media exposure in the prediction of anxiety and anger are both significant and positive ($\beta = .01$,

Table 1. Regression results

	Linear Regression Models			Growth Mixture Models					
					Intercept			Slope	
	(anx)	(sad)	(anger)	(anx)	(sad)	(anger)	(anx)	(sad)	(anger)
followers count	−.001	.002	−.003	.001	.009**	−.003	−.002	−.014**	.002
friends count	−.002	−.002	−.001	.000	.001	.001	−.003	−.004	−.006
gender male	−.01	−.01*	−.002	.000	−.009	−.002	−.005	.018	−.003
prior geographic distance	.000	−.002	.002	−.003	.007	−.007	.005	−.016*	.02**
post geographic distance	−.004	.004	−.004	.000	−.004	.000	−.003	.008	−.006
after the attacks	*aggregated measures*						*repeated measures*		
cognitive complexity	−.002	−.004	−.001				.002	.000	−.001
psychological distancing	−.01**	.001	.001				−.002	.000	−.003
media exposure	.01**	−.004	.01**				.016**	−.002	.013**
communication rate	−.001	−.004	.001				−.004**	−.007**	−.007**
Subjects	1,168	1,168	1,168	1,121	1,121	1,121	1,121	1,121	1,121

Note: the table lists the estimated coefficients. $^*p<.05$; $^{**}p<.01$

$p < .01$, and $\beta = .01$, $p < .01$). Consistently, for the prediction of the recovery process, we also observe significant and positive associations between the media exposure and the level of anxiety and of anger ($\beta = .016$, $p < .01$ and $\beta = .013$, $p < .01$).

In H3a, we expect the ego-network size to positively associate with better distress response. According to Table 1, we find that the size of out-going links (as measured by *friends count*) are insignificant in all models. Rather unexpectedly, followers count has a significant and positive intercept on the initial level of sadness ($\beta = .009$, $p < .01$) while a significant and negative slope in the sadness ($\beta = −.014$, $p < .01$), in the recovery model. H3b predicts that higher rates of interactions with others associate with lower distress responses. The growth mixture model implies there exist negative and significant relations between the communication rate and the degree of anxiety, sadness, and anger ($\beta = −.004$, $p < .01$; $\beta = −.007$, $p < .01$; $\beta = −.007$, $p < .01$).

H4a predicts the positive association between the cognitive complexity and better distress response. We find that the coefficient of cognitive complexity is at the margin of statistical significance and negative in the prediction of sad level but only in the immediate response ($\beta = −.004$, $p = .066$). H4b hypothesizes that a higher psychological distancing corresponds to a lower level of distress response. Our model shows a significant and negative association between psychological distancing and the level of anxiety ($\beta = −.01$, $p < .001$) in the immediate response.

In H5, we hypothesize that female users associate with a higher level of distress response. Our model suggests that gender has a significantly negative coefficient on the degree of sadness in the immediate response model ($\beta = −.01$, $p < .01$). But gender does not seem to have any significant effects on the rate of recovery based on the short-term recovery model since the slope for gender turns out to be insignificant.

6   Discussions

The negative impact of media exposure on the distress response in recovery phase is interesting to us for two reasons. First, we extend the previous finding that a higher TV coverage on 911 attacks correlates with a higher level of distress, to a social media environment. Second, on the two famous-yet-contradictory recovery strategies, "avoidance versus focused attention" [24], we provide the evidence that users who are more towards attentional focus (by reading more information surrounding the attacks), however, appear to have a higher level of distress response.

Echoing with literature in stress recovery after traumatic events, users who have a higher level of social support (as measured by the communication rate) present a reduced level of distress over time. We also find that users with a higher number of followers have a faster drop in the degree of sadness over the study period. In combination, this is consistent with the buffering hypothesis in which social support acts as a buffer against unpleasant effects of the stressors [25]. Interestingly, the number of followers users have on Twitter turns out to be a significant and positive predictor of the initial level of sadness. One possible explanation is that users with more followers probably have a higher tendency of social sharing on the grave. However, further investigation is needed to verify this conjecture.

We find the geographic distance prior the attacks are insignificant for the prediction of overall distress response within four days. One possible explanation is that the acts of terrorism injected a burst of psychological instabilities that spread to areas that are not limited to certain boundaries within the city. Besides, the mixed effects of prior distance in the recovery model suggest a rather interesting story. Users who were further away from the attack sites in the past tended to express more sadness on the first day after the attacks than those who were close. Users who were closer experienced a flatter decrease in the level of sadness but a steeper drop in the degree of anger. Moreover, we find that the geographic distance to the attack sites after the attacks is a quasi-significant predictor of the level of anxiety within four days in the aftermath. One possible explanation is that users who were closer might be anxious about similar events happening again in the focal areas. Combining these results, we argue that geographical proximity produces rather complex effects on the emotional responses. In evaluating its impacts, one needs to consider both prior geographic proximity and continuous geographic exposure to the attack sites.

Limitations: The first limitation is the use of the 2007 LIWC dictionary for sentiment analysis. With many of the more advanced work mainly focus on English text [26], we selected LIWC for its flexibility of dealing with multiple languages. However, the word on Twitter is likely to be different to the one used in written speech, e.g., more abbreviations on Twitter. This fact might result in a lower recall of the affective words. Besides, our results show that the immediate response models have relatively low fit as reflected by the Adjusted R^2. This suggests that there might exist more influential variables yet to be discovered.

The current study aims to discover associations between certain user characteristics with their post-disaster emotional trends. The novel contribution of this study lies in the ability to measure panel users' fine-grained time-varying emotional statuses. We leave it as part of the future work to explore more effective prediction models. Also, we have not discussed in details about the nuances of different type of negative emotional responses. As evidenced by the distinct impacts of geographic distance, we planned to probe further the semantic differences in various types of emotional measures.

7 Conclusion

In this study, we used social media data as an alternative channel to study the disaster response. We collected tweets from users in Paris prior and after the attacks to understand the interplays between the individual differences and the distress immediate outcome as well as its trajectories. Our analysis shows that a higher geographical proximity after the attacks corresponds to a greater level of anxiety. Users' emotional trajectories vary with their level of media exposure and the degree of social interactions. This study provides both theoretical and practical implications for understanding the distress response process, strategic interventions in the aftermath of mass violence events.

References

1. Beevers, C.G., Hixon, J.G.: Individual differences in stress physiology: understanding person by situation influences (2011)
2. Sutton, J., Palen, L., Shklovski, I.: Backchannels on the front lines: emergent uses of social media in the 2007 southern California wildfires. In: Proceedings of ISCRAM (2008)
3. Castillo, C., Mendoza, M., Poblete, B.: Information credibility on twitter. In: WWW (2011)
4. Caragea, C., Squicciarini, A., Stehle, S., et al.: Mapping moods: geo-mapped sentiment analysis during hurricane sandy. In: Proceedings of ISCRAM (2014)
5. Lin, Y.-R., Margolin, D.: The ripple of fear, sympathy and solidarity during the boston bombings. EPJ Data Sci. 3(1), 31 (2014)
6. Panzer, A.M., Butler, A.S., et al.: Preparing for the Psychological Consequences of Terrorism: A Public Health Strategy. National Academies Press, Washington, D.C. (2003)
7. Rodriguez-Llanes, J.M., Vos, F., et al.: Measuring psychological resilience to disasters: are evidence-based indicators an achievable goal. Environ. Health 12, 115 (2013)
8. Bonanno, G.A., Brewin, C.R., Kaniasty, K., La Greca, A.M.: Weighing the costs of disaster consequences, risks, and resilience in individuals, families, and communities. Psychol. Sci. Public Interest 11(1), 1–49 (2010)
9. Silver, R.C., Holman, E.A., McIntosh, D.N., Poulin, M., et al.: Nationwide longitudinal study of psychological responses to September 11. Jama 288(10), 1235–1244 (2002)

10. Schlenger, W.E., Caddell, J.M., Ebert, L., et al.: Psychological reactions to terrorist attacks: findings from the national study of Americans' reactions to September 11. Jama **288**(5), 581–588 (2002)
11. Cohen, S., Wills, T.A.: Stress, social support, and the buffering hypothesis. Psychol. Bull. **98**(2), 310 (1985)
12. Helgeson, V.S.: Social support and quality of life. Qual. Life Res. **12**(1), 25–31 (2003)
13. Thomas, M., Tsai, C.I.: Psychological distance and subjective experience: how distancing reduces the feeling of difficulty. J. Consum. Res. **39**(2), 324–340 (2012)
14. Saslow, L.R., McCoy, S., et al.: Speaking under pressure: low linguistic complexity is linked to high physiological and emotional stress reactivity. Psychophysiology **51**(3), 257–266 (2014)
15. Lin, Y.-R., Margolin, D., Keegan, B., Lazer, D.: Voices of victory: a computational focus group framework for tracking opinion shift in real time. In: WWW (2013)
16. Coppersmith, G., Dredze, M., Harman, C.: Quantifying mental health signals in twitter. In: ACL (2014)
17. Pennebaker, J.W., Francis, M.E., Booth, R.J.: Linguistic Inquiry and Word Count: LIWC 2001. Lawrence Erlbaum Associates, Mahway (2001)
18. De Choudhury, M., Gamon, M., Counts, S., Horvitz, E.: Predicting depression via social media. In: ICWSM (2013)
19. Cohn, M.A., Mehl, M.R., Pennebaker, J.W.: Linguistic markers of psychological change surrounding September 11, 2001. Psychol. Sci **15**(10), 687–693 (2004)
20. Bunney, B.S.: The psychological aftermath of disasters
21. Slatcher, R.B., Chung, C.K., Pennebaker, J.W., Stone, L.D.: Winning words: individual differences in linguistic style among us presidential and vice presidential candidates. J. Res. Pers. **41**(1), 63–75 (2007)
22. Curran, P.J., Obeidat, K., Losardo, D.: Twelve frequently asked questions about growth curve modeling. J. Cogn. Dev. **11**(2), 121–136 (2010)
23. Zheng, Z., Pavlou, P.A., Bin, G.: Latent growth modeling for information systems: theoretical extensions and practical applications. ISR **25**(3), 547–568 (2014)
24. Keogh, E., Hatton, K., Ellery, D.: Avoidance versus focused attention and the perception of pain: differential effects for men and women. Pain **85**(1), 225–230 (2000)
25. Cohen, S., McKay, G.: Social support, stress and the buffering hypothesis: a theoretical analysis. Handb. Psychol. Health **4**, 253–267 (1984)
26. Pak, A., Paroubek, P.: Twitter as a corpus for sentiment analysis and opinion mining (2010)

Conversational Non-Player Characters
for Virtual Training

Dennis M. Buede[1(✉)], Paul J. Sticha[2], and Elise T. Axelrad[3]

[1] Innovative Decisions, Inc., Vienna, VA, USA
dbuede@innovativedecisions.com
[2] Human Resources Research Organization, Alexandria, VA, USA
[3] Sandia National Laboratories, Albuquerque, NM, USA

Abstract. This paper describes an implementation of Dynamic Decision Networks for the creation of an intelligent Non-Player Character (NPC) in a virtual training system. The NPC was required to interact in a text-based system with a trainee to respond as a villager would with helpful information, evasive responses or lies. The NPC was also required to assume a range of personalities that one might find in a foreign village. Our approach proved successful and includes a number of important characteristics that should be considered for future intelligent agents.

Keywords: Behavioral-cultural modeling · Intelligent agents · Dynamic decision networks · Influence diagrams · Bayesian networks

1 Introduction

Solutions for problems of the 21st century are not readily found in the back of a textbook. Unique and complex problems require creative, critical, and analytical problem-solving abilities. 3D Immersive Learning Environments or Virtual World Simulations provide a cost-effective way to simulate these complex problems and learn the skills required to solve them.

To fulfill the requirements to be scalable and extensible as well as to provide realistic complexity, our team developed artificial-intelligence based conversational non-player characters (NPCs) called Adaptive Human Behavior Avatars (AHBAs), which were implemented as part of an Army Research, Development and Engineering Command (RDECOM) activity. These automated, adaptable and unscripted intelligent agents facilitate unpredictable complex human-computer conversational interactions in virtual environments to help learners understand and succeed in complex systems.

The implementation of intelligence within a Dynamic Decision Network (DDN) is justified and described. The DDN approach allows reasoning over time by the NPC so as to adjust its behavior in a reasonable manner as the trainee adjusts.

This paper gives an overview of the AHBA methodology, describes the DDN technology upon which it is based, and summarizes how the AHBA incorporates psychological personality constructs and represents the rapport between the trainees and NPCs. The virtual training scenario involves a military trainee learning how to interact with villagers in a foreign country (in this case, Azerbaijan) in order to obtain useful information about the situation, the local leadership, and the local bad actors.

© Springer International Publishing Switzerland 2016
K.S. Xu et al. (Eds.): SBP-BRiMS 2016, LNCS 9708, pp. 389–399, 2016.
DOI: 10.1007/978-3-319-39931-7_37

2   Overview of Intelligent Agents for Virtual Training

AHBAs use a natural language parser and keyword lookup tables to determine what the trainee is asking, a probability model to track the AHBA's rapport with the trainee, and an expected utility model to determine whether to respond truthfully, evasively, or deceitfully. Included in the AHBA is a model of the rapport between the trainee and itself so that the AHBA can dynamically respond based on the progression of the conversation. Our approach to modeling rapport is an innovative mixture of cultural programming and individual personality characteristics.

Figure 1 defines a spectrum of AI architectures [1] for creating NPCs. The spectrum is not a range from bad to good architectures. Rather it is meant to show that AI architectures differ on qualities such as authorial control and simplicity. Authorial control suggests that the designer of the NPC has a deterministic representation that defines the response of the NPC. AI Autonomy means that the NPC designer has constructed an engine that enables the NPC to respond however the engine dictates. The Utility-based AI approach, as embodied in our AHBA, has a defined set of responses, but there is a probabilistic engine that determines which response is best, based on an expected utility calculation. The best architecture depends largely on the specific situation; each works well in some circumstances and poorly in others [2].

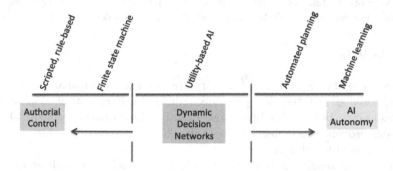

Fig. 1. Spectrum of AI architectures for NPCs

The authors chose the expected utility approach because (1) uncertainty was important to ensuring the Non-Player Character was not predictable (i.e., not going to do the same thing every time the trainee took a certain path), and (2) contradictory evidence was expected for the rapport variable that could not be handled well in a rule-based system. This contradictory evidence occurs in specific statements by the trainee as well as across statements by the trainee. A black-and-white rule based system could not address the subtleties of positive sentence tone with severe cultural insensitivies in the same trainee inquiry.

3   Adaptive Human Behavior Avatar (AHBA) as Dynamic Decision Network

We consider the following scenario: A trainee approaches a simulated villager NPC and engages in a conversation. The words used by the trainee may build, sustain, or diminish the rapport felt by the simulated villager with the trainee. Depending on the likelihood of a positive, neutral or negative rapport, an expected utility is calculated for each potential simulated villager decision: tell the truth, be evasive, or lie.

Our assessment was that the AHBA would have to address the following four issues:

1. Infer the trainee's key message from 1–3 sentences spoken;
2. Determine whether the trainee is saying a greeting, a pleasantry or asking a question;
3. Decide whether the NPC should tell the truth, be evasive or lie; and
4. Once the above three determinations have been made, decide what to respond to the trainee.

To determine the meaning of a message, we choose to select key topic words from the trainee's text. We quickly found that we needed a large thesaurus that was keyed to the topics that a trainee and responsive villager (as played by the NPC) would address in the context of U.S. military forces trying to learn about the local leadership and issues of conflict in a village in Azerbaijan. This worked moderately well but needed improvement so we also implemented the Stanford parser to help adjudicate among multiple possible key topics. This topic tree structure was critical for pleasantries.

We built a statement classifier using a Bayesian network to determine whether a statement was a greeting, pleasantry, or substantive question. The features of this classifier included the presence of words in questions, greetings, pleasantries, farewells, and substance. This classifier was simple and quite accurate.

To determine the NPC actions, we used the conceptual method of Dynamic Decision Networks (DDNs) that was developed by Buede in the early 1990s [3]. DDNs utilize the expected utility approach shown in Fig. 1. They had been implemented successfully as part of the Future Combat System [4], so their capability to address a problem of this complexity was established. For our purpose we considered utility to map onto the concept of rapport, since the trainee would be likely to find that as rapport increases, information sharing increases [5].

To determine the specific NPC response to the trainee's statement, we developed a large table of responses that are keyed to potential topics brought up by the trainee. This effort involved massive research on Azerbaijani life and culture by one of our colleagues (Ms. McCarter) and was also quite successful.

DDNs are designed to solve real time problems in a dynamic environment characterized by one or more variables that evolve with time. A dynamic environment is defined as one in which there are numerous time periods when an individual makes decisions. These decisions not only affect uncertainty in the present but also uncertainty and decisions in the future. In dynamic decision making, the amount of uncertainty can change from period to period; in new periods some uncertainties may be resolved and some new ones may develop. For the AHBA described here, each time period was initiated by a statement from the trainee. The response from the NPC/AHBA ended the time period.

DDNs are implemented as the integration of influence diagrams and Bayesian networks [3, 4]. The influence diagram contains decision, value (utility), and chance nodes and uses expected value (utility) to identify the best alternative for each time period based on the evidence obtained in the current and previous time periods. Embedded in the influence diagram are key chance nodes about which evidence may be obtained in any given time period. A Bayesian network is used in each time period to update these key chance nodes. To simplify computations and remain tractable, the influence diagram does not look ahead to future time periods.

The DDN for time period zero, prior to trainee inputs, is shown in Fig. 2 and includes a utility node, decision node, numerous chance nodes (parchment colored nodes). The personality model is shown in on the right of Fig. 2; this is discussed in the next section. In this case the villager is modeled as having a moderate level for three personality variables. The initial valence of the rapport with the trainee experienced by this villager is roughly one third on positive, neutral and negative. The decision node shows that the expected utility for being evasive (57.19) is higher than for telling a lie (48.43), with telling truth having the lowest value (40.45). Consequently, the trainee must build rapport prior to seeking any information. Besides personality, rapport is affected by the type of sentence, the number of the sentence, and whether the sentence has been repeated. Here a question on the first sentence will drop rapport, and a greeting on the first sentence will increase rapport. Pleasantries for sentences two through four will build rapport as well. Other elements of the sentences that affect rapport are sentence tone (positive versus negative words) and cultural understanding (positive versus negative topics).

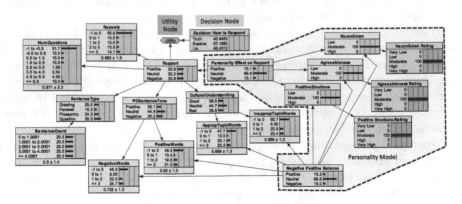

Fig. 2. DDN for time period 0 with no evidence from trainee's statements

Figure 3 shows the updated DDN for Time Period 1 with evidence from the Trainee's first sentence: "Who is Rayhan Karimov?" Since this first sentence is a question, the probability of negative rapport increases to 0.59, which increases the expected value of telling a lie but not enough to overtake being evasive. Note there

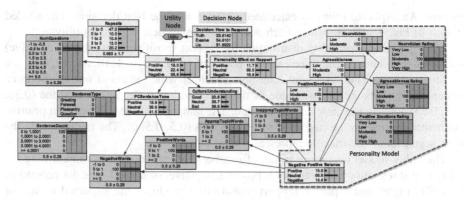

Fig. 3. DDN for time period 1 with evidence from trainee's statements

were no words in this first sentence that favored or disfavored sentence tone or cultural understanding.

Since rapport was the central concept to the decision of telling the truth, being evasive, or telling a lie, we developed a visual representation of the rapport and how it changed over time, see Fig. 4. The triangle in Fig. 4 is a projection of the 3-dimensional space defined by probabilities of positive, negative, and neutral rapport, respectively, onto the plane in which these three values sum to 1.0. The vertices of the triangle are the points where the plane intersects the three axes. Each point in the space depicts the probabilities for the simulated villager's felt rapport after the cumulative effect of the trainee's utterances, that is, the subjective likelihood that the felt rapport is positive, negative or neutral.

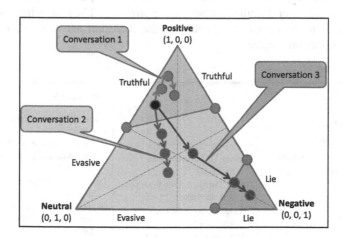

Fig. 4. Rapport probability-decision map

There is a differential utility for each decision under the situation of different levels of rapport. For example, lying has the greatest utility when rapport is substantially negative, while telling the truth has the greatest utility when rapport is substantially positive. An expected utility is calculated to determine the best decision. The shaded regions of Fig. 4 indicate where each action has the highest expected utility.

Every possible probability distribution for rapport (positive, neutral, and negative) fits within the triangle. The three extreme points of (1, 0, 0), (0, 1, 0), and (0, 0, 1) are the three corners. The three faint blue lines, each of which starts at one of the three corners and bisects the opposite side of the equilateral triangle, intersect at the (0.33, 0.33, 0.33) point in the center of the triangle. The midpoint of the line between positive and neutral represents the case where the rapport is (0.5, 0.5, 0). The midpoints on the other segments of the triangle represent similar cases.

The utility model contains a triplet of numbers for each response, representing the utility of that response when rapport is positive, negative, or neutral. The dot product of the utility triplet and triplet of rapport probabilities produces the expected utility for each response. Figure 4 is a probability-decision (or policy) map because it shows how a set of probabilities on rapport affects the decision to be truthful, to be evasive or to lie. The blue dots on each side of the triangle represent the point at which the utility score of each decision (truthful, evasive, or lie) is the same. For example, the blue dot on the left side of the triangle is at (0.5, 0.5, 0), which is where the expected utility for truthful and evasive responses are equal. Connecting these points with straight lines (guaranteed due to the structure of the expected utility calculation) results in three regions. If the set of rapport probabilities fall into the top green region, a truthful response will be given. If they fall into the bottom right reddish region, the NPC will lie. An evasive answer will be given if they fall into the large orange region.

The black dot near the lower left corner of the green (truthful) region is the initial probability for one of the villagers. The green, red and purple arcs emanate from this black dot. The paths for rapport probabilities for the three conversations shown in Fig. 5 allow a training assessor to follow the conversation paths in Fig. 4 to determine why truthful, evasive, or deceitful answers were provided. The text strings provided by the trainee provide the evidence elements for the Bayesian network that explains the path of the rapport probability.

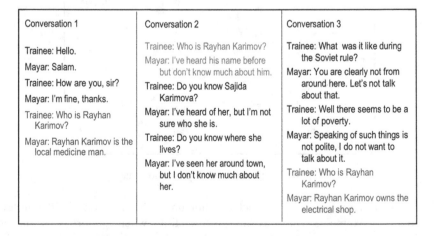

Conversation 1	Conversation 2	Conversation 3
Trainee: Hello. Mayar: Salam. Trainee: How are you, sir? Mayar: I'm fine, thanks. Trainee: Who is Rayhan Karimov? Mayar: Rayhan Karimov is the local medicine man.	Trainee: Who is Rayhan Karimov? Mayar: I've heard his name before but don't know much about him. Trainee: Do you know Sajida Karimova? Mayar: I've heard of her, but I'm not sure who she is. Trainee: Do you know where she lives? Mayar: I've seen her around town, but I don't know much about her.	Trainee: What was it like during the Soviet rule? Mayar: You are clearly not from around here. Let's not talk about that. Trainee: Well there seems to be a lot of poverty. Mayar: Speaking of such things is not polite, I do not want to talk about it. Trainee: Who is Rayhan Karimov? Mayar: Rayhan Karimov owns the electrical shop.

Fig. 5. Sample interactions (Color figure online)

4   AHBA with Personality

Since rapport plays a central role in the DDN, we give a brief description of the theoretical background for this variable and the way that it is implemented in the model. We then describe variables representing the personality of the simulated villager and specify how these variables can affect rapport.

4.1   Theoretical Background to Model Rapport

We considered two constructs that capture the experience of someone responding to a stranger. One construct is the initial impression formed (i.e., by the local of the trainee) and the other construct is the rapport between the local and the trainee that will change over time. While others have described the user experience of flow and immersion in interacting with non-player characters in dialog (see, for example, the work of Serder Sali and colleagues [6, 7], we were interested in the likely reactions of an NPC to the trainee, which we modeled as contingent on the NPC's modeled sense of rapport with the trainee.

Research in social perception focuses on the cognitive processes underlying the integration of data in forming an impression. One finding in this area is that people organize their impressions of others according to two fundamental dimensions: warmth and competence [8]. Another finding is that first impressions and recent impressions matter, but the relative weight of initial and recent information seems to depend on how that information is presented [9]. To reflect these results, our AHBA framework for implementing NPCs enables these variables to be mixed in different proportions, depending upon the training environment.

We consider the warfighter who is trying to elicit social network information from a local in a foreign culture to be in a similar situation to that of the ethnographer who interviews locals to understand their social life. Spradley [5], in a classic paper, characterized ethnographic interviewing as developing rapport and eliciting information. Rapport is a dynamic variable, representing the relationship between the ethnographer and the informant. When rapport breaks down, the local may refuse to share information, or at least stop sharing the kind of information of interest to the ethnographer.

According to Spradley [5], the rapport process has four stages: apprehension, exploration, cooperation, and participation. During apprehension, informants show anxiety and suspicion, feeling uncertain about the motives of the visiting ethnographer. These concerns are best addressed by asking open-ended descriptive questions, letting the informant do most of the talking, listening with interest, and responding in a non-judgmental way. During exploration, the two get to know each other. The shift from apprehension to exploration is marked by each laughing at humorous comments, the informant speaking on an interesting tangent, and the ethnographer dropping prepared questions. The ethnographer must be patient, clarifying the purpose of the conversation, and using key phrases introduced by the informant. Cooperation occurs when there is mutual trust, with neither concerned about giving offence. Expectations for the conversations are clear. The informant may spontaneously correct the ethnographer. Finally, participation occurs when the informant accepts the role of teacher, bringing new unsolicited information to the attention of the ethnographer. We designed our AHBA with consideration of these types of interactions, but the current version only implements some of them.

More recently, Spencer-Oatey [10] has laid out a taxonomy of rapport management as a function of language (in addition to information transmission). The three elements of rapport management are the management of face, sociality rights and obligations, and interactional goals. Culture influences how people interact across these domains.

A number of domains can affect rapport. For example, misuse of speech acts such as orders, requests, apologies and compliments can lead to breakdowns in rapport. In addition, topics can be inappropriate for the conversation. Interruptions can threaten rapport, as can inappropriate use of tone, vocabulary, syntax, terms of address and honorifics.

4.2 Personality

The statements of a warfighter in interaction with a local civilian will have distinct effects on their rapport and the local's willingness to share information, depending on the personality of the local. Some locals may respond charitably to a foreigner's communication error while others might provide false information or turn away in disgust. We expanded the model to represent hypothesized effects of the NPC personality characteristics on the rapport between the NPC and the trainee, based on the nature of trainee statements. The Five-Factor Model (FFM) of personality as assessed by the NEO Personality Inventory [11] considers five general personality traits—neuroticism, extraversion, openness to experience, agreeableness, and conscientiousness. We focused on three characteristics taken from the traits and facets of the FFM. We hypothesized that two traits, neuroticism and agreeableness, and one facet of extraversion, positive emotions, can affect the rapport between the NPC and the trainee. Afonso and Prada [12] used agreeableness in their non-player character models, and suggest that neuroticism and extraversion also matter to human relationships. We narrowed extraversion to the facet within extraversion of positive emotions because we were interested in the emotional reactions of non-player characters. These characteristics are briefly defined and characterized below:

- Neuroticism contrasts negative emotionality, such as feeling anxious, nervous, sad, and tense, with emotional stability and even-temperedness. Individuals high in neuroticism tend to experience more fear, sadness, embarrassment, anger, guilt, and disgust, than more emotionally stable individuals.
- Agreeablenesses contrasts a prosocial and communal orientation toward others with antagonism and includes facets such as altruism, tender-mindedness, and trust. Individuals high in agreeableness tend to be sympathetic, eager to help, and trust others to be helpful. Those low in agreeableness tend to be egocentric, skeptical of others' intentions, and competitive.
- Positive emotions assess the tendency to experience emotions such as joy, happiness, love, and excitement. This variable is one facet of extraversion. Individuals who are high in positive emotions laugh easily and often, are cheerful, optimistic,

enthusiastic, and humorous. Individuals who are low in positive emotions are less likely to express these emotions, though they may be happy.

We hypothesize that neuroticism is inversely related to rapport, while agreeableness and positive emotions are each directly related to rapport.

4.3   Implementation of the Rapport Model

As shown in Fig. 3, rapport is represented as a probability distribution over three states: positive, neutral, and negative. The rapport is based on three primary components: the structure of the conversation (what sentence type is being used and whether this sentence is at the beginning, in the middle or at the end of the conversation; whether the sentence has been used before; and how many questions have been asked in a row), the tone of the input (positive and negative words used), and demonstrated cultural understanding (appropriate and inappropriate topic words). On the right of Fig. 3 are three psychological variables used to characterize the NPC: neuroticism, agreeableness, and positive emotions. These three variables are used to compute an aggregate personality measure called Negative-Positive Balance (bottom right of Fig. 3). An NPC with a negative balance will be initialized as having a more negative rapport and then will respond more strongly to negative words and inappropriate topics than to positive word and appropriate topics. Conversely an NPC with a positive balance will be initialized as having a more positive rapport and will respond more strongly to positive words and appropriate topics.

There are two mechanisms by which the personality variables affect rapport. One mechanism proceeds fairly directly from the personality variables, to the node named 'Personality Effect on Rapport,' to rapport itself. This mechanism represents the first impression that the NPC has of the warfighter trainee before they interact. An NPC that is high in agreeableness and positive emotions and low in neuroticism will have a more positive initial view of the trainee, thus a higher likelihood of positive rapport.

The second mechanism has a more indirect effect that includes the node named 'Negative Positive Balance,' as well as several nodes that evaluate features of the trainee's communications. This mechanism has no effect when there is no communication to interpret. However, when there is a communication, the mechanism represents a positive or negative bias in how it is interpreted. Those who are high in positive emotions and agreeableness and low in neuroticism will place more emphasis on positive words and appropriate topics, to produce a more favorable interpretation of the trainee's sentence tone and cultural understanding. Those with the opposite values on these variables will tend to produce more negative interpretation.

5   Conclusions

This paper has presented an innovative solution for intelligent agents based on DDNs with a personality module. The DDN approach allows the intelligent agent to receive inputs from the environment and respond periodically to the inputs in a way that adapts over time.

The DDN-based AHBA described here has the ability to extract evidence about what the trainee is asking, evidence for the type of response (e.g., answer to a question, response to a greeting) the trainee is expecting, and the politeness and sensitivity of the trainee in the conduct of the conversation. In addition, the AHBA can respond appropriately to these text strings to keep the trainee engaged and provided with cues about how well the trainee is performing. Our approach to visualizing the course of the interaction between the trainee and the NPC over time in the response space of the NPC telling the truth, being evasive or lying enables the NPC designers to debug and validate the system. This visualization also provides important cues to the trainee either in real time or after the training event is completed.

Our approach to integrating an array of NPC personalities into the DDN is also described. Our approach built upon recent research regarding rapport between people and human personality. Specific personality traits were selected from the FFM of personality for this particular application. This personality model not only makes adjustments to the starting position of the NPC in the three element rapport space of positive, neutral and negative, but it also adjusts how the NPC will respond to positive and negative statements from the trainee.

References

1. Dill, K.: Introducing GAIA: A Reusable, Extensible Architecture for AI Behavior. Simulation Interoperability Standards Organization, 2012 Spring Simulation Interoperability Workshop. Paper number 12S-SIW-046 (2012)
2. Dawe, M., Champandard, A., Mark, D., Rich, C., Rabin, S.: Deciding on an AI architecture: which tool for the job. In: Game Developer's Conference (2010)
3. Buede, D.M.: Dynamic Decision Networks: An Approach for Solving the Dual Control Problem, 1999 Spring INFORMS, Cincinnati, OH (1999)
4. Kobylski, G.C., Buede, D.M.: Dynamic decision networks: an alternative to dynamic programming. Int. J. Inf. Decis. Sci. 3(3), 203–227 (2011)
5. Spradley, J.P.: Asking descriptive questions. In: Spradley, J.P. (ed.) The Ethnographic Interview, 1st edn. Wasworth, Belmont (1979)
6. Sali, S., Wardrip-Fruin, N., Dow, S., Mateas, M., Kurniawan, S., Reed, A.A., Liu, R.: Playing with words: from intuition to evaluation of game dialogue interfaces. In: Proceedings of the Fifth International Conference on the Foundations of Digital Games, pp. 179–186. ACM (2010)
7. Sali, S.: Playing with words: From intuition to evaluation of game dialogue interfaces. Univ. Calif., Santa Cruz (2012)
8. Cuddy, A.J., Fiske, S.T., Glick, P.: Warmth and competence as universal dimensions of social perception: The stereotype content model and the BIAS map. Adv. Exp. Soc. Psychol. 40, 61–149 (2008)
9. Dreben, E.K., Fiske, S.T., Hastie, R.: The independence of item and evaluative information: Impression and recall order effects in behavior-based impression formation. J. Pers. Soc. Psychol. 37, 1758–1768 (1979)
10. Spencer-Oatey, H.: Face, (im)politeness, and rapport. In: Spencer-Oatey, H. (ed.) Culturally Speaking: Managing Rapport through Talk across Cultures, pp. 11–47 (2000)

11. Costa, P.T., McCrae, R.R.: Revised NOE Personality Inventory (NEO PI-R) and NEO Five-Factor Inventory (NEO-FFI) Professional Manual. Psychological Assessment Resources, Inc., Lutz (1992)
12. Afonso, N., Prada, R.: Agents that relate: improving the social believability of non-player characters in role-playing games. In: Stevens, S.M., Saldamarco, S.J. (eds.) ICEC 2008. LNCS, vol. 5309, pp. 34–45. Springer, Heidelberg (2008)

Holy Mackerel! an Exploratory Agent-Based Model of Illicit Fishing and Forced Labor

Kyle M. Ballard$^{(\boxtimes)}$

U.S. Department of State,
Office to Monitor and Combat Trafficking in Persons, Washington, D.C., USA
kyllard@gmail.com

Abstract. This paper introduces an agent-based model to explore the existence of positive feedback loops related to illegal, unregulated, unreported (IUU) fishing; the use of forced labor; and the depletion of fish populations due to commercial fishing. The author hypothesizes the use of forced labor adversely impacts economic activity, provides incentive for illicit activity, and depletes the population of fish. Left unchecked, such a dynamic may lead to irreversible environmental impacts, exacerbate international tensions, and yield significant economic losses. The lack of reliable data on human trafficking and global fisheries makes statistical analysis extremely difficult. This model serves to consolidate several behavioral and impact assumptions into a single exploratory model in order to test these assumptions and establish a proof of concept to guide future research.

Keywords: Trafficking in persons · Forced labor · IUU fishing · Agent-Based modeling · Overfishing · International conflict

1 Introduction

Human trafficking is a global scourge from which no country is immune. The use of force, fraud, or coercion to compel individuals into sex trafficking or forced labor happens everywhere – from the least developed to the most advanced; from kleptocracies to democracies. The U.S. Department of State's 2015 Trafficking in Persons Report documents 188 countries in which victims of modern slavery are found, and the International Labor Organization estimates 21 million victims worldwide and an illicit economy of $150 billion per year attributed to human trafficking [1, 2].

While these statistics are staggering, efforts to quantify human trafficking are riddled with challenges. Despite an international legal definition of "trafficking in persons," jurisdictions within the international system operate under drastically different definitions of the crime. This complicates efforts to collect, validate, and aggregate data. In addition, human trafficking is a hidden crime in that victims are often unwilling to self-identify, and instances are obscured by the presence of other crimes such as

Kyle M. Ballard—The views expressed here are solely of the author and do not necessarily reflect the opinions or policies of the U.S. Department of State or any other U.S. government agency.

K.S. Xu et al. (Eds.): SBP-BRiMS 2016, LNCS 9708, pp. 400–409, 2016.
DOI: 10.1007/978-3-319-39931-7_38

other forms of illicit trafficking (e.g. drugs, weapons), unlawful fishing, prostitution, and immigration violations. Ultimately, reliable data simply does not exist. This has not prevented advocates, activists, and policymakers from citing such data as they craft approaches to combating human trafficking.

One area of human trafficking that has garnered public and policymaker attention is the use of forced labor in supply chains, particularly in the seafood industry [3, 4]. Despite increased anecdotal evidence and a nascent understanding of actor behavior, concrete data on forced labor within the seafood industry remains scant. As a result, there is a significant gap in the ability of policymakers to make evidence-based policy, as well as a strong need for analytical techniques that thrive in the absence of "big data." To fill this gap, this paper uses exploratory modeling – an approach that allows for the testing of various assumptions and hypothesis when confronted with "insufficient knowledge or unresolvable uncertainty preclude[ing] building a surrogate for the target system" [5]. Rather than drilling down to increasingly granular micro-levels of the system, this paper attempts to consolidate broader system-level information and assumptions to test hypotheses and reveal deeper insight into the overall system.

As such, this paper introduces a simple agent-based model that explores the existence of a positive feedback loop between forced labor; illegal, unregulated, unreported (IUU) fishing; and overfishing. This author hypothesizes that IUU fishing and labor exploitation serve in conjunction to drive down labor costs, and therefore market prices for fish. Furthermore, the author postulates lower market prices push fishermen to catch and sell more fish to sustain their businesses and to compete in the market. This hypothesis holds that the use of forced labor hurts legitimate businesses, incentivizes illicit activity, and depletes the population of fish. Left unchecked, such a dynamic may lead to irreversible environmental impacts, exacerbate international tensions, and yield significant economic losses.

Policymakers are currently pursuing separate policies to address forced labor in the fishing sector and IUU fishing. There are efforts to incentivize more transparent and lawful supply chains, including eliminating forced labor from seafood supply. At the same time, policies on IUU fishing generally focus on environmental impacts such as depleted fish stock and economic losses [6]. There are signs that the two policy interests are ripe for convergence, suggesting that this is an opportune time to leverage modeling to inform policy on these issues. The new Trans-Pacific Partnership trade deal addresses both forced labor and environmental issues, and interagency efforts on IUU are increasingly aware of forced labor in the fishing industry [7, 8]. Agent-based modeling may prove a useful tool for providing policymakers deeper insights into the fishing industry, potentially resulting in diplomatic solutions, more exacting policy, and innovative program development such as tailored public-private partnerships.

2   Workings of the Agent-Based Model

In addition to the aforementioned challenges of quantifying human trafficking and the extent of forced labor in the fishing industry, data on fish stocks and the prevalence of IUU fishing are also difficult to find. This is due to several factors, including the vastness of the ocean; territorial disputes in places like the South China Sea; unclear jurisdiction in international waters; and the economic interests of sovereign countries that maintain secrecy around natural resources they seek to exploit. Therefore, this agent-based model was built in NetLogo [9] with adjustable attributes to accommodate available data, test values where data is unavailable, and explore various assumptions. Various scenarios can be subject to experiment in order to learn more about the interplay of forced labor, IUU fishing, and overfishing. The model space is a 100 × 100 Cartesian plane, which roughly scales to the portion of the South China Sea with Hong Kong to the north, Brunei and Malaysia to the south, Vietnam to the west, and the Philippines to the east. This is only a rough fit and is not meant to be a reliable geographical representation.

2.1   Model States

Each run of the model can be run in one of three states. The first is the absence of IUU fishing in which boats pursue fish according to standard rules and adhere to common labor practices. In this state, boats only fish during "on season." The second state is the presence of IUU fishing but the absence of forced labor. The prevalence of IUU fishermen as a percentage of total boat population is adjustable for each run. IUU boats in this state engage in fishing activity during the "off-season" but adhere to wage standards. The third state is the presence of both IUU fishing and forced labor. In this state, IUU boats engage in substandard labor practices and fish during the off-season, while the other boats do not. Having three states allows the observer to compare outcomes of the various states to determine the impact of IUU and forced labor.

Global Attributes.

Attribute	Description
Number of boats	Number of boats
Fish population	Number of fish at the outset of a model run
Fish per boat	Number of fish a boat should carry at a given time (serves as a safety standard or fishing quota). Law-abiding boats adhere while IUU boats exceed this number, according to their own "Fish capacity" attribute described below.
IUU fishing	True/false statement that determines whether a subset of boats engage in IUU
IUU prevalence	Percentage of total boat population engaged in IUU
Forced labor	True/false statement that determines whether IUU boats engage in substandard labor practices
Minimum wage	Standard wage all law-abiding boats pay workers. IUU boats pay workers a random wage below this number (0 to n-1)

Seasons. The model has two seasons: on-season and off-season. On-season lasts 36 time steps and off-season lasts 16 time steps. Together, this constitutes a full calendar year, broken down into 52 time steps (1 time step = 1 week). The 16-week off-season represents the real-world off-season imposed on catching Grouper fish such as that imposed by the Thai government and others around the world. This off-season coincides with the annual period during which Grouper fish lay eggs and those eggs hatch (usually February through May). Fish reproduction in the model occurs during the off-season. Law-abiding boats in the model only fish during on-season. Boats that engage in IUU and forced labor (if these features are activated) also fish during off-season. Boats that adhere to the season schedule return to port and take their fish to market during off-season.

2.2   Agents

Boats. Boats represent fishermen and their economic interests, as they troll the ocean in pursuit of fish. Each boat starts at a port at the edge of the model space. These ports also serve as local markets where boats sell their catch. As boats pursue fish, they accrue operating costs and labor costs. The operating costs are the same for all boats, representing standard costs such as fuel and food. Labor costs vary from boat to boat, depending on whether the boat adheres to wage standards (legitimate fishermen) or engage in illicit labor practices (IUU fisherman) such as substandard wages or forced labor. In addition to illicit labor practices, IUU fisherman also engage in fishing activities during "off-season" and exceed any quotas or safety standards regarding the number of fish that can be at a given time.

Boat Attributes.

Attribute	Description
Fish capacity	Number of caught fish a boat is willing to hold at any given time
Worker capacity	Number of fisherman a boat can hold
Fish total	Actual number of fish on board
Worker pay rate	Rate at which each worker is payed
Cost	Boat's total operating cost
Vision	Radius within which a boat can see fish
IUU?	True/false statement indicates whether boat engages in IUU/forced labor

Fish. The number of fish in the model can be adjusted to accommodate various fishery estimates, geographical regions, and species. Each fish starts at a random x and y coordinate in the model space and schools according to the rules of Wilensky's NetLogo Flocking model [10]. Wilensky's model provides each fish a standard set of rules from which schools emerge endogenously. This ensures schools remain in the model regardless of population shifts, reducing the possibility of unintended outputs of the model. Half the fish population are female and can reproduce at an adjustable rate of reproduction (See "Annual eggs" below). The female fish carry a "maturity" attribute by which female fish grow capable of reproduction. The default maturity value and rules are set to represent actual average age at which Grouper fish in the South China Sea mature and become fertile, estimated to be roughly 5 years (or 260 time steps in this model). This attribute is adjustable. The number of offspring is also adjustable to allow for the replication of real-world fish behavior such as growth/ decline rates according to environmental issues, climate change, predation, disease, and other factors that are not hardcoded into this model. All eggs hatch new fish and the live fish population increases accordingly. Mature female fish procreate the same time each year, deemed "off-season" during which law-abiding fishermen refrain from fishing activity. IUU fishermen do not adhere to this rule.

Fish Attributes.

Attribute	Description
Maturity	Numerical value (0–260 time steps) to emulate 5 year maturity period before female can reproduce
Annual eggs	Number of offspring each mature female fish produces each off season

2.3 Model Behaviors

During each run of the model, several coordinates on the edge of the model space are deemed "port." Boats return to ports in the off-season and when they take fish to market. Each boat trolls the ocean seeking fish within its "vision" range. It then catches fish within a radius of two lattice squares ("patches" in NetLogo parlance), which is equivalent to roughly two miles (remember, each time step is equivalent to a week). Once a boat reaches its "fish-capacity," it take its fish to market. Two behavior rules of marketing are possible in this model that must be selected before each model run. The first rule has boats return to the nearest port to their location once they reach fish capacity. The second rule sets their port of origin as their home to which they always return. This simulates the difference between a truly international market and localized markets.

Local Markets. Once a boat arrives at a port, it sets it's per fish asking price by dividing its "cost" by the total number of fish it caught. This value is added to that particular port's running list of fish prices, the average of which determines the local market price. Boats then sell their catch at the local market price, adding the value of their catch to their revenue. This is a rudimentary implementation of the cost-of-production theory of value and Gordon's Economic Theory of a Common-Property Resource [11]. In addition to being relatively simple organizing principles for economic activity, this construct provides realistic stability in market prices by ensuring single transactions do not drastically swing prices. This is important because the number of boats in this model is proportionate but significantly fewer in number than real-world observations. Such a small number of agents raises the possibility of skewed data outputs.

Global Market. Each time a boat sells its catch at a port, its fish total is added to the global "market" which counts the number of fish caught and sold over the entire run of the model. At each time step, ports also send a mean of their local price to a global list of fish-prices. The average of the global list serves as the global market price of fish.

It is by comparing these outputs and the number of live fish at various time steps that the observer can experiment with variable values and model states (i.e. normal, IUU, forced labor).

IUU and Forced Labor. In model states that include IUU fishing, IUU boats continue to fish in the off-season while law-abiding boats return to port. IUU boats also set their "fish capacity" at a random number between the global value of the "fish-per-boat" attribute and 125 % of "fish-per-boat." This represents the fact that IUU fishermen do not, by definition, adhere to established quotas, and often disregard accepted safety or labor standards that may dictate the amount of weight allowed on a fishing vessel. If "forced labor" is in effect, IUU boats engage in labor exploitation and their worker-pay-rate is set at a random float between zero and the minimum wage. This represents substandard wages, to include slavery (NOTE: in some cases, labor recruiters are used by fishermen who then are indebted to said recruiters. In such cases, they still accrue a small wage, although that wage is passed on to pay their debts. This is a practice called "debt bondage," and is considered a form of forced labor. This explains why forced labor victims may still be paid in this model) [12].

3 Initial Findings

After several thousand runs of the model in various model states, initial findings suggest a clear decline of global revenue with the presence of IUU fishing and further decline with forced labor. For the purposes of this paper, the author maintained a fish-to-boat ratio of 667:1. This was largely due to limits in computation power; something future work will seek to address. Runs in various model states and at several values for IUU prevalence indicate that economic losses increase along with the prevalence of IUU fishing and labor exploitation.

The model was runs for several thousand iterations for 260 time steps (equivalent to five years) in which boats sell fish at any port in the model space. Total global revenues from fish sales was measured at each time step. The resulting data shows progressive decline in economic output as the prevalence of IUU fishing increases. Further declines results from the introduction of forced labor. This was measured by comparing outputs from runs of the model without IUU fishing, runs with IUU fishing but no forced labor, and runs with IUU fishing and forced labor. With an IUU prevalence of 7 % the total global revenue over the several thousand runs declined by 6.6 % without forced labor compared to runs without IUU, and 6.9 % with forced labor compared to runs without IUU. These figures jumped dramatically as prevalence increased. At 27 % IUU prevalence, economic activity declined by 7 % without forced labor and 19 % with forced labor; at 34 % prevalence, 19 % and 26 %; and at 67 % prevalence, 33 % and 45 %, respectively. If boats are restricted to selling their catch only at their home ports, economic activity declined even further in some cases, but revenue in the model is not particularly sensitive to this parameter. In these scenarios, a 7 % IUU prevalence yields a decline of 7.1 % without forced labor and 7.4 % with forced labor. At 27 %

prevalence, there is an 18 % decline without forced labor and a 16 % decline with forced labor; at 34 % prevalence, 23 % and 25 %; and at 67 % prevalence, 33 % and 45 %, respectively.

The presence of IUU fishing also had a drastic impact on fish population. Over the aforementioned model runs, total fish population at the end of 260 time steps dropped as the prevalence of IUU increased. First, boats sell fish only at their home port. With 7 % IUU prevalence, the total fish population dropped 12 % without forced labor and 27 % with forced labor. With 27 % prevalence, the population dropped 65 % without forced labor and 43 % with forced labor; with 34% prevalence 79 % and 66 %; and with 67 % prevalence, 89 % and 85 %, respectively. Interestingly, the presence of forced labor generally does not result in further decline, and some cases show the opposite effect.

However, runs in which boats sell fish globally (at the nearest port) show a clear negative impact on total fish population. In such scenarios, a 7 % IUU prevalence yields a 24 % population decline without forced labor but a 37 % decline with forced labor. With a 27 % IUU prevalence, the population declines by 39 % without forced labor and 58 % with forced labor; with 34 % prevalence, 64 % and 73 %; and with 67 % prevalence, 95 % and 96 %, respectively.

Unlike revenue, total fish population appears to be very sensitive to whether boats can sell only at their home ports or globally. Further study of this phenomenon is required, but this likely results from boats that sell globally being able to more quickly take their catch to market and return to sea. This means each boat spends more time catching fish and less time travel to and from the ports of sale since they simply move to the nearest port rather than back to their home port.

Additionally, initial study of each time step finds that IUU fishing leads to female fish being caught a higher rate before they are able to reproduce. Note that the maturity rate of female fish in this model is much longer than a single fishing season. The presence of IUU fishing also means that fertile fish are removed from the population at a faster rate. This results in a more rapid depletion of fish stock. Figure 1 shows the decline of total fish population by counting total population immediately after each reproduction season. The fish population is less likely to replace caught fish as prevalence of IUU increases. In fact, the slope of these lines becomes much steeper as IUU prevalence increases. Stated differently, Fig. 2 shows the declining average catch rate of each boat over time and at various levels of IUU prevalence. As prevalence of IUU increases, the number of fish a boat catches per time step declines. This is a result of fish stock depletion but also highlights the economic implications as boats have fewer fish to take to market and require more time at seas to reach fish capacity (i.e. fewer fish at increased overhead costs). In addition to the clear environmental and economic implications, the potential for significant international conflict is already manifest. For example, Erickson and Kennedy suggest one role for China's "Maritime Militias" – local militias supported by the People's Liberation Army often drawing from local fishing communities – is to undertake "confrontations with other states' fishing and naval vessels, due to the depletion of fishery resources and the need to fish farther from shore" [13].

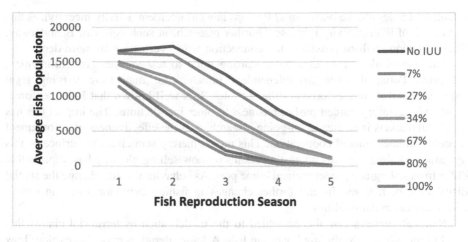

Fig. 1. Fish population after each reproduction season

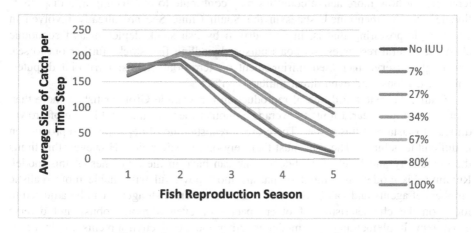

Fig. 2. Average catch rate per boat per time step

4 Future Work

This model set out to test a connection between IUU fishing, forced labor, and over-fishing. It is a highly-abstracted proof-of-concept. Much work is required to match the existing model with available data on fisheries and law enforcement activities. Moreover, the behavior and attributes of the model as described in this paper do not yet truly explore the behavior that may lead to overfishing. Future iterations of this model will attempt to translate real-world economic incentives into computer code. In other words, business decisions of various agents are not currently represented. Will law-abiding boats continue fishing if IUU boats continue to drive down market prices? Or, will some law-abiding boats find incentive to engage in illicit activity themselves? Similarly, as IUU boats generate significant revenue, will they seek to become more established as legitimate businesses? Perhaps law enforcement activity incentivizes this as the risk of illicit activity increases. Further research on such behavior is underway. These additions will help explore the connection with overfishing in more depth.

This model also provides a basic starting point to test various law enforcement strategies. Ports, if taken as independent jurisdictions, may have varying legal requirements that they enforce at different rates. It is well known that heavy enforcement and regulatory burden pushes business to other jurisdictions. The impact this has on global markets is an area of on-going research, as is the effectiveness of enforcement models and international cooperation. This model merely scratches the surface of this dynamic and already yields interesting results on how selling globally has impacts that differ from selling at a predetermined home port. As behavioral shifts change the spatial distribution of fish and inspire further changes in fishing behavior, one can expect additional emergent qualities.

National interests can also be added to the model, such as territorial claims that would limit where law-abiding boats can fish. A future iteration may also explore how overfishing might lead to more confrontation between national vessels in disputed territories or how more active countries may contribute to overfishing at a disproportionate rate. As mentioned, states in the South China Sea are already involved in geostrategic posturing that is, in part, driven by fish stock depletion and economic claims to natural resources. As boats pursue dwindling fish stocks further out to sea, one might observe increased infringement on territorial claims, territorial disputes, political tensions, and even direct military conflict.

Finally, the real-world rate of reproduction for a single Grouper fish is more than 1 million fish per year. Limits to average computer memory makes this rate of growth difficult to sustain in this model. Therefore, a realistic rate of depletion or reproduction is difficult to achieve. The user must then rely on realistic ratios. However, this limits the granularity and level of confidence one can have in the outcomes of this model. Running this model on computers that are more powerful will enable more realistic numbers of agents and values for various attributes. The fish agents can be adjusted to focus on the characteristics of other species or even a more robust and diverse ecosystem. Implementing this model in other modeling environments might better accommodate higher volumes of agents, more robust verification and validation efforts, as well as an ability to add parameters to this proof-of-concept.

References

1. U.S. Department of State: Trafficking in Persons Report 2015. U.S. Department of State, Washington, D.C., July 2015. http://www.state.gov/j/tip/rls/tiprpt/2015/
2. International Labour Organization: Forced labour, human trafficking and slavery. ILO, Geneva, Switzerland (2016). http://www.ilo.org/global/topics/forced-labour/lang–en/index.htm
3. Pope, A.: Combating Human Trafficking in Supply Chains. White House, Washington, D.C., 29 January 2015. https://www.whitehouse.gov/blog/2015/01/29/combating-human-trafficking-supply-chains

4. Urbina, I.: The outlaw ocean. In: The New York Times. New York, 25 July 2015. http://www.nytimes.com/interactive/2015/07/24/world/the-outlaw-ocean.html
5. Bankes, S.: Exploratory modeling for policy analysis. Oper. Res. **41**, 435–449 (1993). INFORMS, Cantonville, MD
6. Deese, B. et al.: How to Tackle Illegal Fishing. White House, Washington D.C., 15 March 2015. https://www.whitehouse.gov/blog/2015/03/15/how-tackle-illegal-fishing
7. Somanader, Y.: See What the Most Progressive Trade Agreement in History Looks Like. White House, Washington, D.C., 4 March 2015. https://www.whitehouse.gov/blog/2015/03/04/see-what-most-progressive-trade-agreement-history-looks
8. Presidential Task Force on Combating IUU Fishing and Seafood Fraud: Action Plan for Implementing the Task Force Recommendations. U.S. National Oceanic and Atmospheric Administration (NOAA), Washington, D.C., 15 March 2015. http://www.nmfs.noaa.gov/ia/iuu/noaa_taskforce_report_final.pdf
9. Wilensky, U.: NetLogo. Center for Connected Learning and Computer-Based Modeling, Northwestern University, Evanston, IL (1999). http://ccl.northwestern.edu/netlogo/
10. Wilensky, U.: NetLogo Flocking model. Center for Connected Learning and Computer-Based Modeling, Northwestern University, Evanston, IL (1998). http://ccl.northwestern.edu/netlogo/models/Flocking
11. Gordon, H.S.: The economic theory of a common-property resource: the fishery. J. Polit. Econ. **62**(2), 124–142 (1954)
12. U.S. Department of State: Trafficking in Persons Report 2015, pp. 7–9
13. Erickson, A.S., Kennedy, C.A.: China's Maritime Militias. CNA (Center for Naval Analysis) Corporation, Arlington, VA, 7 March 2016. https://www.cna.org/cna_files/pdf/Chinas-Maritime-Militia.pdf

Author Index

Printed in the United States
By Bookmasters